Fiber Bragg Grating Based Sensors and Systems

Fiber Bragg Grating Based Sensors and Systems

Editor

Oleg Morozov

MDPI • Basel • Beijing • Wuhan • Barcelona • Belgrade • Manchester • Tokyo • Cluj • Tianjin

Editor
Oleg Morozov
Department of Radiophotonics
and Microwave Technologies, 10,
Karl Marx st., 420111 Kazan,
Tatarstan, Russia
Kazan National Research
Technical University named after
A.N. Tupolev-KAI
Kazan
Russia

Editorial Office
MDPI
St. Alban-Anlage 66
4052 Basel, Switzerland

This is a reprint of articles from the Special Issue published online in the open access journal *Sensors* (ISSN 1424-8220) (available at: www.mdpi.com/journal/sensors/special_issues/FBGSS).

For citation purposes, cite each article independently as indicated on the article page online and as indicated below:

LastName, A.A.; LastName, B.B.; LastName, C.C. Article Title. *Journal Name* **Year**, *Volume Number*, Page Range.

ISBN 978-3-0365-1286-0 (Hbk)
ISBN 978-3-0365-1906-7 (PDF)

Contents

About the Editor

Oleg Morozov

Oleg G. Morozov (D. of Tech. Sc.) is a Full Professor and the Head of the RadioPhotonics and Microwave Technologies Department of the Institute of RadioElectronics, Photonics and Digital Technologies of the Kazan National Research Technical University - KAI (KNRTU-KAI) n.a. A.N. Tupolev, Russia. He is also the Director of the R&D Institute in Applied Electrodynamics, Photonics and Living Systems of the KNRTU-KAI. Since 2001 he is a member of Russian Federal Register of Experts in Science and Technology, since 2017 he is a member of the expert council in Electronics, Photonics, Instrumentation And Communications of the Supreme Attestation Commission of Russia. Over the past five years he has been awarded the title of Senior member of SPIE, OSA and IEEE. His main research interests include microwave photonics; fiber-optic sensors and inerrogation systems; infocommunication systems of optical, microwave and terahertz ranges.

Preface to "Fiber Bragg Grating Based Sensors and Systems"

Today, no one doubts that fiber Bragg gratings (FBGs) have become the most used tool for measuring various physical parameters, the structural integrity of engineering systems, and biological activity of living systems. Classical approaches to measurements based on temperature and mechanical deformations and changes in the refractive index of the surrounding sensor environment are actively developing. New measurement principles are emerging based, for example, on physical changes in the length of the grating. The search for ways to simplify and reduce the price of FBG interrogation systems on the one hand, and improve their metrological characteristics on the other, is ongoing. One of the winning directions of these studies is the transition to microwave photonics measurement systems, which have been developed on the basis of schemes of optoelectronic generators, frequency mapping, probing using multi-frequency laser radiation with difference frequencies lying in the microwave range, and comb generators. The second promising direction is the development and creation of addressable FBGs, the use of which makes it possible to increase the efficiency of processing measurement data and provides the ability to visualize quasi-distributed sensors and map their readings.

This volume is a collection of papers that originated as a Special Issue, focused on some recent advances related to Fiber Bragg Grating Based Sensors and Systems. Conventionally, the book can be divided into three parts: intelligent systems, new type of sensors and original interrogators. Intelligent systems including evaluation of strain transition properties between cast-in FBGs and cast aluminum during uniaxial straining, multi-point strain measurements on a containment vessel, damage detection methods based on long-gauge FBG for highway bridges, evaluation of a coupled sequential approach for rotorcraft landing simulation, wearable hand module and real-time tracking algorithms for measuring finger joint angles of different hand sizes, and glaze icing detection of 110 kV composite insulators are presented. New type sensors are reflected on multi-addressed fiber Bragg structures for microwave-photonic sensor systems, its applications in load-sensing wheel hub bearings and more complex influence in problems of generation of vortex optical beams based on chiral fiber-optic periodic structures. Original interrogators include researches in optical designs with curved detectors for FBG interrogation monitors, demonstration of a filter less, multi-point, and temperature-independent FBG dynamical demodulator using pulse-width modulation, and dual wavelength differential detection of FBG sensors with a pulsed DFB laser. The authors of this book are grateful to all the contributing authors, journal editors, reviewers and the production team.

Oleg Morozov
Editor

sensors

MDPI

Article

Evaluation of Strain Transition Properties between Cast-In Fibre Bragg Gratings and Cast Aluminium during Uniaxial Straining

Florian Heilmeier [1,*], Robert Koos [2], Michael Singer [1], Constantin Bauer [1], Peter Hornberger [3], Jochen Hiller [4] and Wolfram Volk [1]

1 Chair of Metal Forming and Casting, Technical University of Munich (TUM), 85748 Garching, Germany; michael.singer@singeroelk.de (M.S.); cob@utg.de (C.B.); wv@utg.de (W.V.)
2 Research Neutron Source Heinz Maier-Leibnitz, TUM, 85748 Garching, Germany; robert.koos@frm2.tum.de
3 Application Center for CT in Metrology, Fraunhofer Institute for Integrated Circuits, IIS, 94469 Deggendorf, Germany; peter.hornberger@th-deg.de
4 Department of Mechanical Engineering and Mechatronics, Deggendorf Institute of Technology, 94469 Deggendorf, Germany; jochen.hiller@th-deg.de
* Correspondence: fhe@utg.de; Tel.: +49-89-289-13988

Received: 30 September 2020; Accepted: 29 October 2020; Published: 4 November 2020

Abstract: Current testing methods are capable of measuring strain near the surface on structural parts, for example by using strain gauges. However, stress peaks often occur within the material and can only be approximated. An alternative strain measurement incorporates fibre-optical strain sensors (Fiber Bragg Gratings, FBG) which are able to determine strains within the material. The principle has already been verified by using embedded FBGs in tensile specimens. The transition area between fibre and aluminium, however, is not yet properly investigated. Therefore, strains in tensile specimens containing FBGs were measured by neutron diffraction in gauge volumes of two different sizes around the Bragg grating. As a result, it is possible to identify and decouple elastic and plastic strains affecting the FBGs and to transfer the findings into a fully descriptive FE-model of the strain transition area.We thus accomplished closing the gap between the external load and internal straining obtained from cast-in FBG and generating valuable information about the mechanisms within the strain transition area.It was found that the porosity within the casting has a significant impact on the stiffness of the tensile specimen, the generation of excess microscopic tensions and thus the formation of permanent plastic strains, which are well recognized by the FBG. The knowledge that FBG as internal strain sensors function just as well as common external strain sensors will now allow for the application of FBG in actual structural parts and measurements under real load conditions. In the future, applications for long-term monitoring of cast parts will also be enabled and are currently under development.

Keywords: Fibre Bragg Gratings; neutron diffraction; X-ray tomography; tensile test

1. Introduction

A precise understanding of material behaviour is essential for the load-specific design of structural components. Although there is much effort involved in designing structural parts, exact data for local strains and stresses under load are often unavailable. Consequently, the compounds are oversized by design, in order to ensure that they do not fail under normal load conditions [1]. The use of advanced internal measurement methods now provides a means to obtain valid strain information in structural parts under operating conditions and is therefore a promising approach to address this problem. Fibre-optical strain sensors (Fibre Bragg Gratings, FBGs) are strain sensors which can be cast into aluminium parts, as already shown by Weraneck et al. [2]. At the current state of development, structural parts are commonly monitored by the use of FBGs. In addition to using FBG as a substitute for strain gauges, the trend in research and development is to use component-integrated sensors based on FBG. There are several interesting applications of this method. Due to the low diameter of the glass fibre FBGs are being put into fibre reinforced plastics [3]. One application of embedded FBGs is the possibility to record strains during the curing of epoxy resin matrix. The associated knowledge of process-related residual stresses has particular advantages in manufacturing fibre-reinforced laminated metals [4]. FBG sensors integrated into fibre reinforced plastics are often used to examine impairments and delaminations by Low-Velocity-Impacts [5]. This principle is also applicable to steel cables with cores consisting of fibre reinforced plastics used for bridge constructions. This has the advantage of long-term strain measurements due to the inserted FBG [6]. Thus, FBG sensors can also be used in the construction industry. A further application proposes the use of FBG as a humidity sensor in composites of wooden bridges. Since wood swells under humidity, FBG strain measurements allow for conclusions about the humidity value and an assessment of the damages in the outer structure of the bridge [7]. FBGs are also applied to monitor the progression of corrosion of construction steels [8]. This is possible due to the resistance of glass fibres against corrosive. This quality is also beneficial for applications in the oil industry, where Zhou et al. describes FBGs as being used as pressure sensors within boreholes [9]. To summarize, FBGs have a lot of advantages and thus can be applied to a vast variety of measurement applications.

In this work, we utilized the small diameter and the resistance of glass fibres against the corrosive effect of aluminium melts to cast FBGs into aluminium parts, see Figure 1a. By doing this, they function as internal strain sensors (Figure 1b). This measurement principle has been applied to cast tensile specimens made from the hypoeutectic cast alloy AlSi9Cu3 at utg. First calibration efforts, conducted by comparing strain measurements on the inside as well as on the outside of tensile specimens [10], showed that the calibration factor of the FBG differs from the factor obtained from the free fibres as determined by Jülich et al. [11].

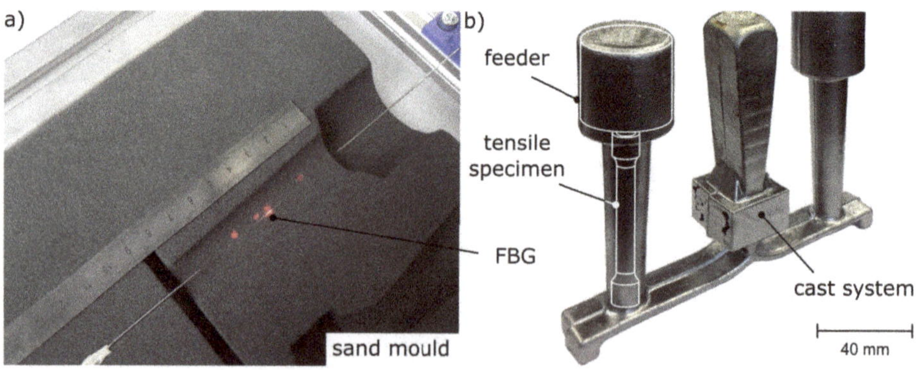

Figure 1. (**a**) One half of the instrumented mould with highlighted FBG. (**b**) Unfinished casting with cast system and two specimens each. After machining, the standard tensile specimen according to [12] contains a fully functional FBG. According to [13].

In the effort to calibrate cast-in FBGs during casting [13] and uniaxial straining [10], Heilmeier et al. shows a wide dispersion of calibration factors. Obviously, cast-in FBGs behave differently compared to free FBGs, which can be precisely calibrated. In general, embedded FBGs can be loaded with axial and transversal strains according to [14]. Thus, for a better understanding of the strain transfer mechanisms from aluminium to fibre, the transition area between FBG and surrounding aluminium needs to be investigated in axial and transversal direction. To achieve this, the local microstrains within the contributing phases need to be obtained by neutron diffraction on two different volume scales to identify their influence on the strain measurement by FBGs. As a result, we will be able to identify and decouple elastic and plastic strains affecting the FBGs. This will allow the application of internal strain sensors in actual structural parts and subsequently, measurements under real load conditions.

2. Materials and Methods

2.1. Fibre Bragg Gratings

The following section describes the measurement method that was used. An FBG is a periodic change in the refraction index within the core of a glass fibre. Thus, a strain sensitive area is given along the fibre's axis, which is measured by an optical interrogator. For the internal strain measurements during tensile testing we used a 3 mm long femtosecond grating within an SMF28 glass fibre, which can be cast into aluminium alloys [2]. According to the findings of Heilmeier et al., the fibre has a force-locked connection to the surrounding casting [10]. Figure 2a shows a sketch of a single mode fibre with a core diameter of 8 μm. The coating is removed before any further processing leading to an effective fibre diameter of 125 μm. The grating within the fibre's core is depicted in Figure 2b. This figure shows that the grating is affected by both the axial strain ϵ_z and the transversal strains ϵ_x and ϵ_y.

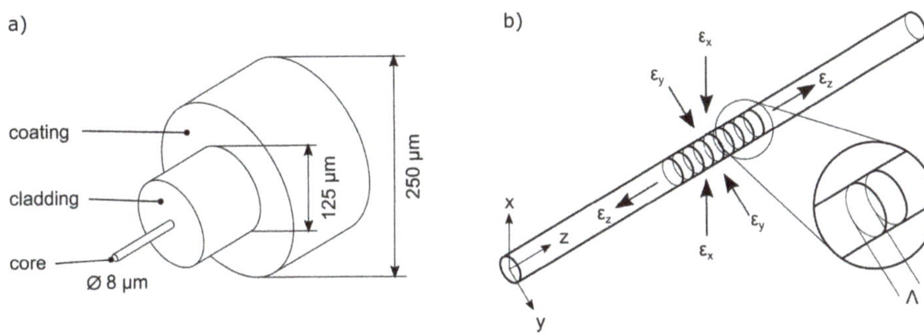

Figure 2. (**a**) Structure of a single-mode SMF28 glass fibre. The resulting diameter after decoating is 125 μm. (**b**) Inner structure of a Bragg grating within the fibre's core. The embedded fibre can be impinged by longitudinal (ϵ_z) and transversal strains (ϵ_x, ϵ_y).

The absolute Bragg wavelength λ_B of the reflected spectrum is given by

$$\lambda_B = 2n_{eff}\Lambda \tag{1}$$

and depends on the grating period Λ as well as the effective refractive index n_{eff} [15]. Changes in Λ, by external straining, lead to a shift $\Delta\lambda_B$ of the peak wavelength according to [16]:

$$\Delta\lambda_B = \lambda_{B,0}(1 - p_e)\epsilon_z \tag{2}$$

Here, $\lambda_{B,0}$ is the initial peak wavelength of the free fibre without external straining and p_e is the effective photoelastic constant. For embedded FBGs, the strain ϵ_z can cause transversal strains by the transversal contraction of the surrounding material. In this case, Equation (2) expands to

$$\frac{\Delta\lambda_{B,x}}{\lambda_{B,0}} = \epsilon_z - \frac{n_0^2}{2}[p_{11}\epsilon_x + p_{12}(\epsilon_y + \epsilon_z)] \tag{3}$$

$$\frac{\Delta\lambda_{B,y}}{\lambda_{B,0}} = \epsilon_z - \frac{n_0^2}{2}[p_{11}\epsilon_y + p_{12}(\epsilon_x + \epsilon_z)] \tag{4}$$

for the x- and y-direction [17]. Here, local photoelastic constants p_{11} and p_{12} depend on the direction which is currently referred to. Due to birefringence, there may be more than one distinct peak within the FBG's spectrum. This is why we use a peakfinding algorithm which tracks the primary peak according to Heilmeier et al. The primary peak originates from the initial peak of the FBG and represents the axial strain ϵ_z during tensile testing [10].

We chose femtosecond FBGs because of their thermal stability. In our recent work, we found that the gratings withstand cast temperatures up to 750 °C without an excess degradation of their spectra. The resulting reflectivity of at least 50% of the initial intensity grants robust measurements during and after casting [14]. Table 1 shows the main FBG properties, including the conversion factor $k = 0.795$ of the free fibre, which we used for the evaluation of ϵ_z according to Equation (5) [18]:

$$\frac{\Delta\lambda}{\lambda_B} = k \cdot \epsilon_z \tag{5}$$

We found that this approach is valid, if only the primary peak is evaluated.

Table 1. Fibre properties.

Fibre Type	Single Mode SMF28
fibre diameter	125 μm
grating type	femtosecond FBG
grating length	3 mm
initial wavelength $\lambda_{B,0}$	1550 nm
k-factor of the free fibre	0.795 [11]

2.2. Cast Materials

The standardized hypoeutectic cast alloy AlSi9Cu3(Fe) [12] was used to cast the specimens for this investigation. It is commonly used for the production of structural parts using sand moulds, die casting and high pressure die casting. For grain refinement we used an aluminium-titanium boride (Al-TiB_2) master alloy [19] to ensure better grain statistics during neutron diffraction [20]. The actual composition of the cast material is shown in Table 2. The characterization using the specimens after testing was conducted by spark emission spectroscopy.

Table 2. Standardized composition of AlSi9Cu3(Fe) and measured composition of grain refined AlSi9Cu3(Fe) obtained by spark emission spectroscopy.

(wt.%) Type	Si	Cu	Fe	Mn	Mg	Ti
AlSi9Cu3(Fe) standardized [12]	8.0–11.0	2.0–4.0	1.3	0.55	0.05–0.55	<0.20
AlSi9Cu3(Fe) as-cast and grain refined	9.1	3.1	0.74	0.28	0.21	0.010

For casting we used 3D-printed furan resin-bound silica sand moulds, from which one half is shown in Figure 3a. The melt is poured into the inlet at a cast temperature of 700 °C and is split by a runner after passing through the filter. This way, two specimens, each containing an FBG, can be cast. During machining, the feeder remains on the specimen. This way, the fibre is not harmed during machining and the specimen contains a fully functional FBG for tensile testing—see Figure 3b.

The effect of grain refinement is shown in Figure 4. The dark field microscopy shows the texture of the etched micrographs with distinct areas of different colours. Each area represents one grain of the cast alloy, which is measured in directions of the large and small half axis. The mean values of (a) 3260 μm versus 2040 μm and (b) 2200 μm versus 1180 μm, respectively, show a significant reduction in the grain sizes, which leads to a stiffer material. This effect is described in Section 3.2.

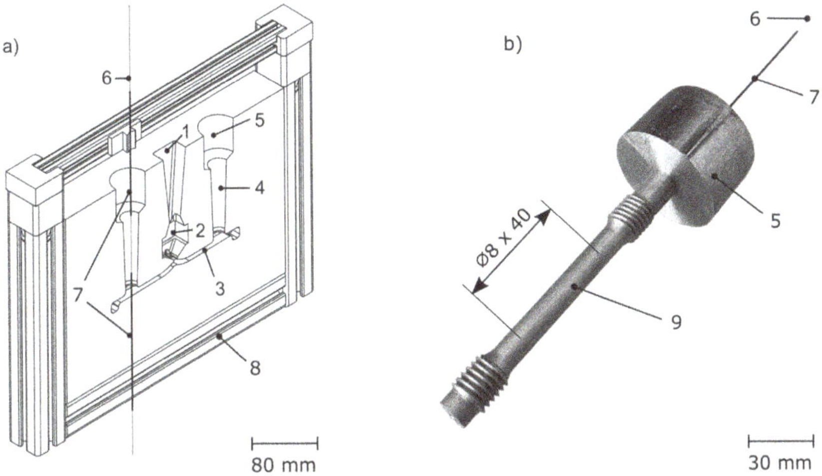

Figure 3. (**a**) One half of the cast mould. The melt is poured into the inlet (1), passes the filter (2) and is split by the runner (3). This way, two specimens (4) with feeders (5) on top can be cast simultaneously. The fibre (6) is protected by steel capillaries (7), which are supported by a frame (8). (**b**) Tensile specimen (9) after machining, containing a fully functional internal FBG.

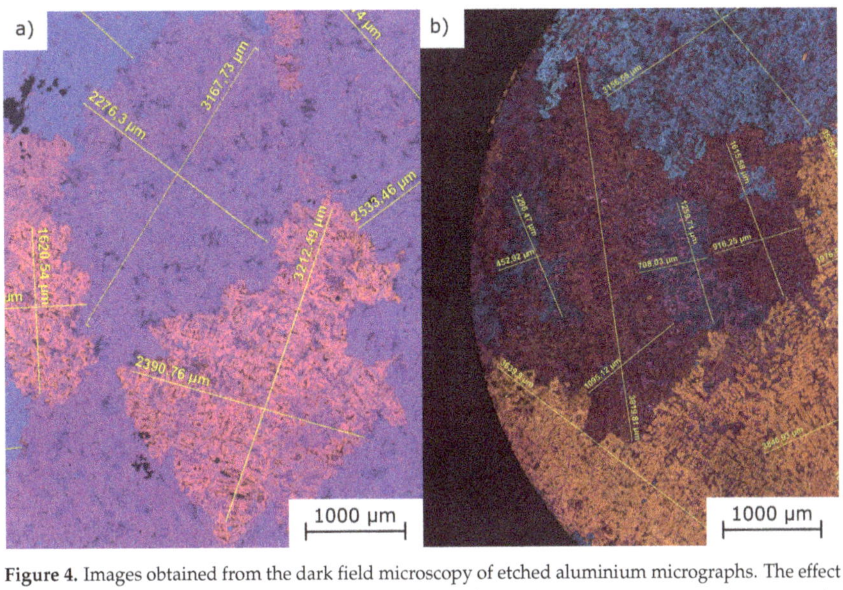

Figure 4. Images obtained from the dark field microscopy of etched aluminium micrographs. The effect of grain refinement of titanium boride in (**b**) leads to a finer microstructure compared to non-grain refined aluminium in (**a**) and thus to a higher stiffness of the tensile specimens.

2.3. Neutron Diffraction

Neutron diffraction is a common method for destruction-free measurement of the internal straining of crystalline materials. At the research neutron source FRM2 (TUM) in Garching [21], the neutron flux is generated by nuclear fission of ^{235}U in a water-moderated chain reaction which emits white neutron radiation [22]. After being reflected by a Si400 monochromator at a wavelength of 1.67 $\AA$, the monochromatic neutron beam passes through the primary slit and penetrates the specimen, where it is diffracted by the hkl-lattice planes. After passing through a secondary slit, followed by a radial collimator, the diffracted beam is detected—see Figure 5. The gauge volume inside the specimen is defined by the height and width of the primary beam in combination with the width of the secondary slit [23].

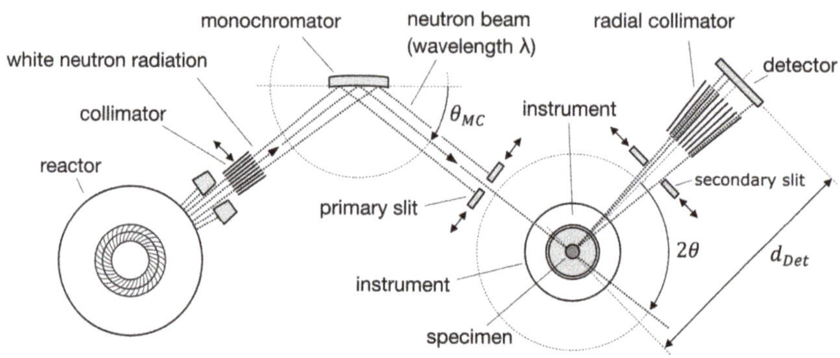

Figure 5. Fundamental setup of the neutron diffractometer STRESS-SPEC [23].

Neutron diffraction as a measurement technique is based on the scattering of neutrons by the lattice planes of crystallines materials. The resulting path difference causes an interference, which was first defined by W. H. Bragg as a fundamental equation [24] as follows:

$$2d_{hkl}\ sin\theta_{hkl} = \lambda \tag{6}$$

The measurement principle is illustrated in Figure 6. The gauge volume exactly matches the middle of the tensile specimen, where the internal FBG is situated. The diffraction angle $2\theta_{hkl}$ is extracted from the detector images. The change in $2\theta_{hkl}$ results from the change in d_{hkl} during external straining and serves as basis for the calculation of the lattice spacing by using Equation (7) [25].

$$\epsilon_{hkl} = \frac{d_{\sigma,hkl} - d_{0,hkl}}{d_{0,hkl}} = \frac{sin\Theta_{0,hkl}}{sin\Theta_{\sigma,hkl}} - 1. \tag{7}$$

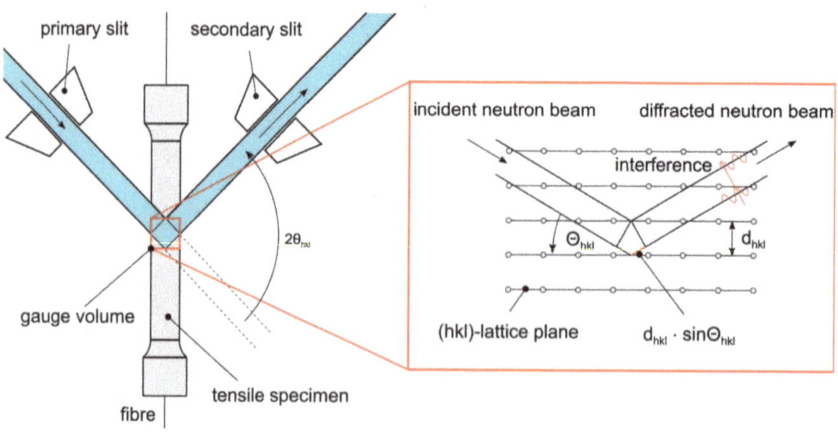

Figure 6. Principle of neutron diffraction, illustrated by the gauge volume within a tensile specimen. The incident neutron beam is diffracted by the hkl-lattice planes of the α-crystallites, according to Equation (6).

2.4. X-ray Computed Tomography

X-ray computed tomography (CT) is an imaging technique, which is increasingly used as a powerful, non destructive tool for visualizing 3D micro-structures. It enjoys enormous popularity in research and development, especially for material science applications due to its ability to achieve volumetric image resolutions at micrometer scale [26].

The cone-beam CT represents a state of the art scanning principle [27] which is depicted in Figure 7. Multiple 2D X-ray projection images are taken from different angles, enabled by a rotation of the scanned object within the X-ray cone beam. The flat photon detector is used to digitize the projections in form of grey value coded images for further processing.

The principle is based on the partial attenuation of the X-ray beam by matter, following an exponential law. Longer penetration depths lead to darker areas on the projected image. The absorption also increases with both higher density and atomic number, while it decreases with higher photon energies.

With the total amount of images taken, the inner structure of the sample can be determined by mathematical 3D reconstruction methods [28]. For this reason, we used the cone-beam algorithm by Feldkamp, Davis and Kress [29].

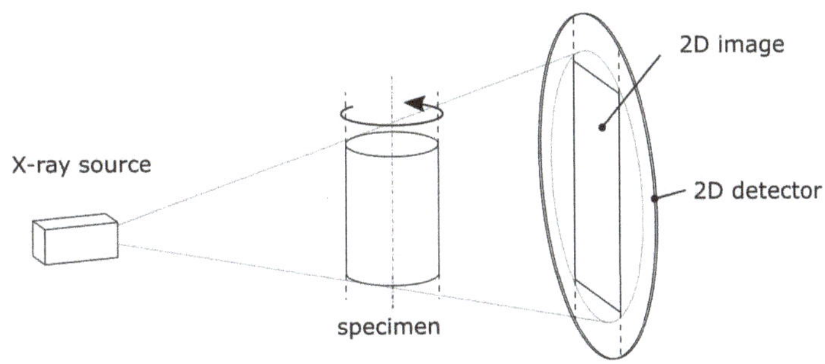

Figure 7. Schematic representation of the cone-beam CT imaging technique which gives evidence of the inner structure of the specimen, such as porosity. According to [29].

2.5. Mathematical Operations

All measured strains depicted in Section 4 are fitted by using the empirical, mathematical description of flow curves by Ludwik-Hollomon, according to [30]:

$$\sigma_\phi = k \cdot \phi^n + \sigma_0 \quad (8)$$

where σ_ϕ is the yield stress and ϕ the degree of deformation. σ_0 considers a possible prestress on the specimen. This basic exponentional term is used to smooth the measured data points for a direct comparability of the experimental variants. The root mean squared error obtained by fitting Equation (8) to the measured data is propagated as an additional error value in combination with the errors of the measured data. An estimation of the resulting errors is given by the square root of the summed squared individual errors $s_{\epsilon,i}$, according to Gauss [31]

$$s_{\epsilon,res} = \sqrt{\sum_{i=1}^{n} s_{\epsilon,i}^2} \quad (9)$$

3. Experimental Setup and Simulation Model

In order to evaluate the strain transition between cast-in FBGs and surrounding aluminium, we simultaneously measured the internal straining obtained from FBG and neutron diffraction as well as the external strain with an extensometer during unixaial tensile testing. The measurements are accompanied by a simulation, which models the cast-in FBG under external loads. The employed methods are described below.

3.1. Experimental Setup

The experimental setup for internal strain measurements using FBGs has been established onsite the STRESS-SPEC instrument [32] at the research neutron source FRM2 (TUM) in Garching [21] as shown in Figure 8. The setup incorporates a mounting system, which allows the feeder to remain on the tensile specimen. This is of particular importance because the FBG requires a functional connection to the measurement equipment.

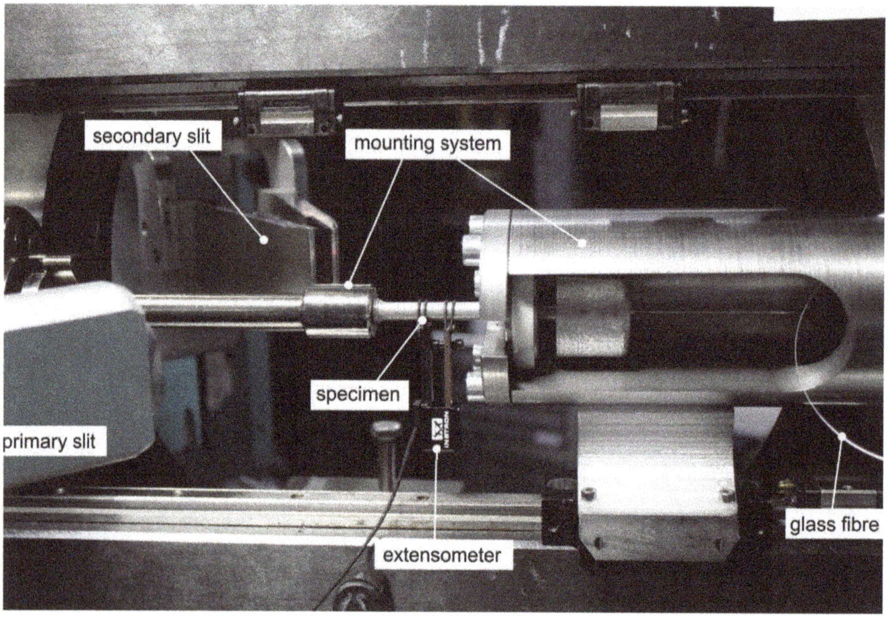

Figure 8. Experimental setup on site for the STRESS-SPEC instrument for simultaneous strain measurements using FBG and neutron diffraction. The primary and secondary slits are adjusted automatically to enable two different gauge volumes during testing. There is an additional extensometer which is situated directly on the specimen.

The strain of the specimens was measured in situ and ex situ by extensometer, FBGs and neutron diffraction simultaneously using the tensile rig at STRESS-SPEC [33]. The neutron beam covered two differently sized gauge volumes ($0.5 \times 8 \times 8$ mm^3 and $0.5 \times 0.5 \times 8$ mm^3) formed by a 0.5 mm collimator and an automatically adjustable primary slit. A wavelength of $\lambda = 1.67$ Å was used for acquiring the diffraction peaks of Al(311) at $2\theta = 86°$ and a scan time of 300 s. The specimens rotated around the vertical axis to improve grain statistics. The primary and secondary slit are directed to the middle of the specimen and form a gauge volume which is congruent to the FBG and its measurement direction along the axis of the specimen. Figure 6 shows the measurement principle on a microscopic scale. The incident neutron beam is diffracted by the lattices planes of the α-crystallites within the casting, which meet the Bragg relation in Equation (6) at an angle of $2\theta \sim 86°$.

During the experiment, the specimens were loaded with increasing load steps (in situ), subsequently followed by a release of the force (ex situ). The load steps are shown in Table 3. For a sufficient neutron count rate by the detector, each step with big volume lasts 480 s, whereas the small volumes have a scan time of 900 s.

Table 3. Increasing load steps during the tensile test. These represent the in situ load steps whereas the ex situ load steps are given by each subsequent force relief with identical measuring durations.

	Step 1	Step 2	Step 3	Step 4	Step 5	Step 6	Step 7	Step 8	Step 9
tension/MPa	5	7.5	10	20	40	60	80	120	160
time (big volume)/s	480	480	480	480	480	480	480	480	480
time (small volume)/s	900	900	900	900	900	900	900	900	900

3.2. Simulation

For a distinct evaluation of the strain transition from aluminium to fibre, we used the implicit finite element (FE) simulation given in Figure 9, as already presented in [13]. In order to model the interaction properly, the simulation starts with the cooling of the aluminium body, which forces compression strains onto the fibre—see Figure 10. The resulting force-locked connection is used as a start condition for the subsequent simulation of the tensile test with increasing load steps, as listed in Table 3. We extended the model by defining a step-wise increasing load σ onto the upper surface of the aluminium body as depicted in Figure 9a.

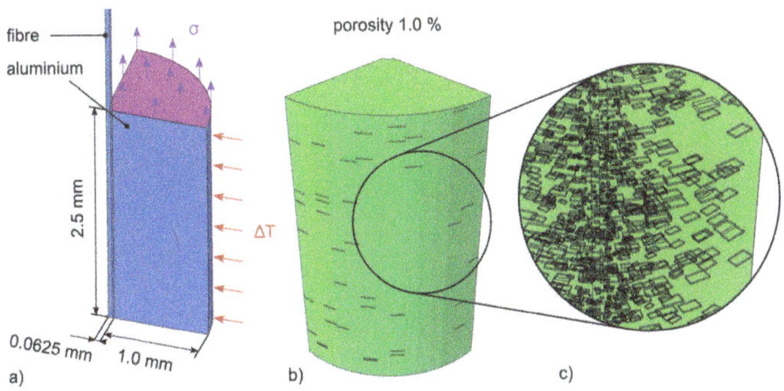

Figure 9. (**a**) Sketch of the simulation model consisting of fibre with surrounding aluminium. Due to symmetry, the model represents one eighth of the whole body. The two boundary conditions ΔT and σ represent two consecutive steps of the straining depicted in Figure 10. (**b**) Mesh of the model with a porosity of 1.0 vol.-%. (**c**) Detailed view on the void distribution within the model.

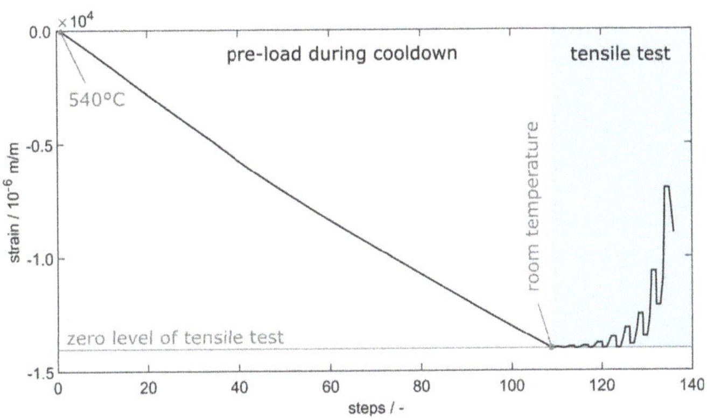

Figure 10. Definition of the simulation steps with explicit differentiation between the pre-load and tensile test. The pre-load according to [13] leads to a realistic force-locked connection between aluminium and fibre as a start condition for the subsequent tensile test at room temperature.

For an additional evaluation on how the occurring porosity within the specimen affects the strain transition, the aluminium body contains voids, which are simulated by a random deletion of nodes in the FE-mesh, as already described by Heilmeier et al. [13]. The mesh with deleted nodes is shown in Figure 9b, while Figure 9c shows a detailed view of the voids on the model's surface.

The material model we used is based on a series of hot tensile tests, which were conducted by Reihle [34]. The resulting temperature-resolved Mises-yield surface represents the macroscopic material behaviour. The cast material tested by Reihle was not grain refined like the one in this survey, which is why we calibrated the simulation using the in situ strain response measured by FBG. The yield strength at room temperature has been extended to 104% of its original stiffness to match load step 9 in Figure 14b.

3.3. *Porosity Evaluation by Computed Tomography*

In order to perform a porosity analysis, high-resolution 3D-image data were obtained from the tensile specimen using industrial computed tomography. The corresponding scanner is equipped with a micro-focus X-ray source, which illuminates a 2k-detector with a pixel pitch of 200 μm. The scanning region in the middle of the sample comprises the complete diameter by a scan height of 10 mm. Due to this small region, a high magnification (compare with Figure 11) and thus a high volume resolution of 8 μm is achieved at a scanning time of 90 min and a maximum X-ray energy of 170 keV. In order to avoid artefacts arising from the polychromatic characteristic of the spectrum, the spectrum is pre-filtered by a copper plate of 1 mm thickness [35].

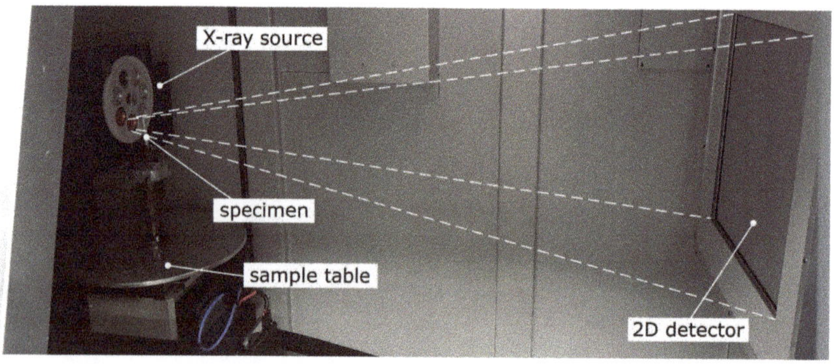

Figure 11. The micro-focus CT-scanner Tomoscope HV 500 from Werth Messtechnik GmbH with the tensile specimen, positioned directly in front of the X-ray source for high image magnification.

Since materials can be distinguished in the volume, a porosity evaluation is possible, as long as they have different absorption coefficients. For evaluation, the VGDefX algorithm was used, which is part of the porosity/inclusion analysis module of VG Studio Max [36]. This algorithm offers a reliable pore detection and indicates the probability of occurrence for a statistic evaluation. It was used because it takes grey value variations into account and applies noise reduction. In the analysis, an automatic surface determination with local thresholding is used.

4. Results

In this investigation, we conducted tensile tests with three independent strain measurement techniques. One of them is the new approach of optical strain measurements with cast-in FBGs. The CT-scan gives evidence of the porosity within the specimens with an overall evaluation using 3D-image-reconstruction methods. We will present specimens FS54 and FS55 as well as ZS25 and ZS26, which show less and more porosity, respectively. Both variants are fitted by Equation (8) for a generalization of the strain progressions. All tests are accompanied by an FE-simulation, which is used for the final evaluation of the strain transition from aluminium to fibre.

4.1. Porosity Evaluation

Figure 12 shows the micro CT-scan of specimens FS54 and FS55 with porosity values of 0.15% and 0.06%, which are rather small for AlSi9Cu3(Fe) cast into sand moulds. The majority of the occurring voids is formed by microscopic blowholes with occasional bubbles of entrained gas held onto the glass fibres, as can be seen marked by the red arrows. In contrast, ZS25 and ZS26 show higher porosity values of 0.63% and 0.73%, which are formed by the very prominent micro blowholes and the more frequent occurrence of entrained gas bubbles on the glass fibres. The air bubbles are marked by red arrows in Figure 13. These specimens were cast on a different day, leading to different cast conditions, such as ambient temperature and humidity.

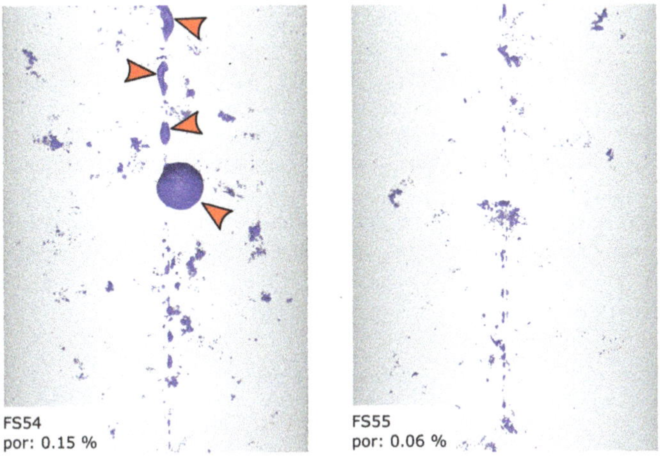

Figure 12. CT scans of specimens FS54 and FS55. The microstructure shows only a small amount of micro blowholes and only occasional entrained gas bubbles (marked by arrows). This leads to an overall porosity values of 0.15 vol.-% and 0.06 vol.-%.

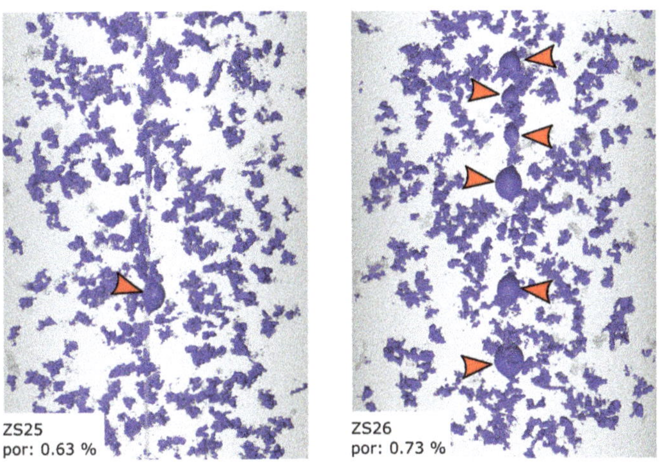

Figure 13. CT scans of specimens ZS25 and ZS26. These specimens were cast on another day leading to a much higher amount of micro blowholes as well as more entrained gas bubbles on the fibres marked by arrows. The resulting overall porosity values of 0.63 vol.-% and 0.73 vol.-% lead to a weaker microstructure.

4.2. Strain Evaluation during Tensile Testing

The tensile tests enable the comparison of three independent strain measurement techniques. The extensometer data represent the macroscopic behaviour of the specimens under external load. The cast-in FBGs provide strain information from within the specimens. The phase-specific straining is given by neutron diffraction with two different gauge volumes around the fibre. Each strain evolution consists of two data sets measured by specimens obtained from one casting, which were unified using a fitting curve given by Equation (8).

The tensile test routine presented in Table 3 leads to strain reactions, which are shown in Figure 14 for FS54 and FS55. Figure 14a depicts the phase-specific strains within the Al311-crystallites. AlSi alloys commonly show heterogeneous microstructures, which combines stiff Si particles with ductile α-aluminium into a composite with a combined strength, see Schöbel et al. [37]. Due to the ductility of the α-aluminium, which is measured specifically by neutron diffraction in form of the Al311-reflection, the measured strains in Figure 14a are smaller than the macroscopic strains depicted in Figure 14b. The big and small gauge volumes both show similar in situ strain reactions, whereby the small volume generates slightly smaller values. The similar strains can be explained by the small amount of porosity as depicted by Figure 12, leading to an even strain distribution over the gauge volumes. The simulated strain data show very good agreement with the measured data, where only the last two load steps are overvalued by up to 340×10^{-6} m/m. The ex situ data of Al311 show compressive strains, which is a microscopic reaction to the presence of silicon-rich precipitations within the alloy. Due to the homogeneous material model within the simulation, this effect cannot be accurately recreated and thus the elastic strains take on a value of zero when the force is relieved.

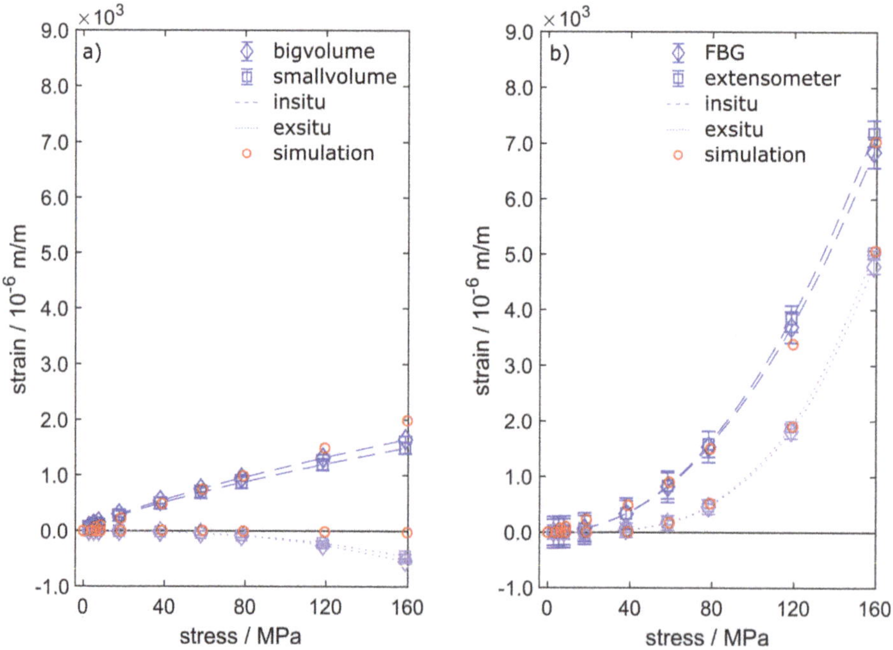

Figure 14. Straining of specimens FS54 and FS55 under tensile load steps until fatigue. The figure shows both the comparison of in situ and ex situ data obtained by (**a**) neutron diffraction and (**b**) by extensometer and FBG. The simulation does not consider the microstructure of the aluminium and thus the microscopic straining does not show any permanent strains. Besides that, the calculated data in (**b**) match the measured strains well. The measured strain reactions have been fitted by Equation (8).

The straining obtained from FBG and extensometer is depicted in Figure 14b. The data show good agreement between FBG and extensometer, meaning that cast-in FBG are perfectly capable of measuring precise strain data conforming to standards [38]. The maximum in situ straining shows values as high as 7000×10^{-6} m/m. This data point is used to calibrate the simulation model, from which all other calculated strain values arise.

The ex-situ straining shows the plastification of the specimens, beginning at tensions higher than 40 MPa. Both extensometer and FBG recognize plastic strains in the form of a permanent deformation, which remains after strain relief as ex situ straining. Here, again, the calculated strain data perfectly match the measured data. Due to the small amount of porosity within specimens FS54 and FS55, the strain calculations were obtained from the simulation model without porosity.

Concerning specimens ZS25 and ZS26, the best fitting results were calculated by adding 1.0% porosity to the model as described in Section 3.2. The resulting quality was evaluated with respect to the highest in situ strain value in Figure 15b. Again, the model does not take the microstructure of the alloy into account. Thus, the phase-specific straining of Al311 is overrated for the in situ steps and takes on zero-values for the ex situ steps. The in situ and ex situ straining of the fibre and the extensometer, which again show very good agreement to each other, are perfectly matched by the simulation.

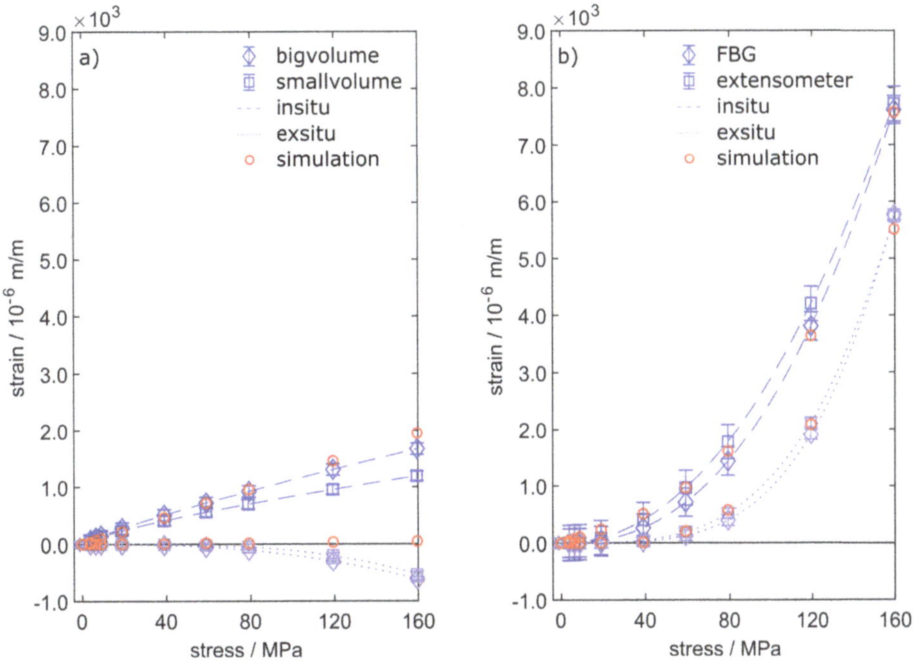

Figure 15. Straining of specimens ZS25 and ZS26 under tensile load steps until fatigue. The figure shows both the comparison of in situ and ex situ data obtained by (**a**) neutron diffraction and (**b**) by extensometer and FBG. The calculated data include 1.0 vol.-% porosity and match the measured strains well. The measured strain reactions have been fitted by Equation (8).

In conclusion, the results of the simulation model presented in Section 3.2 could be verified by all the different strain measurement techniques we used during tensile testing. Especially the phase-specific strain data obtained by neutron diffraction provides valuable support, leading to a fully descriptive model of the interaction between glass fibre and cast aluminium. Based on these findings, we are now able to examine the strain transition from aluminium to fibre as a function of the radius on a microscopic scale.

4.3. Evaluation of the Strain Transition Area

The simulation model now allows for a closer look at the strain transition area as a function of the model's radius. Therefore, Figure 16 shows the strain distribution for load step 9 at 160 MPa, according to Table 3. The radius is given by normalized values beginning from zero in the middle, where the fibre is indicated, to the fringe of the aluminium.

In Figure 16 the non-porous material to the left is compared to the material with 1.0 vol.-% porosity to the right. The in situ elastic straining shows 2000×10^{-6} m/m for both models, which is in compliance to the results in Figures 14a and 15a. In addition, the plastic straining is given, which is generated during the in situ steps and remains as permanent straining when the force is relieved.

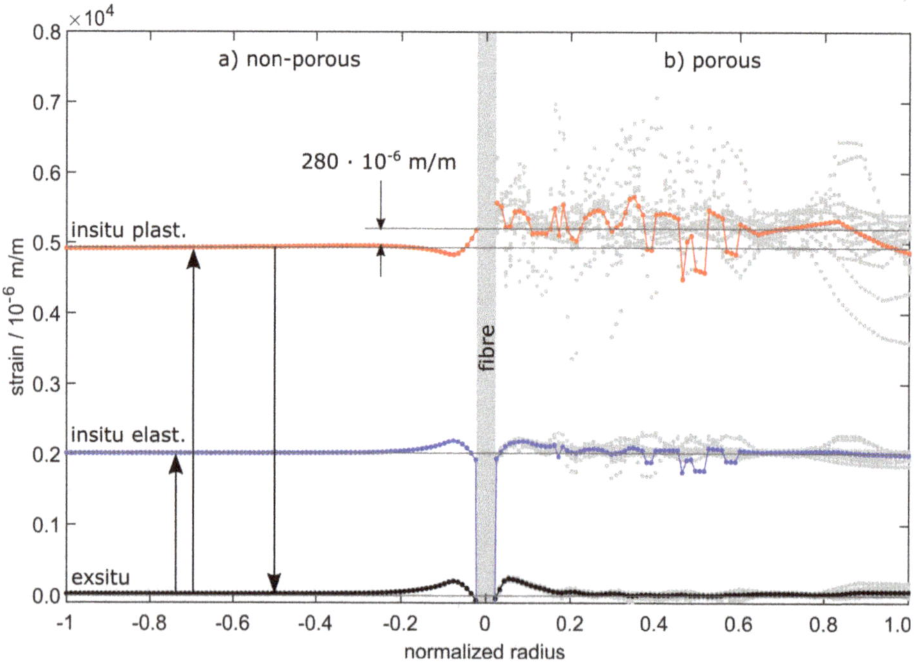

Figure 16. Strain transition area as obtained from the calibrated simulation model. The straining versus normalized radius shows the in situ elastic and plastic strains as well as the ex situ strains for (**a**) the non-porous model and (**b**) the model with 1.0 vol.-% porosity.

After relief, the elastic ex situ strains return to zero except for the immediate surrounding of the fibre. Note that the data points deposited in grey colour are the direct calculations of the porous model. As the model is speckled with voids, the results are not radially symmetrical any more, which is why we calculated the mean values versus radius.

Naturally, the voids within the aluminium are not able to transfer stresses, which weakens the whole material by the generation of local excessive tensions. This leads to increased overall plastic strains beginning at tensions higher than 40 MPa. The plastic strains sum up over the entirety of all in situ load steps leading to higher measured values in Figure 15b than in Figure 14b. This does not affect the elastic share of the overall straining, which solely depends on the yield stress given by the material model. This is why the elastic strain progressions for both models share a common mean value in Figure 16.

Due to the voids, local excessive tensions are forming within the aluminium. The related plastic strains are depicted in Figure 17b in direct comparison to the plastic strains in the absence of the voids (Figure 17a). The voids are statistically allocated, which is why the effect of excessive tensions averages itself. The glass fibre is only affected by changing transversal tensions, which can cause a change in the FBG's spectrum [17]. According to Heilmeier et al. [10], this has no influence on the FBG's axial straining as long as only the primary peak is evaluated.

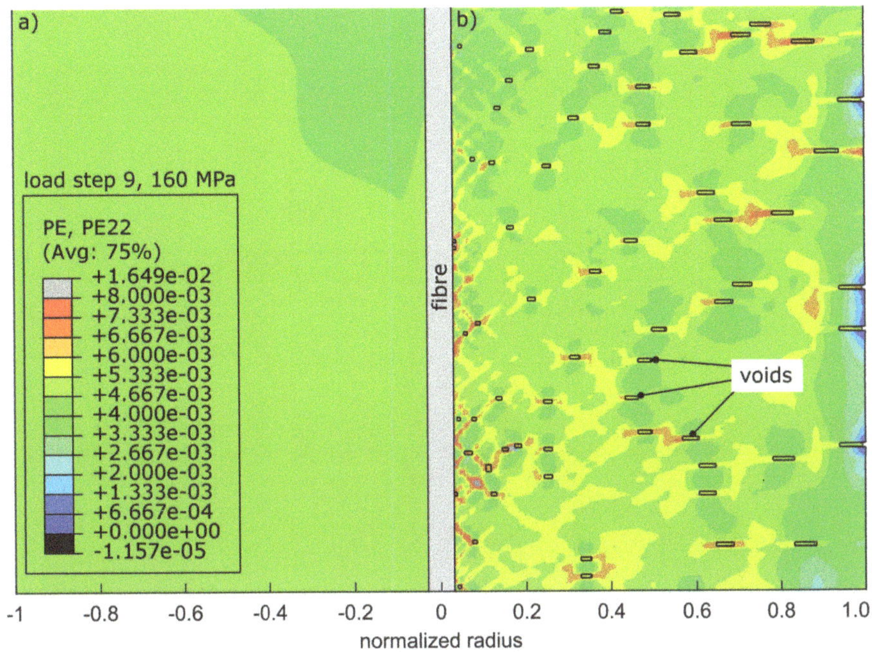

Figure 17. Microscopic plastic straining in the transition area around the glass fibre (**a**) without and (**b**) with 1.0 vol.-% porosity. By adding voids to the model, local excessive tensions are formed, weakening the structure of the specimen.

4.4. Spectra Analysis

Figure 18 elucidates the alteration of the FBG's spectrum. The position of the primary peak determined by the peakfinding algorithm is given by the black arrows. We chose specimen FS55 due to the small defect content (see Figure 12) in order to keep the influence of porosities on the spectrum at a minimum. All spectra are normalized using the initial peak, which is the given by the free fibre before casting. The peak intensity after casting is beneath 40% of its original intensity. The spectrum of the FBG after machining shows a relief of the force on the fibre, leading to a moderate rise in the peak intensity and a shift back to higher wavelengths. This is the initial state of the FBG before tensile testing.

In order to show the effect of the external force during tensile testing, the figure shows the spectrum which forms during the last load step and after relief. During in situ step 9, the spectrum shows distinct secondary peaks, which may be caused by transversal loads, strain gradients along the Bragg grating [17] or changes in the fibre's local photoelastic constants—see Equations (3) and (4).

The peakfinding algorithm is designed to track the primary peak, which represents the axial straining ϵ_z of the grating. This was already proven by Heilmeier et al. [10] and can be directly seen by the agreement of the strain measurements in Figures 14b and 15b. The cause of the occurring secondary peaks, however, has to be determined by further experiments. After relief, the spectrum obviously reshapes to a less distorted peak. This may be a proof for the secondary peaks to be caused by transversal strains, as the lateral contraction of the tensile specimen reduces after relief.

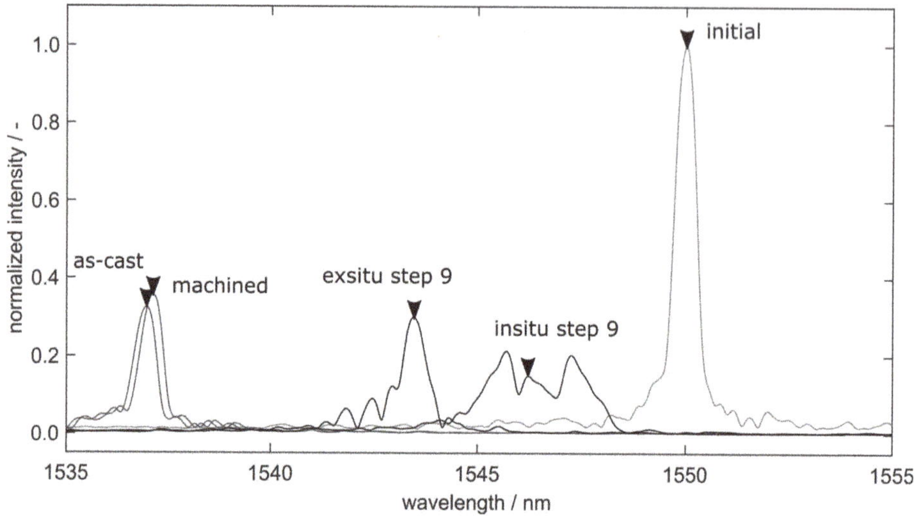

Figure 18. FBG-spectra obtained from every intermediate condition of the specimen before and during testing. The spectrum shows a decrease in reflectivity and degenerates with an increased load subsequently, where secondary peaks are formed beside the primary peak (marked by a black arrow for each step).

5. Discussion

The calculations show that increased plastic strains are generated by the voids within the aluminium due to local excessive tensions. This greatly affects the macroscopic behaviour of the specimens and thus the strain measurements by extensiometer and FBG—see Figures 14b and 15b. Apparently, the microscopic strain distribution only affects the aluminium, as the glass fibre only shows changes in macroscopic strains. Obviously, the casting's supporting effect is still granted in the presence of pores, as the glass fibre would otherwise have breached.

Above a tension of 40 MPa, plastic strains are generated, which build up during the in situ steps. The total plastic strain is then retained as total strain information within the glass fibre. Obviously, the strain transition between aluminium and fibre is not affected by the porosity. There is rather the effect of increased plastification of the specimens which, of course, cannot be directly detected during continuous tensile testing. Regarding [10], only in situ data were generated during testing, making it impossible to decouple plastic and elastic strains afterwards.

The wide range of porosity values between 0 vol.-% and 2 vol.-% can now explain the variation of calibration factors. This claim is sustained by [13], which states that the overall defect volume has a greater influence on the strain response of the cast-in FBG than the defect area directly on the fibre.

Nevertheless, the difference between strains measured by FBG and extensometer has not yet been clarified. This may be due to stick-slip effects, which we could not verify in our simulation.

Although the statement of the simulation is valid and provides quantitative results, the microscopic strain distribution highly depends on the shape and distribution of the voids within the model. This may have a small impact on the local microscopic strains, but sums up to a macroscopic difference in the overall straining and stiffness of the specimen. The validation in Figures 14 and 15 shows excellent agreement between measurement and simulation and yields valuable information about the strain transition area. Thus, the FE-model closes the gap between microscopic and macroscopic straining. Of particular significance is the statement that cast-in FBG behaves like the external strain measurement method given by extensometer. The microstructure of the surrounding aluminium and the external load condition do affect the reflected spectra obtained from the Bragg grating. However, the strain measurement by FBG is not compromised if the primary peak is reliably tracked during the spectra analysis. Other than Lammens et al., who investigated transversal strains during the curation of cross-ply composites [39], we only focus on the distinct detection of the primary peak in this work, which represents the axial straining ϵ_z. Nevertheless, if the occurring secondary peaks can be reliably related to transversal strains, the peak-finding method we used could be of certain interest for the evaluation of embedded FBGs.

6. Conclusions

In recent research, cast-in FBG showed a wide dispersion of calibration factors. In this research, we examined the question on how the strain transition from aluminium to fibre forms during tensile testing. Microscopic porosity has a great impact on the local microstrains generated. Both effects were evaluated by a calibrated finite element simulation, which is based on both the measurement of phase-specific strains using neutron diffraction and macrostrains by the extensometer, as well as FBG. As a result, we were able to identify and decouple elastic and plastic strains affecting the FBGs.

The experiments show that plastic strains develop at tensions higher than 40 MPa. Plastic strains remain permanently within the specimen and add up over all load steps. Because the fibre has a force-locked connection to the surrounding casting due to shrinkage after solidification, the FBG shows the same permanent straining in the plastic deformation regime of the specimen as the external strain measurement by extensometer.

Because of local excessive tensions in presence of pores regions more plastic straining is transferred into the FBG. This is shown by the FE simulation, which gives evidence of the local stress and strain distribution in the transition area, especially in the direct surroundings of the glass fibre. The content of porosity within the casting turns out to be the main influence on the straining of the fibre, which is then measured by the Bragg grating.

The porosity affects neither the calculated nor the measured phase-specific elastic strains of the aluminium crystallites. The mechanisms of plastification do not occur within the aluminium grains, but in between. This is why neutron diffraction is only able to measure elastic strains, which turn out to be independent from the pore content. This is substantiated by simulation, which is not able to show the microscopic, inter-granular ex situ strains within the aluminium, though. An interesting extension of the simulation model would hence consider a heterogeneous microstructure of the aluminium.

In conclusion, we accomplished to close the gap between the external load and internal straining obtained from cast-in FBG by development of a fully descriptive FE-model considering the contact between casting and glass fibre. This enables the generation of valuable information about the mechanisms within the strain transition obtained from the strain evolutions directly around the fibre. We see that the porosity has a significant impact on the stiffness of the tensile specimen, the generation of excess tensions and thus the formation of permanent plastic strains, which are well recognized by the FBG. The knowledge that FBG as internal strain sensors function just as common external strain sensors will allow the application of FBG in actual structural parts and measurements under real load conditions. In future, applications for long-term monitoring of cast parts will also be enabled and are currently under development.

Author Contributions: F.H. and R.K. conceived and designed the experiments. F.H., R.K. and M.S. performed the experiments. M.S. and F.H. performed the simulation. P.H. and J.H. provided the CT scans as well as the porosity analysis. F.H., R.K. and C.B. analyzed the data. W.V. coordinated the project and discussed the theory. All authors have read and agreed to the published version of the manuscript.

Funding: This work was funded by the Deutsche Forschungsgemeinschaft (DFG) [Grant Nos. VO 1487/11-1 and VO 1487/11-2].

Conflicts of Interest: The authors declare that they have no known competing financial interests or personal relationships that could have appeared to influence the work reported in this paper.

References

1. Warnke, E. Mit Eigenspannungen leben: Entstehung, Auswirkungen, Messung, Berechnung und Vermeidung von Eigenspannungen, Ausblick. In *Konstruieren und Giessen*; Bundesverband der Deutschen Gießerei-Industrie: Düsseldorf, Germany, 2008; pp. 37–40.
2. Weraneck, K.; Heilmeier, F.; Lindner, M.; Graf, M.; Jakobi, M.; Volk, W.; Roths, J.; Koch, A.W. Strain Measurement in Aluminium Alloy during the Solidification Process Using Embedded Fibre Bragg Gratings. *Sensors* **2016**, *16*. [CrossRef]
3. Luyckx, G.; Voet, E.; Lammens, N.; Degrieck, J. Strain measurements of composite laminates with embedded fibre bragg gratings: Criticism and opportunities for research. *Sensors* **2011**, *11*, 384–408. [CrossRef]
4. Prussak, R.; Stefaniak, D.; Kappel, E.; Hühne, C.; Sinapius, M. Smart cure cycles for fiber metal laminates using embedded fiber Bragg grating sensors. *Compos. Struct.* **2019**, *213*, 252–260. [CrossRef]
5. Mulle, M.; Yudhanto, A.; Lubineau, G.; Yaldiz, R.; Schijve, W.; Verghese, N. Internal strain assessment using FBGs in a thermoplastic composite subjected to quasi-static indentation and low-velocity impact. *Compos. Struct.* **2019**, *215*, 305–316. [CrossRef]
6. Li, F.; Du, Y.; Sun, X.; Zhao, W. Sensing performance assessment of twisted CFRP with embedded fiber Bragg grating sensors subjected to monotonic and fatigue loading. *Sens. Actuators A Phys.* **2018**, *271*, 153–161. [CrossRef]
7. Nair, A.; Cai, C.S.; Kong, X.; Hou, S. Bridge Retrofitting Using FRP-Wrapped Balsa Wood Deck: Experimental Study and Field Evaluation. *J. Aerosp. Eng.* **2019**, *32*, 04019065. [CrossRef]
8. Cinitha, A.; Sampath, V.; Kesavan, K. Strain monitoring of low carbon steel in a corrosive environment using fiber Bragg technology. *Constr. Build. Mater.* **2019**, *217*, 265–272. [CrossRef]
9. Zhou, X.; Yu, Q.; Peng, W. Fiber-optic Fabry—Perot pressure sensor for down-hole application. *Opt. Lasers Eng.* **2019**, *121*, 289–299. [CrossRef]
10. Heilmeier, F.; Koos, R.; Weraneck, K.; Lindner, M.; Jakobi, M.; Roths, J.; Koch, A.W.; Volk, W. In-situ strain measurements in the plastic deformation regime inside casted parts using fibre-optical strain sensors. *Prod. Eng.* **2019**, *618*, 271. [CrossRef]
11. Jülich, F.; Aulbach, L.; Wilfert, A.; Kratzer, P.; Kuttler, R.; Roths, J. Gauge factors of fibre Bragg grating strain sensors in different types of optical fibres. *Meas. Sci. Technol.* **2013**, *24*, 094007. [CrossRef]
12. Deutsches Institut für Normung e.V. *Aluminium und Aluminiumlegierungen—Chemische Zusammensetzung und Mechanische Eigenschaften*; Beuth Verlag GmbH: Berlin, Germany, 2019.
13. Heilmeier, F.; Koos, R.; Hornberger, P.; Hiller, J.; Weraneck, K.; Jakobi, M.; Koch, A.W.; Volk, W. Calibration of cast-in fibre Bragg gratings for internal strain measurements in cast aluminium by using neutron diffraction. *Measurement* **2020**, *163*, 107939. [CrossRef]
14. Weraneck, K. *Strukturüberwachung Mittels Eingebetteter Faser-Bragg-Gitter*; Universitätsbibliothek der TU München: München, Germany, 2018.
15. Erdogan, T. Fiber grating spectra. *J. Light. Technol.* **1997**, *15*, 1277–1294. [CrossRef]
16. Rao, Y.J. In-fibre Bragg grating sensors. *Meas. Sci. Technol.* **1997**, *8*, 355–375. [CrossRef]
17. Wagreich, R.B.; Sirkis, J.S. Distinguishing Fiber Bragg Grating Strain Effects. In Proceedings of the 12th International Conference on Optical Fiber Sensors, Williamsburg, VA, USA, 28–31 October 1997; pp. 20–23. [CrossRef]
18. Werneck, M.; Allil, R.C.S.B.; Ribeiro, B.A.; de Nazaré, F.V.B. A Guide to Fiber Bragg Grating Sensors. In *Current Trends in Short- and Long-Period Fiber Gratings*; Cuadrado-Laborde, C., Ed.; IntechOpen: London, UK, 2013. [CrossRef]
19. Mondal, D.P.; Jha, N.; Badkul, A.; Das, S. Effect of Al—TiB master alloy addition on microstructure, wear and compressive deformation behaviour of aluminum alloys. *Trans. Nonferrous Met. Soc. China* **2012**, *22*, 1001–1011. [CrossRef]
20. Randau, C. Entwicklungen am Neutronendiffraktometer STRESS-SPEC für Schnelle und Lokale Polfigurmessungen zur Bestimmung Ortsaufgelöster Texturen: Dissertation. Ph.D. Thesis, Technische Universität Clausthal, Clausthal, Germany, 2012.
21. Röhrmoser, A. Core model of new German neutron source FRM II. *Nucl. Eng. Des.* **2010**, *240*, 1417–1432. [CrossRef]

22. Gabrys, B.J. *Applications of Neutron Scattering to Soft Condensed Matter*; Chapman and Hall/CRC: Boca Raton, FL, USA, 2014.
23. Meier, L. *In-Situ-Messung der Phasenumwandlungskinetik von Ausferritischem Guasseisen*; utg-Forschungsberichte; Shaker Verlag: Aachen, Germany, 2017; Volume 76.
24. Bragg, W.H.; Bragg, W.L. The Reflection of X-rays by Crystals. *Proc. R. Soc. A Math. Phys. Eng. Sci.* **1913**, *88*, 428–438. [CrossRef]
25. ISO International Organization for Standardization. *Non-Destructive Testing: Standard Test Method for Determining Residual Stresses by Neutron Diffraction*; ISO: Geneva, Switzerland, 2005.
26. Hanke, R.; Fuchs, T.; Salamon, M.; Zabler, S. X-ray microtomography for materials characterization. In *Materials Characterization Using Nondestructive Evaluation (NDE) Methods*; Woodhead Publishingr: Sawston, Cambridge, 2016; pp. 45–79. [CrossRef]
27. Buzug, T. *Computed Tomography: From Photon Statistics to Modern Cone-Beam CT*; Springer: Berlin/Heidelberg, Germany, 2008. [CrossRef]
28. Rodet, T.; Noo, F.; Defrise, M. The cone-beam algorithm of Feldkamp, Davis, and Kress preserves oblique line integrals. *Med. Phys.* **2004**, *31*, 1972–1975. [CrossRef]
29. Feldkamp, L.A.; Davis, L.C.; Kress, J.W. Practical cone-beam algorithm. *J. Opt. Soc. Am. A* **1984**, *1*, 612. [CrossRef]
30. Lange, K.; Pöhlandt, K. *Vergleichende Betrachtung der Verfahren zur Prüfung der Plastischen Eigenschaften Metallischer Werkstoffe*; Springer: Berlin/Heidelberg, Germany, 1984; Volume 80. [CrossRef]
31. Papula, L. *Mathematik für Ingenieure und Naturwissenschaftler*; Vieweg+Teubner Verlag: Wiesbaden, Germany, 2011. [CrossRef]
32. Hofmann, M.; Schneider, R.; Seidl, G.A.; Rebelo-Kornmeier, J.; Wimpory, R.C.; Garbe, U.; Brokmeier, H.G. The new materials science diffractometer STRESS-SPEC at FRM-II. *Phys. B Condens. Matter* **2006**, *385–386*, 1035–1037. [CrossRef]
33. Hoelzel, M.; Gan, W.M.; Hofmann, M.; Randau, C.; Seidl, G.; Jüttner, P.; Schmahl, W.W. Rotatable multifunctional load frames for neutron diffractometers at FRM II—Design, specifications and applications. *Nucl. Instrum. Methods Phys. Res. Sect. A Accel. Spectrometers Detect. Assoc. Equip.* **2013**, *711*, 101–105. [CrossRef]
34. Reihle, M.M. Entstehung und Ausprägung von Eigenspannungen in Verbundgussteilen. Ph.D. Thesis, Technische Universität München, Munich, Germany, 2016.
35. Firsching, M.; Salamon, M.; Keßling, P.M.; Nachtrab, F.; Krumm, M.; Uhlmann, N.; Hanke, R. *Micro Structural Analysis of AlSi6Cu4 Using Quantitative Computed Tomography Methods*; Fraunhofer IIS: Erlangen, Germany, 2010.
36. Reinhart, C. *VGStudio MAX Referenzhandbuch*; Volume Graphics GmbH: Heidelberg, Germany, 2018.
37. Schöbel, M.; Baumgartner, G.; Gerth, S.; Bernardi, J.; Hofmann, M. Microstresses and crack formation in AlSi7MgCu and AlSi17Cu4 alloys for engine components. *Acta Mater.* **2014**, *81*, 401–408. [CrossRef]
38. Deutsches Institut für Normung e.V. *Metallische Werkstoffe—Zugversuch: Teil 1: Prüfverfahren bei Raumtemperatur*; Beuth-Verlag GmbH: Berlin, Germany, 2019.
39. Lammens, N.; Kinet, D.; Chah, K.; Luyckx, G.; Caucheteur, C.; Degrieck, J.; Mégret, P. Residual strain monitoring of out-of-autoclave cured parts by use of polarization dependent loss measurements in embedded optical fiber Bragg gratings. *Compos. Part A Appl. Sci. Manuf.* **2013**, *52*, 38–44. [CrossRef]

sensors

MDPI

Article

On the Evaluation of a Coupled Sequential Approach for Rotorcraft Landing Simulation

Demetrio Cristiani [1], Luca Colombo [1], Wojciech Zielinski [2], Claudio Sbarufatti [1,*], Francesco Cadini [1], Michal Dziendzikowski [2] and Marco Giglio [1]

1 Dipartimento di Meccanica, Politecnico di Milano, 20156 Milano, Italy; demetrioluigi.cristiani@polimi.it (D.C.); luca1.colombo@polimi.it (L.C.); francesco.cadini@polimi.it (F.C.); marco.giglio@polimi.it (M.G.)

2 Air Force Institute of Technology, Airworth, Division, 01-494 Warsaw, Poland; wojciech.zielinski@itwl.pl (W.Z.); michal.dziendzikowski@itwl.pl (M.D.)

* Correspondence: claudio.sbarufatti@polimi.it; Tel.: +39-022399-8213

Received: 31 March 2020; Accepted: 26 April 2020; Published: 29 April 2020

check for updates

Abstract: Maximum loads acting on aircraft structures generally arise when the aircraft is undergoing some form of acceleration, such as during landing. Landing, especially when considering rotorcrafts, is thus crucial in determining the operational load spectrum, and accurate predictions on the actual health/load level of the rotorcraft structure cannot be achieved unless a database comprising the structural response in various landing conditions is available. An effective means to create a structural response database relies on the modeling and simulation of the items and phenomena of concern. The structural response to rotorcraft landing is an underrated topic in the open scientific literature, and tools for the landing event simulation are lacking. In the present work, a coupled sequential simulation strategy is proposed and experimentally verified. This approach divides the complex landing problem into two separate domains, namely a dynamic domain, which is ruled by a multibody model, and a structural domain, which relies on a finite element model (FEM). The dynamic analysis is performed first, calculating a set of intermediate parameters that are provided as input to the subsequent structural analysis. Two approaches are compared, using displacements and forces at specific airframe locations, respectively, as the link between the dynamic and structural domains.

Keywords: fiber Bragg gratings; landing simulation; rotorcraft; coupled sequential method; landing structural response; finite element analysis (FEA)

1. Introduction

Landing event characterization is fundamental when aircraft structural assessment has to be performed. Load spectra are inherently affected by landings, and their accurate determination cannot be pursued without considering the landing event itself. Indeed, depending on the landing severity and occurrence, the structure might be subjected to high stress levels and non-negligible performance degradation caused by material damage, especially when the landing ranks in the harsh landing regime. The definition of harsh landing is still neither clearly defined nor well-established and is generally considered as a phenomenon that occurs whenever the landing event induces abnormal operational conditions. Whether these conditions are related to the aircraft's structural response only or also involve the passengers' comfort is a matter of debate. Harsh landing can thus be generally located in between normal landing operations and crash events. More specifically, harsh landing is defined by the regulatory authorities in EASA Certification Specification (CS) 25 and Federal Aviation Regulations (FAR) 25 as a landing with a vertical descent velocity exceeding 3 m/s [1,2]. The same threshold is used in the Aircraft Crash Survival Design Guide (Volume 3—Aircraft Structural Crash Resistance) [3].

However, this definition is not sufficient when the landing structural response is concerned, since the vertical descent rate alone is not the unique parameter influencing the landing severity. The structural response to landing depends on multiple variables; the total aircraft weight, the mass distribution, the landing attitude (i.e., the pitch and roll angles), the weight to lift ratio, the forward and lateral landing velocities (generally null or negligible when rotorcrafts are considered), the pitch, roll and yaw rates (if non-negligible), and other parameters related to the environment, have proven to be crucial for the landing assessment.

Nowadays, the flight crew judges the classification of the landing, and eventually establishes if the landing was harsh or not [4,5]. This judgment is generally based on subjective perception, and even when it is supported by objective data, these are insufficient to assess the structural response to landing accurately. This can lead to a biased classification of the landing conditions, thus affecting the safety of the system or leading to unnecessary time and money consuming maintenance procedures. Hence the need to be fully aware of the landing structural response. Precise knowledge of the landing event structural consequences might result in considerable economic savings and increased system safety, since maintenance would be undertaken only when necessary, and based on objective data evaluation, allowing the current maintenance philosophies to evolve into potentially more cost-effective condition-based maintenance philosophies. Understanding the relationship between the landing characteristics and their structural consequences will enable not only the estimation of the actual aircraft aging, as well as an objective classification of the landing, but furthermore the potential occurrence of structural failures, thus paving the way for the implementation of health and usage monitoring systems (HUMS).

In general, any monitoring system entails the observation of a structure over time using periodical measurements, the extraction of proper features from these measurements and the analysis and interpretation of these features to determine the current state of the system, whether the focus is on its health state or its load level [6,7]. The actual state of the investigated system can often result from a statistical comparison of the current data with a database of experimental data. The latter, depending on the examined system and on the variety of circumstances it operates in, might be economically unsustainable, as well as time-consuming to be obtained. An efficient solution to this problem lies in mathematical modeling. Indeed, using numerical models and proper simulation strategies, it is possible to virtually reproduce the phenomena of concern, developing a database of the structural response based on the most influential landing features, consistently limiting the costs of an experimental campaign. The modeling and simulation of the investigated system also allow to (i) select the most suited sensor technology, (ii) predict, in the modeling space, the sensitivity of some parameters to the features of interest, (iii) design and optimize the monitoring system.

This work aims to provide an analysis method for simulating, investigating and characterizing the landing event of a medium-size rotorcraft (Mil Mi-8 helicopter), both from a dynamic and structural point of view, thus leading the way for future implementation of load monitoring (LM) and structural health monitoring (SHM) systems. Specifically, the present work addresses the landing operational range which spans between the nominal landing operations and the harsh landings. These are determined based on the structural consequences the landing has on the airframe. The present framework is enhanced by full-scale experimental tests carried out on a Mil Mi-8 helicopter, which allowed the numerical models of the examined aircraft to be evaluated and verified, proving the legitimacy of the hereafter described method.

More specifically, the landing problem field is divided into two separate domains, namely the dynamic and the structural domain. According to the pursued strategy, the dynamic domain, which relies on a multibody model of the aircraft, takes as an input the main landing characterizing features, and (i) it returns the dynamic response to the landing event; (ii) it provides the structural domain with the proper parameters to assess the landing event structural response. This strategy provides a tool that can predict the landing structural response based on some characteristic landing parameters.

The present work is structured as follows: first, the state of the art of the subject and the related topics is given, then the landing event analysis method is presented, followed by the illustration of an insight of the experimental activity. Subsequently, the modeling of the aircraft landing is examined, and the dynamic and structural domains are investigated. Finally, a conclusive section is provided.

2. State of the Art

The rotorcraft open literature lacks a comprehensive method able to describe and simulate the landing event from both a dynamic and structural point of view. Furthermore, general studies on the simulation strategies of unconstrained aircraft structures subjected to landing loads are lacking. Concerning rotorcraft landing simulation related topics, the open literature mainly covers two types of research lines: (i) crashworthiness related research activities, which usually consist of experimental activities (e.g., full-scale crash tests), modeling (FE models) and simulation of the crash event; (ii) research activities related to the dynamics of the rotorcraft during the landing, such as the rotorcraft-ground interaction and the development and optimization of landing gears design.

Crashworthiness related research activities date back to the pioneering work of Hugh De Haven in the 1940s [8], which provided design guidelines that are still pertinent [9,10]. Aviation crash dynamics research, besides the experimental crash assessment of already existing aircraft, is aimed at providing meaningful design guidelines to make the aircraft structures crashworthy, minimizing the harm to the passengers [11,12], as illustrated in the Aircraft Crash Survival Design Guide [3], which is considered to be a milestone in its field [9]. Apart from full-scale crash experimental tests, in recent decades, research activity has been mainly oriented towards the development of analytical/computational tools for the accurate simulation of an airframe structural response to crash loads [13–15]. Resources have been invested in the validation of numerical simulations and models [14,16,17], which are nowadays supported by powerful, efficient and relatively economic calculators. Virtual modeling and simulation allow us to evaluate numerous situations that are economically unfeasible with full-scale crash testing. However, full-scale experimental tests are still the unique procedure which allows the numerical models to be validated. The National Aeronautics and Space Administration (NASA) has been one of the major players in the crashworthiness related research activities. The Impact Dynamics Research Facility (IDRF) located at NASA Langley Research Center in Hampton, Virginia, witnessed decades of full-scale crash experiments (it was converted into a full-scale crash test facility for light aircraft and rotorcraft in the early 1970s) [18]. A significant example of FE code validation and simulation procedures is represented by a research project which was initiated to demonstrate the capabilities of the state-of-the-art commercial crash simulation codes in predicting the dynamic structural response of the Sikorsky ACAP (Advanced Composite Airframe Program) helicopter, during a full-scale crash test, in 1998. The objective of the crash simulation was to evaluate the capabilities of the code (MSC.Dytran) in predicting the response of a composite airframe subjected to impact loading. The numerical results were correlated with the experimental data to validate the simulation [19]. The level of agreement obtained between the experimental and analytical data ensured the numerical modeling to be an effective tool for the design and certification of crashworthy aircrafts structures. More focused activities have investigated the crashworthiness of single helicopter components or sections, such as in [17,20–22]. Landing gears—mainly skid landing gears—also have been included in the crashworthiness research programs, aiming at providing and optimizing effective structures for safe landings [23–27].

Research activities related to the dynamic response of the rotorcraft during the landing usually focus on (i) the helicopter terrain interaction, (ii) the landing gear design and optimization, whether the landing gear is wheel or skid equipped. These two topics are generally linked and usually rely on models describing the rigid dynamics of the aircraft, neglecting the deformability of structures. Concerning the landing gear related topics, in 1981 Bell Helicopters Textron published a technical report which presented the results of an investigation summarizing the landing gear criteria for helicopters [28]. The investigation was conducted in two phases: the first phase constituted a summary of a literature survey and the second phase consists of a design study of various landing gear configurations. Useful information on the landing gear characteristics identification is also present in [3,29]. Skid landing gear operations were investigated in [25], where a multibody model of a skid landing gear is presented; plastic bending deformations of structural members, damper behavior, and the characteristics of the attachments with fuselage are reproduced. Simulations in different landing conditions are carried out, and the outcomes are compared with experimental results. The investigation is performed in various attitudes and soil conditions, and the sensitivity to soil friction factor is investigated. Another work concerning a skid landing gear was published in 2007 by the NASA, describing an experimental program to assess the impact performance of a skid gear for the use on the Wasp kit-built helicopter [23]. In recent years the application of automated robotic landing gears has seen the light. For instance, in [30] a novel solution for the hard landing mitigation is proposed: the implementation of a robotic legged landing gear system, which aims at softening the hard landings by acting as a shock absorber with a relatively large stroke, thus allowing the aircraft to decelerate over a much larger distance compared with a traditional landing gear system. The investigation of such a system was explored using a multibody dynamics simulation tool. An exhaustive study on the modeling of an articulated robot landing gear was given in [31]. Concerning the dynamic landing response, a study dealing with the analysis of the helicopter-terrain interaction is carried out in [32]; this work analyzed the significant aspects related to the interaction between rotorcrafts and the terrain using a typical medium weight helicopter, with detailed wheel landing gear and full rotor dynamics, during significant maneuvers. That work aimed to investigate the interaction between the landing devices and the terrain under realistic conditions, and analyzing the effects of realistic ground loads on the rotor components. Other works specifically focused on the rotorcraft harsh landing proposed interesting landing simulation techniques [33,34], introducing on-board monitoring techniques (e.g., health monitoring systems) able to assess the structural integrity of the helicopter fuselage [33–37]. In [38,39] a tool for aircraft hard landing detection is suggested, being able to reveal landing gear overloads.

The present work aims at filling a void in the aircraft landing simulation open literature, proposing an effective and efficient multidomain strategy for the landing simulation and providing an objective evaluation of the proposed methods.

3. Method Overview

Sequential field-coupling is the combination of analyses from different engineering disciplines that interact to solve a multidisciplinary engineering problem. When the input of one analysis belonging to one field depends on the results from another analysis belonging to another domain, the analyses are coupled. Sequential refers to the fact that the simulations are solved one after the other; the results from one analysis become the input parameters (e.g., loads and boundary conditions) for the next analysis. In the present case, the coupling is unidirectional, since data are flowing only forward, and no backpropagation nor iteration procedures take place. The pursued approach to the landing problem investigation can be thus defined as a "hybrid" or "sequentially-coupled" analysis approach, since its development involves the exploitation of two different interacting fields, resulting in a solution depending on a two-domain stepwise simulation procedure. Indeed, the complex landing problem is divided into two separate domains, which are ruled by their respective numerical models. The former domain is the dynamic one, which is governed by the multibody model of the rotorcraft. The multibody model takes as an input the landing event/aircraft characterizing features (i.e., the rotorcraft vertical touchdown velocity, mass distribution, landing attitude and weight to lift ratio), providing (i) the dynamic response to the landing and (ii) the proper parameters to feed the structural domain with, to perform the landing structural response. The latter domain is the structural one, which is ruled by a finite element model, whose loads and boundary conditions are derived directly from the dynamic analysis, as Figure 1 shows. Therefore, the described method works according to a feedforward scheme, in which the dynamic domain precedes the structural domain, providing this last domain the proper parameters to assess the landing from a structural point of view. According to the presented approach, the two domains are related in both the chronological and causal planes. Indeed, there is a clearly defined time-based relation between the two, since the dynamic domain always precedes the structural domain, besides an evident cause-effect relationship, in which the dynamic domain provides the structural domain with the parameters to perform the structural response. This strategy allows a complex multidisciplinary problem to be decoupled into two simpler monodisciplinary subproblems, making its understanding and management more effortless.

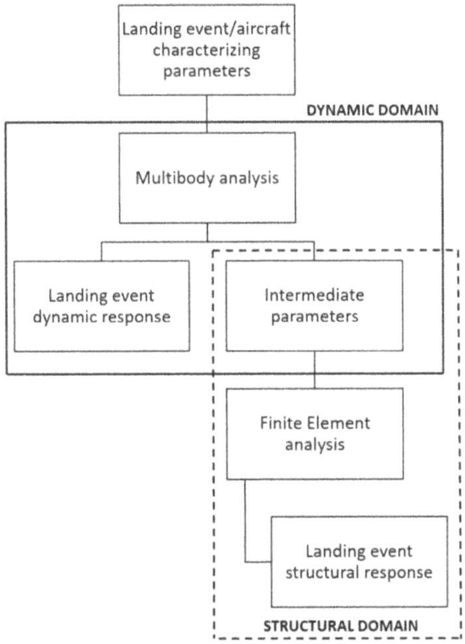

Figure 1. The "sequentially-coupled" method schematics.

The following advantages are achieved: (i) the construction of a single overcomplex multidomain model is avoided; (ii) the two domains are fully exploited and faithfully developed, taking advantage of the specific features of each numerical model and related simulation software; (iii) detailed understanding of the dynamic response is available at a low cost, both in terms of computational power and time consumption, allowing sensitivity analyses based on the landing characterizing parameters. However, these advantages come at a cost. Indeed, the main problem associated with the present approach lies in the fact that the assumptions made in the dynamic analysis (i.e., the multibody model), which provides the structural domain (i.e., the finite element model) with the proper parameters to perform the structural response to the landing, inherently affect the structural response. Moreover, the selection of the appropriate parameters linking the two domains is not trivial, and great care has to be taken in the choice.

Concerning the intermediate parameters relating the two domains, two different approaches were followed; the former relies on a displacement control strategy, in such a way that the multibody analysis provides the finite element model with the displacements in time of some characteristic points of the airframe, namely the landing gears-fuselage attachment points. These displacements represent the boundary conditions governing the finite element analysis, which is thus time-dependent, and was performed via an implicit dynamic algorithm. The main difficulty associated with this approach lies in the fact that the multibody model is composed of rigid bodies. Therefore, enforcing these displacements in the finite element model, results in a fictitious stiffness increase of the airframe.

To overcome the issues related to the displacement control strategy, a second approach was developed, which relies on a load control strategy rather than on a displacement control strategy.

Therefore, the intermediate parameters are represented by the forces exchanged by the landing gears and the fuselage. The structure being loaded and unconstrained, the inertia relief method was used [40]. In finite element analysis, the inertia relief method provides the response of an unconstrained structure subjected to constant or relatively slowly varying external loads, with static analysis computational costs. According to the inertia relief method, the unconstrained structure is assumed to be in a state of static equilibrium; indeed, an acceleration field is computed to counterbalance the applied loads. A set of translational and rotational accelerations provide distributed body forces over the structure in such a way that the sum of applied forces and the sum of moments are zero. To obtain accurate inertia relief calculation, the periods of applied loads should be much higher than the periods of rigid body modes restrained [40], so that the dynamic structural effects can be legitimately neglected. If this condition is not met, the inertia relief method might provide not meaningful results, since the stress component associated with the dynamic structural response is unaccounted for. This second approach is not exempt from issues; indeed, it does not account for the dynamic structural effects, and, unlike dynamic analyses, it is not time-dependent.

4. Experimental Activity

The experimental activity consists of a set of full-scale drop tests of a medium-size Mil Mi-8 helicopter (Figure 2), in the fully-equipped full-load configuration, with gradually increasing drop heights. The Mil Mi-8 helicopter is a medium-size twin-turbine helicopter with a semi-monocoque airframe. The structure consists of an external aluminum alloy skin, stiffened with frames and stringers. The rotorcraft is provided with a wheeled landing gear structure equipped with oleo-pneumatic shock absorbers. The main landing gear (MLG) is equipped with a two-stage oleo-pneumatic absorber system, i.e., two oleo-pneumatic shock absorbers placed in series. The lower cylinder has a low-pressure low-velocity gas chamber while the upper cylinder has a high-pressure high-velocity chamber. The nose landing gear (NLG) is composed of a single oleo-pneumatic shock absorber. Since the test required the helicopter to be in a configuration as realistic as possible, the major subcomponents of the helicopter were installed on the airframe, e.g., engines, gearbox, fuel equivalent mass, etc., and additional weight representing the helicopter freight was added in the form of sandbags in the cargo compartment. The main and tail rotor blades were substituted with equivalent lumped masses. A standard gantry crane was used to perform the set of drops. The helicopter was connected to the gantry crane release mechanism employing a special lug linked to the main rotor shaft-end.

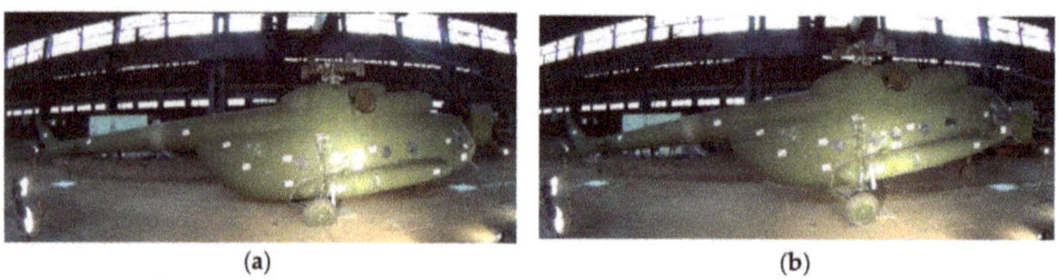

(a) (b)

Figure 2. Experimental activity. Helicopter during: maximum strut compression (**a**) and front-wheel spring back (**b**).

The drop height was gradually increased to raise the touchdown vertical velocity. The test was executed with the helicopter in the full weight configuration, with a total weight approximately equal to 11,000 kg.

The experimental test consisted of a set of six drops (see Table 1), with an increased drop height, starting from a 0 m height, in which the main landing gears wheels barely touched the ground and the shock absorbers were fully extended, up to a 0.78 m height, which is characterized by a vertical touchdown velocity slightly exceeding the critical one, which identifies the onset of the harsh landing regime, as stated in the introduction, according to [1–3]. The test configuration provided a vertical touchdown velocity direction, as for the majority of rotary-wing landings, in which the forward speed was either null or negligible. The test configuration was characterized by a zero-roll angle and a positive pitch angle equal to approximately 4 degrees, which caused the main landing gear wheels to impact the ground slightly before the nose landing gear wheels.

Table 1. List of drop experiments carried out.

Drop Height (m)	Freefall Touch Down Velocity (m/s)		
	Theoretical	Marker on Wheel	Error (%)
0.00	0.00	0.00	0
0.10	1.40	1.39	−0.71
0.25	2.21	2.19	−0.90
0.35	2.62	2.61	−0.38
0.48	3.07	3.01	−1.95
0.78	3.84	3.81	−0.78

To acquire data from the experimental activity, a sensor network was designed and installed on the helicopter.

The experimental measurements, besides displaying the dynamic and structural response of the aircraft to the drop experiment, also allowed to link together the drop landing parameters with (i) the parameters characterizing the dynamic response of the airframe and (ii) the stress level of the structure. Furthermore, experimental data are essential to assess and validate the numerical models of the aircraft itself and are also crucial to identify unknown or uncertainty affected parameters, as it is evidenced in the following sections. Two main categories of sensors were installed: (i) dynamic response acquisition sensors, i.e., fast camera markers, laser distance meters and shock absorbers length sensors, and (ii) structural response acquisition sensors, e.g., fiber Bragg grating (FBG) and strain gauges (SG) sensors. The former class of sensors was aimed at providing a collection of data that illustrate the dynamic behavior of the aircraft during the experimental activity. In contrast, the latter class provided data revealing the structural strains resulting from the landing event loads. Laser distance meters were fixed on the outside of the fuselage, measuring the distance between the sensor and the ground; they were located in three different points: in front of the nose landing gear, on the fuel tank in the proximity of the main landing gear, and on the tail boom extremity (Table 2). The landing event was also recorded with fast camera instrumentation (1000 frames per second); using this equipment, the displacement in time of some markers placed on the fuselage was derived. The fast camera markers were located near the main landing gear, more precisely at the fuel tank level, one forward and one rearward with respect to the main landing gear. The last camera marker was placed on the main landing gear wheel (Table 2). Shock absorbers length sensors were placed across the main landing gear shock absorbers (two shock absorbers in series configuration), measuring the stroke of the double stage suspension system (Table 2). The FBG sensors were placed in the most loaded areas, namely the frames 7 and 10 and the tail boom (Figure 3, Table 3); thus, three arrays of strain measuring sensors were available, for a total of 48 sensors, of which two were designed for temperature compensation. The FBG sensors located on the frames measured the strain along the frame circumferential direction; the ones located on the tail boom, instead, measured the axial strain with respect to the tail boom geometry. Finally, SG sensors were located side by side with some FBG sensors, to cross-check their measurements. Further experimental measurements consisted of optical scanning of the skin surface in some specific areas, performed after each drop event. Structural deformation analysis was achieved by comparing the same scanned surface morphology before and after the landing event, to detect eventual permanent (i.e., plastic) deformations [34,36].

Table 2. Dynamic response sensors.

Sensor Type	Sensor Location	Sensor Number
Laser distance meter	Nose landing gear (NLG)	1
Laser distance meter	Tail boom free-end	1
Laser distance meter	Lateral fuel tank (MLG) (right and left sides)	2
Fast camera marker	Wheel center (right side)	1
Fast camera marker	Fuselage (MLG) (right side)	1
Fast camera marker	Auxiliary fuel tank (right side)	1
Shock absorbers length sensors	MLG absorbers	4 (2 per side)

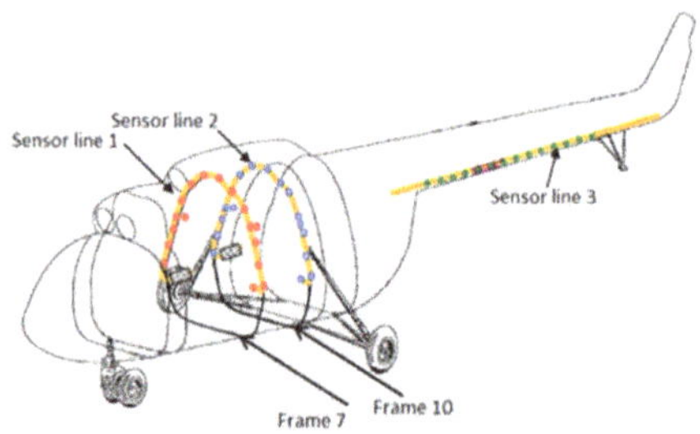

Figure 3. Fiber Bragg grating (FBG) sensor lines location.

Table 3. Structural response sensors.

Sensor Type	Sensor Location	Sensor Number
FBG	Frame 7	17 (included T sensor)
FBG	Frame 10	17 (included T sensor)
FBG	Tail boom	14

5. The Dynamic Domain

The multibody model, being part of the dynamic domain, is aimed at providing the dynamic response of the rotorcraft to the landing event.

5.1. The Multibody Model

The model is made up of rigid interacting bodies representing the key features of the aircraft (Figure 4). The semi-monocoque structure is described as a rigid body to which the computed inertial properties of the real fuselage structure were assigned. The landing gears were faithfully modeled, including the tire behavior. The rigid body assumption made for the fuselage structure arises from the fact that the landing gear performance, and, consequently, the general aircraft behavior, appears to be relatively unaffected by the elastic deformation of the structure [29,41], also because the landing gears fuselage attachments were located in strengthened zones, whose stiffness was thus increased. The present assumption legitimacy depends on the investigated landing regime; for instance, the more the vertical touchdown velocity is increased, the more influential the elastic structural deformation becomes. To faithfully reproduce the experimental drop test via the multibody model, this model has to devotedly reproduce the real aircraft features characterizing its dynamic behavior, i.e., its inertial properties and shock absorbers characteristics.

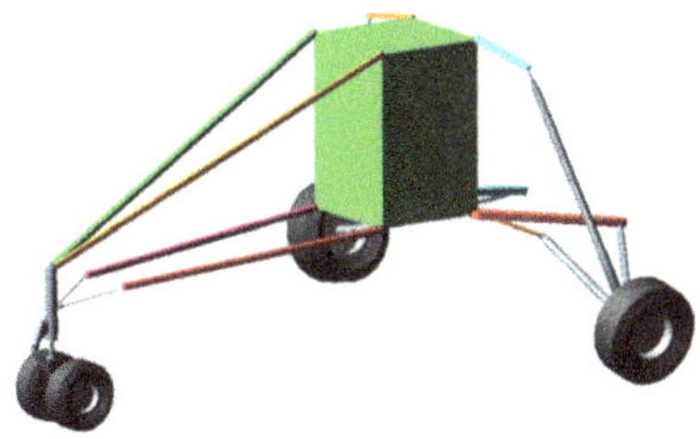

Figure 4. Mil Mi-8 helicopter multibody model.

5.1.1. The Inertial Properties

The multibody model inertial properties were extrapolated from the finite element model, which is characterized by having an accurate representation of the fuselage geometry, which was obtained via reverse engineering techniques. The semi-monocoque structure alone, including the tail boom, stabilizers, empty fuel tanks, etc., weighs approximately 1450 kg; to reach the experimental aircraft drop test weight of 11,000 kg, which is the in-service total full-load weight, a set of feature-representative masses was added to the FE model of the airframe. These masses include both the helicopter equipment and the payload. Some of the masses were modelled with solid elements, and shaped similarly to the real ones, to devotedly reproduce their inertial properties, e.g., the engines, the main rotor, the gearbox, etc., while others were modelled by one-dimensional elements, e.g., the payload, the tail rotor transmission shaft, the on-board instrumentation, etc. This process is by its very nature relatively inaccurate, and potentially introduced some errors in the modeling of the aircraft. However, the computed inertial properties are assumed to be estimated sufficiently well for the objective of this work.

5.1.2. The Shock Absorbers Characteristics

The shock absorbers characteristics were calculated according to theoretical formulas and practical considerations [29] and were then entered in the multibody model software (MSC ADAMS). The elastic and damping characteristics were displayed as a set of force vs. displacement and force vs. rate of displacement, respectively, as illustrated in Figure 5. The forces acting on the strut are here illustrated:

$$F_S = F_h + F_a + F_f \tag{1}$$

where: F_h is the hydraulic force, F_a is the pneumatic force and F_f is the friction force. The hydraulic force in the shock strut, which is responsible for the damping characteristic, results from the pressure difference associated with the oil flow through the orifice in the oil chamber. In a landing gear, the orifice area is usually small enough in relation to the diameter of the strut so that the jet velocities and Reynolds numbers are sufficiently large that the flow is fully turbulent. As a consequence, the damping force varies as the square of the displacement rate rather than linearly. Therefore, through the Bernoulli's principle and considering the continuity equation, the hydraulic force can be derived:

$$F_h = \frac{\dot{s}}{|\dot{s}|}(p_h - p_a)\,A_h = \frac{\dot{s}}{|\dot{s}|}\,\frac{A_h^3 \dot{s}^2 \rho}{2(\,C_d\,A_n)^2} \tag{2}$$

where: C_d is the coefficient of discharge through the orifice, A_n is the net orifice area (assumed here to be constant), ρ is the hydraulic fluid density, A_h is the hydraulic area, i.e., the net area on which the pressure difference is acting, and $\dot{s}$ is the displacement rate of the shock strut. The net orifice area A_n may be either a constant value (as assumed here) or, when a metering pin is used, can vary with the strut stroke:

$$A_n(s) = A_0 - A_p(s) \tag{3}$$

where A_0 is the area of the opening in the orifice plate and A_p is the metering pin cross section area in the orifice plate plane. Generally, the coefficient of discharge could vary because of the (i) variable net orifice area, (ii) any change in the exit and entry conditions due to variations in the length of the chambers, and (iii) because of variations in the Reynolds number of the flow. These considerations are usually neglected when dealing with constant orifice area, under the assumption of constant discharge coefficient.

The pneumatic force gives to the strut an elastic contribution, being the force dependent on the stroke. The force is determined by the initial chamber inflation pressure, the area subjected to the air pressure, and the instantaneous compression ratio. According to the polytropic law for compression of gases:

$$P_a\, V_a^n = P_{a0}\, V_{a0}^n = constant \tag{4}$$

where: P_a is the air pressure in the upper chamber of the strut, P_{a0} is the air pressure in the upper chamber of the strut when completely extend, V_a the air volume of the shock strut, V_{a0} the air volume of the shock strut when wholly extended and n is the gas polytropic exponent. Since the instantaneous volume of the shock strut V_a is equal to the difference between the initial volume V_{a0} and the stroke s times the pneumatic area A_a, it is possible to write:

$$P_a = P_{a0}\left(\frac{V_{a0}}{V_{a0} - s\, A_a}\right)^n \tag{5}$$

$$F_a = P_a A_2 \tag{6}$$

where: F_a is the pneumatic force and A_2 is the external cross-sectional area of the inner cylinder. The gas polytropic exponent n depends on the compression rate and the rate of heat transfer from the gas to the surrounding environment. Low compression rate can be associated to isothermal compression, resulting in values of n approaching the value of 1; higher values of n, limited by the adiabatic compression value of 1.4, are typical of higher stroke rates. Here the compression was considered to be adiabatic ($n = 1.4$), since the stroke rates were relatively high, and the heat transfer phenomena could be neglected. The force term associated to the friction was neglected.

Since not every landing gear parameter was accurately known or easily identifiable and some simplifying assumptions were inevitably made in Equations (1)–(6), a parameter identification procedure was carried out to adjust the multibody model shock absorbers characteristics to mimic the experimental dynamic behavior. This adjustment process was performed on the 0.48 m drop, which was selected as the reference drop, since it exhibited the touch down critical velocity that identifies the onset of the harsh landing regime, thus obtaining the elastic and damping characteristics in Figure 5. Then, a comparison check was carried out for the whole experimental drop sequence, as shown in the next section.

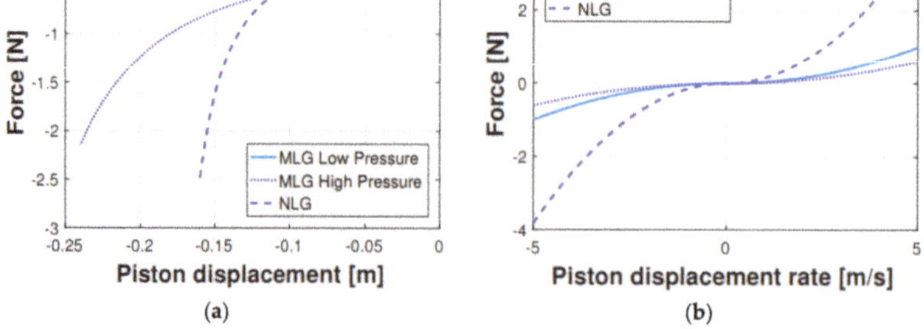

Figure 5. Shock absorbers elastic (**a**) and damping (**b**) characteristics.

5.1.3. The Model Verification

The vertical displacement vs. time of the characteristic points listed in Table 4 is a good indicator of the whole aircraft dynamic behavior. The experimental drops measurements come from the laser distance meters and the fast camera markers. These measurements were taken as comparison parameters. The fuselage was assumed to be rigid, and the fuselage motion was considered symmetrical to the plane containing the roll and yaw axes, which is reasonable given the fact that the aircraft is almost symmetric to the roll-yaw plane and considering that the landing was initialized with a zero-roll angle. A plane roto-translation can thus approximate the motion of the aircraft in the roll-yaw axes plane.

Table 4. Comparison sensors.

Sensor Type	Sensor Location
Laser distance meter	Nose Landing Gear (NLG)
Fast camera marker	Wheel center
Fast camera marker	Fuselage (MLG)
Fast camera marker	Auxiliary fuel tank

Figure 6 is showing the vertical displacement vs. time of the sensors listed in Table 4, specifically for the 0.48 m drop. According to Figure 6 the multibody model satisfactorily reproduced the experimental dynamic response of the helicopter. Finally, a quantitative evaluation of the comparison is presented in Figure 7, where the root mean square error (RMSE) concerning the experimental and numerical results is shown. The RMSE was computed for each investigated sensor and including each drop height. It is useful to notice that even if the shock absorbers set-up procedure was carried out on the 0.48 m drop, the multibody model satisfactorily matched the experimentally measured behavior in the whole drop range.

Apart from the displacement-based comparison (Figure 6; Figure 7), another parameter was used as a reference to prove the legitimacy of the multibody model output parameters. This further comparison parameter is represented by the main landing gear strut loads (Figures 8 and 9). The experimental forces acting on the main landing gear struts were computed via the shock absorbers stroke measurements, which were then converted—using the shock absorbers elastic and damping characteristics illustrated in Figure 5—in their related forces. In Figure 8, the comparison regarding the main landing gear forces—experimental and numerical—is shown for both the right and left side forces. The comparison refers to the 0.48 m drop. Subsequently, in Figure 9, the percentage error between the experimental and numerical peak values of the said forces was computed and illustrated.

The comparison shows that the multibody model is reproducing the real helicopter landing behavior quite accurately, both with regards to the displacements and the loads governing the structure dynamics. This entails that the intermediate parameters, whether they are the displacements or the loads acting on the fuselage, provide the finite element model with proper boundary conditions. It is finally worth noticing that further error reduction can be obtained by adopting more sophisticated models for the landing gear strut load prediction and more advanced optimization strategies, which are, however, outside the scope of the present activity.

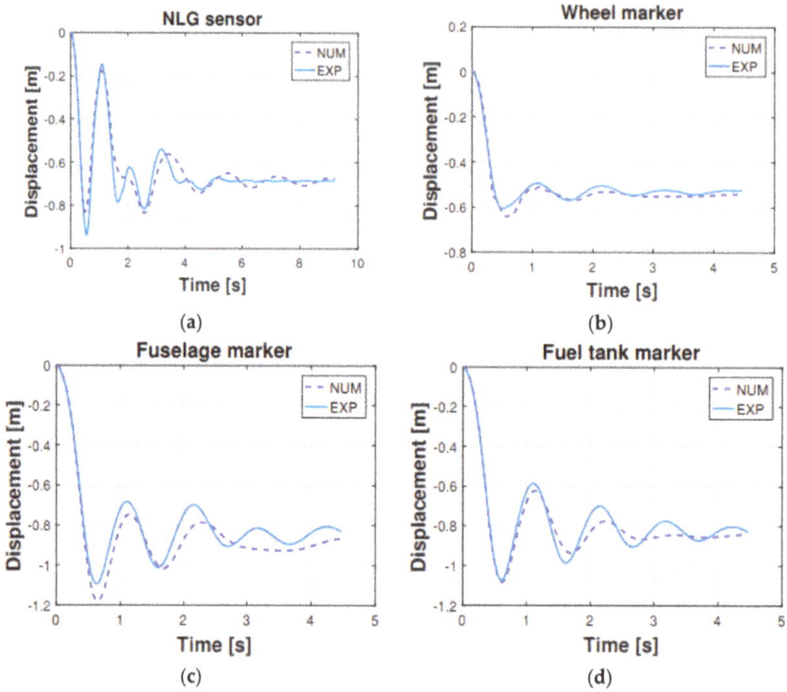

Figure 6. Numerical model (multibody) vs. experimental measurements of the vertical displacement for the 0.48 m drop. Displacement refers to sensors in Table 4.

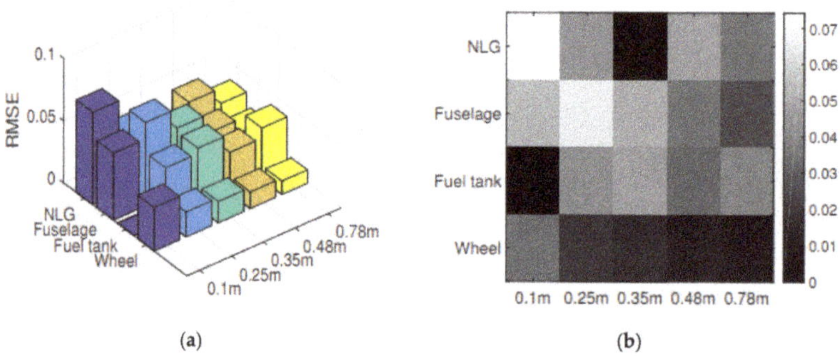

Figure 7. Root mean square error (RMSE) bar plot (**a**) and colormap (**b**).

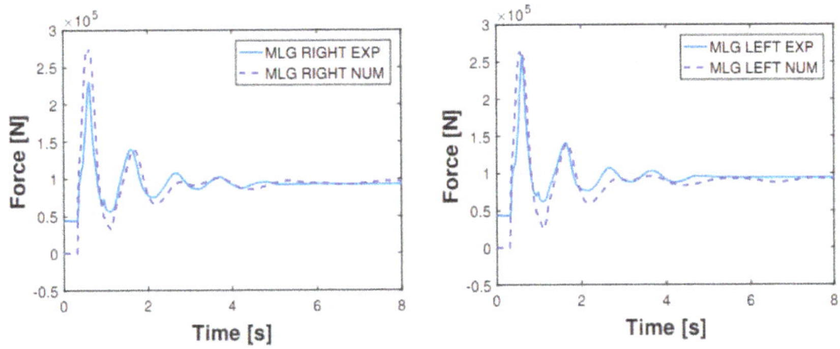

Figure 8. Main landing gear (MLG) shock absorbers total force for the 0.48 m drop: (**a**) right side; (**b**) left side.

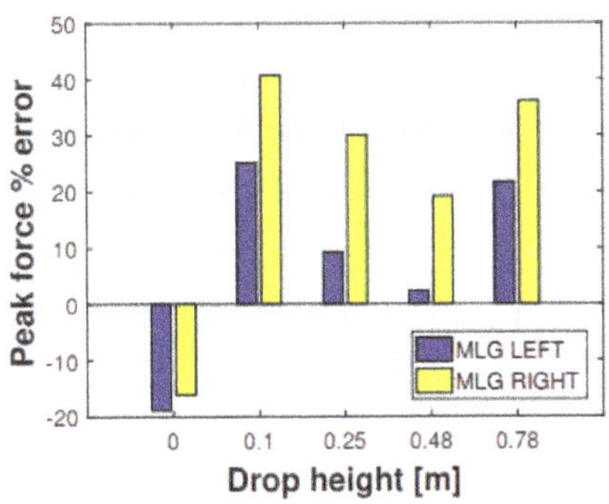

Figure 9. Percentage error between the experimental and numerical (multibody) MLG (right and left) peak forces for each drop height.

6. The Structural Domain

The structural domain, which relies on finite element analysis procedures, is meant to provide the rotorcraft structural response to the landing event. A clear and comprehensive view of the structural response is fundamental when landing is assessed or classified, whether the dynamic failure criteria (e.g., fatigue criteria) is concerned or the static damage related to overload and immediately following the landing itself, enabling a reliable diagnosis of the event and providing data for the residual life prognosis.

6.1. The Finite Element Model

The FE model was developed starting from the helicopter fuselage geometry, which was obtained employing reverse engineering techniques, and then imported in the Abaqus CAE software (Figure 10). The model consists mainly of four-node shell elements with reduced integration (S4R) and an hourglass control, characterized by a generic edge length equal to 30 mm. Some additional characteristic features which were not present in the preliminary model were added, e.g., the gearbox and its sustaining frame structure, the engines, openings in the fuselage, etc. Finally, a set of distributed masses was annexed to the fuselage to take into account the items which do not directly belong to the fuselage but were present in the experimental test, e.g., the aircraft equipment and the payload. These masses are crucial in determining both the helicopter dynamic and structural performances since they represent a large share of the total helicopter mass (87%). These additional features were modeled as lumped masses or via solid elements faithfully reproducing their key features, depending on their relevance on the global inertial properties. The fuselage is mainly made of aluminum alloy (2024 T3 alloy), whose mechanical properties are listed in Table 5.

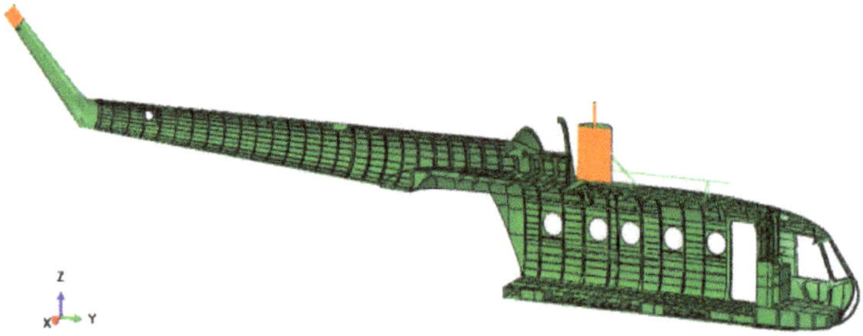

Figure 10. Mil Mi-8 helicopter FE model cutaway view.

Table 5. Fuselage material properties.

Parameter	Value
Yielding stress	335 MPa
Ultimate tensile stress	512 MPa
Young's modulus	71,000 MPa
Poisson's modulus	0.33

6.2. The FE Model Static Verification

Preliminary verification of the FE model was carried out comparing the strain measurements retrieved from the experimental activity with the corresponding numerical strains. The comparison refers to the static configuration of the helicopter resting on the ground after the landing event. Thus, the helicopter is laying on its landing gears in a static equilibrium configuration. The strain sensors (FBG sensors) location was qualitatively shown in Figure 3 and more accurately in Figure 11. The whole experimental drop data set was used, meaning that the experimental strain database includes the static "rest on-ground" configuration for each drop test, thus providing six strain samples referring to each drop experiment for each strain sensor. The relative data dispersion is shown in Figure 12.

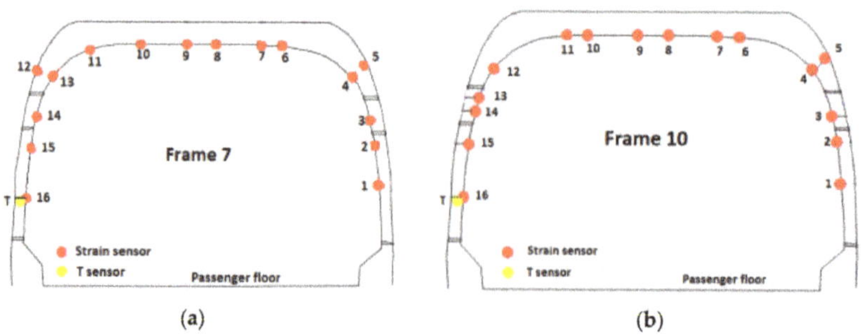

Figure 11. *Cont.*

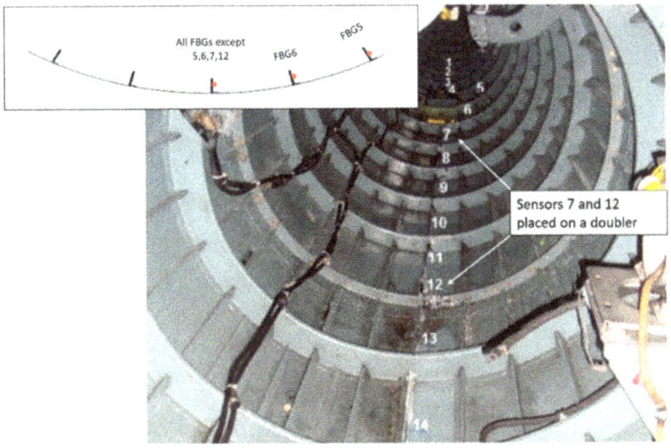

(c)

Figure 11. Sensor location for: (**a**) frame 7, (**b**) frame 10 and (**c**) tail boom.

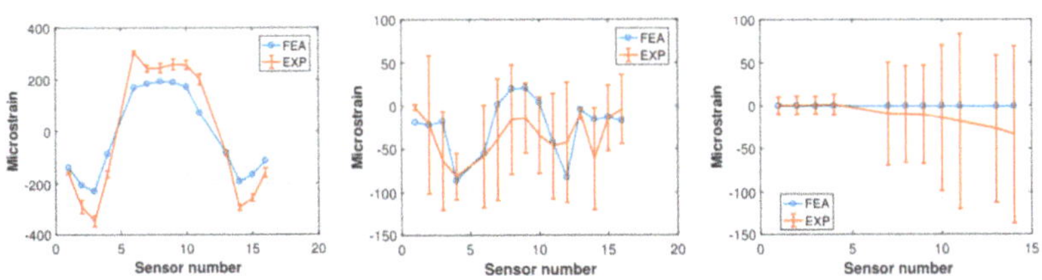

(a) (b) (c)

Figure 12. Experimental (EXP) vs. finite element analysis (FEA) strain comparison for (**a**) frame 7, (**b**) frame 10, and (**c**) tail boom.

The measured strains were calculated as perturbations with respect to the helicopter in the hung configuration (prior to the drop); as a result, the measured strain field is null in the hung configuration. The measured strain tensor can be thus formally defined as:

$$[\varepsilon_M(t,x)] = [\varepsilon_A(t,x)] - [\varepsilon_H(t,x)] \tag{7}$$

The strain tensor is generally time and location dependent. As Figure 12 illustrates, frame 7 shows the best match between the FE analysis outcomes and the experimental results. In contrast, frame 10 seems to be highly sensitive to the environmental noise, having a higher data dispersion. Still, the experimental data dispersion envelope of frame 10 includes the FE strain values. The tail boom numerical strains get more accurate as the tail rotor is approached (the tail boom sensor numbering increases from the tail boom free end to the fuselage connection). In contrast, in the fuselage attachment zone the bias and the data dispersion are higher. Again, the FE strain values were included in the experimental envelope.

To objectively comment on the already presented results alone, some considerations about the FE model have to be pointed out. The finite element model suffers from having a mass distribution that is not supported by accurate data or measurements. This, in turn, affects the numerically computed structural response of the rotorcraft, since the mass distribution is crucial in determining the stress partition on the structure. Furthermore, the real aircraft structure is made up of riveted or bolted connected subcomponents; however, the FE model fuselage does not display such features. The FE model fuselage is made as a unique body, and there are no joints of any kind. Thus, the FE model exhibits an increased structural stiffness with respect to the real riveted structure, which might cause the load paths to deviate from the real ones. This increased stiffness also affects the dynamic structural behavior, since, besides having a higher stiffness, even the characteristic damping phenomena of the riveted structures are not considered. In the FE model, the modeling of substructures like the one supporting the main rotor and its gearbox, or the main landing gear strut, was proven to considerably affect the strain measurements on the frames 7 and 10. The said structures are directly in contact with frames 7 and 10, respectively, thus their influence on the investigated frames is more significant. The connection between the substructures mentioned above and the fuselage was revealed to be a central parameter, although its modeling, based on the real structure, is not straightforward.

6.3. *The Dynamic Structural Analysis*

The maximum loads acting on an aircraft structure generally occur when the aircraft is undergoing some form of acceleration, such as during landing, take-off, and other common maneuvers. From a structural point of view, the dynamic landing phase is crucial, especially when the landing is harsh, since it generates the highest loads; it is thus fundamental to focus on it. According to the present analysis approach, the loads and boundary conditions defining the structural domain are the results of the multibody analysis, while the structural strength assessment was achieved using the finite element analysis. As anticipated in Section 3, in this work, two different finite element analysis approaches were used; the former is based on a displacement control strategy, while the latter operates in the load control domain.

6.3.1. The Displacement Control Strategy

The dynamic domain includes all the landing related phases in which the aircraft is not in a static equilibrium configuration. In the dynamic phase, unlike in the static one, the rotorcraft body is not constrained, and the acting loads are balanced by the inertial reaction of the structure, according to d'Alembert's principle.

The finite element model is the same as in the static finite element analysis; however, as the present finite element analysis was performed in the dynamic domain, a different approach was required. The basic idea behind the dynamic finite element analysis is to provide the model with the landing event loads and the boundary conditions, i.e., the displacements in some characteristic points, which result from the multibody analysis. Thus, the finite element model was provided with the displacement in time of the nine landing gears-fuselage attachment points, which are illustrated in Figure 13. The unique external load acting on the structure was the gravity load. The step was defined as implicit dynamic, neglecting the nonlinear geometric effects of large displacements, which can be ignored in the investigated scenario, especially with reference to the fuselage.

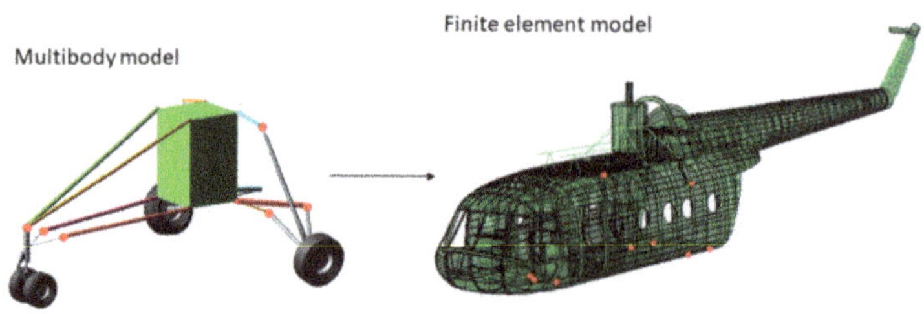

Figure 13. The displacement control approach schematics. Red dots represent the control points, i.e., the landing gear-fuselage attachment points.

The results of the displacement control strategy are shown in Figure 14; the most critical time frame for the 0.78 m drop was shown to be 0.67 s after the release of the helicopter. This approach was affected by some relevant drawbacks:

1. The imposed motion of the landing gear-fuselage attachment points was derived from the multibody analysis. As a matter of fact, with the control point displacements derived from a rigid body model, their relative displacements were null. This, in turn, increased the finite element model structure stiffness above the actual stiffness by having a completely rigid frame connecting the control points. The abruptly increased stiffness of the structure heavily influences the loads, which travel along the stiffest pathways, which were here strongly affected by the imposed constraints. This, in turn, also affected the stress distribution.
2. Imposing the motion of the fuselage control points (and considering the rigid body assumption) made the landing gear structure geometry irrelevant. The fuselage was rigid in the multibody model, and therefore in the imposed displacements derived after the multibody analysis no evidence of the forces not parallel to the above displacements was identified.

These considerations were corroborated by the experimental activity findings, which derive from (i) the structure examination after each drop event and (ii) the FBG strain measurements. Concerning (i), some selected external skin regions were optically scanned after each drop attempt (an example is illustrated in Figure 15), to record detectable eventual permanent skin deformations. Structural deformation analysis was achieved by comparing the same scanned surface morphology before and after the landing event. The comparison procedure resulted in a color map of the scanned regions, which indicates the level of permanent deformation of the surface itself. The scan procedure was repeated after each drop event; thus, for each drop event, the progressive damage with respect to the previous conditions was recorded. The result of the optical scanning procedure outcomes was that only the 0.78 m drop produced any permanent structural deformation, which, however, was very localized and was not captured by the global finite element model (Figure 15); a dedicated sub modeling strategy could potentially resolve this issue. According to the specified inspection campaign, the zones prone to yielding were the following: (i) tail boom-main fuselage connection, (ii) main landing gear strut attachment area. The displacement control analysis, on the contrary, showed large yielded areas.

Regarding the FBG strain measurements, these were compared with the corresponding strain values derived from the FE analysis. The comparison is illustrated in Figure 16. As expected, the comparison reflects the considerations already revealed by the qualitative observation of the FE analysis outcomes and the comparison carried out with the optical scans of the fuselage skin, highlighting the limits of the displacement control strategy. Indeed, the experimental measurements and the numerical values were inconsistent, especially when considering frame 7 and frame 10. The tail boom, instead, as already mentioned above, was less affected by the constraints imposed to the fuselage visbile in the comparison (Figure 16), which clearly showed that the tail boom FE analysis strains were much more consistent with the experimental ones.

The experimental inspection procedures suggest that there is further space for the structural model improvement with respect to the approach based on the displacement control. In the present environment, the displacement control approach was demonstrated to be inaccurate, since it largely overestimated the structural stress level, and also displayed a stress distribution which was biased due to the control points enforced rigid body motion.

Although the above considerations depict the displacement control approach as faulty, this is not necessarily always the case. Indeed, the bias of this approach with respect to the ideal structural behavior was firstly caused by the assumptions forming the basis of its implementation. Therefore, the displacement control approach might be appropriate in other contexts, since its legitimacy depends on:

1. The investigated area. If the area of interest is distant enough from the displacement control points, or if the structure is not affected by the said constraints, the bias might be tolerable, e.g., the tail boom displacement field is not affected by the main body features.
2. The multibody rigid body assumption. This hypothesis might be accurate in a wide range of scenarios. For instance, concerning the helicopter landing event, one may state that the rigid body hypothesis is less valuable with increasing landing velocity.

3. The connectivity of the displacement control points and their number, i.e., how the control region is connected to the rest of the structure. The connectivity of the control points with the rest of the structure is crucial in determining their influence. Additionally, their number is fundamental: the more there are, the higher their impact.

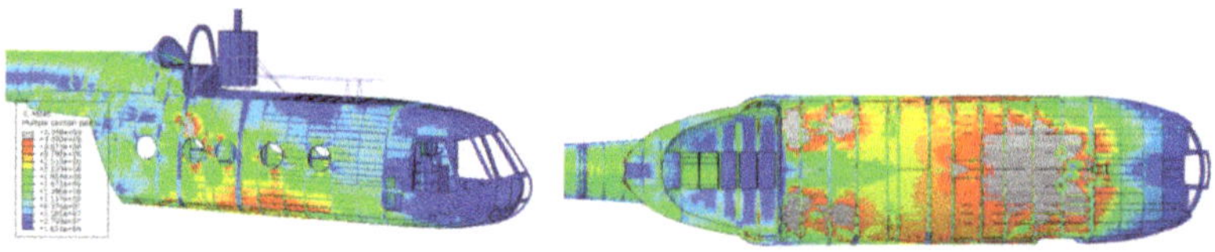

(**a**) (**b**)

Figure 14. Equivalent Von Mises stress field for the 0.78 m drop. Perspective view (**a**), bottom view (**b**).

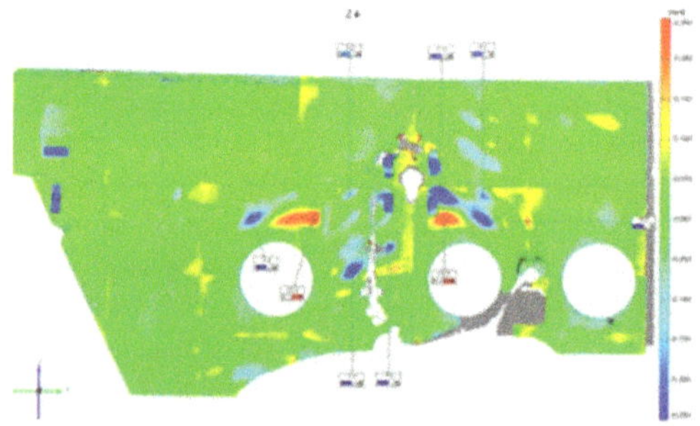

(**a**)

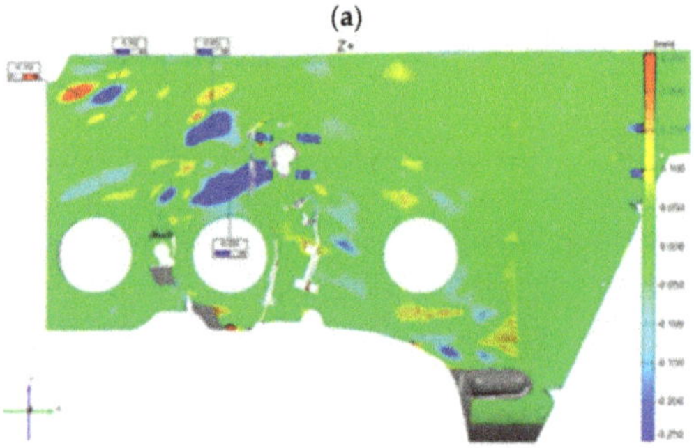

(**b**)

Figure 15. *Cont.*

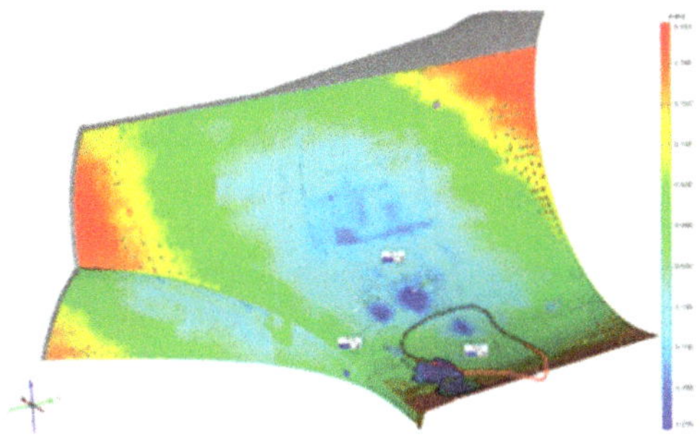

(c)

Figure 15. Optical scans of the fuselage skin after the 0.78 m drop. MLG attachment zone, right (**a**) and left (**b**); tail boom-fuselage connection (**c**).

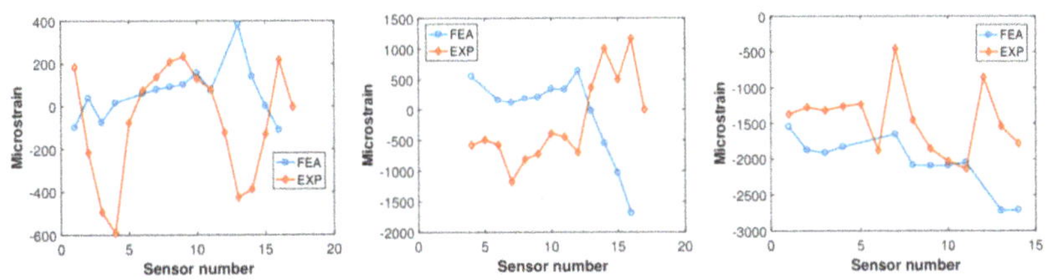

(a) (b) (c)

Figure 16. Experimental (EXP) vs. FEA strain comparison for (**a**) frame 7, (**b**) frame 10, and (**c**) tail boom.

6.3.2. The Inertia Relief Method

Aiming at providing an alternative analysis approach to the displacement control strategy, the inertia relief method (which is supported by Abaqus CAE) was applied to the landing problem. This method enables the analysis of the rotorcraft structure without the need to perform a dynamic analysis, thus permitting the examination of unsupported structures as if they were in a static equilibrium condition, as described in Section 3. Consequently, the inertia relief analysis is not time-dependent, but it investigates a single time instant. The loads are enforced to the landing gear-fuselage attachment points, i.e., the same control points used in the displacement control strategy (Figure 13). In the present analysis, the loads were selected from the multibody analysis load time history, aiming at analyzing the structure under the most critical conditions, which are likely to belong to the time window immediately following the ground impact, namely when the sum of the elastic and damping forces related to the shock absorber is at its highest. Another valuable information which has to be provided to the model is the rotorcraft attitude corresponding to the investigated time history instant. The structure attitude is fundamental, since the aircraft is inserted into the gravitational field, and the direction of the same with respect to the structure coordinate directions is essential. The rotorcraft attitude is controlled by the gravity load vector, whose components are adjusted depending on the required attitude. As stated in Section 3, to obtain accurate inertia relief analyses, the periods of the applied loads should be much greater than the periods of the rigid body modes restrained [40], so that the dynamic structural effects can be legitimately neglected. Inertia relief analysis does not account for the structure dynamic response. Depending on the structure and on the spectrum of the applied loads, the stress/strain contribution related to the structure dynamics might be relevant. Thus, the inertia relief analysis has to be applied cautiously. Here, the analysis was focused on the first instants after the ground impact (i.e., when the main landing gear loads on the structure reach their maximum value); thus, the low frequency-high displacements (i.e., high strain/stresses) contribution on the fuselage was expected not to be yet settled, while the contribution due to the elastic return subsequent the helicopter release was supposed to be irrelevant for the fuselage, while it might affect the tail boom. Ergo, the structural dynamic response does not consistently affect the overall structural response, at least as far as the fuselage is concerned; it is shown in the following section that the same might not apply for the tail boom, which is more prone to dynamic excitation.

The outcomes of the inertia relief analysis showed a completely different scenario with respect to the displacement control strategy results; firstly, the resulting stresses were far lower, and secondly, the stress distribution was rather different (Figure 17). The difference between the displacement control strategy and the inertia relief method is visible upon a comparison of Figures 14 and 17. The stress magnitude overestimation is evident—the 0.78 m drop analysis conducted with the displacement control strategy displays much larger stress magnitudes than the 0.78 m drop analysis performed via the inertia relief analysis—and also the stress distribution is divergent.

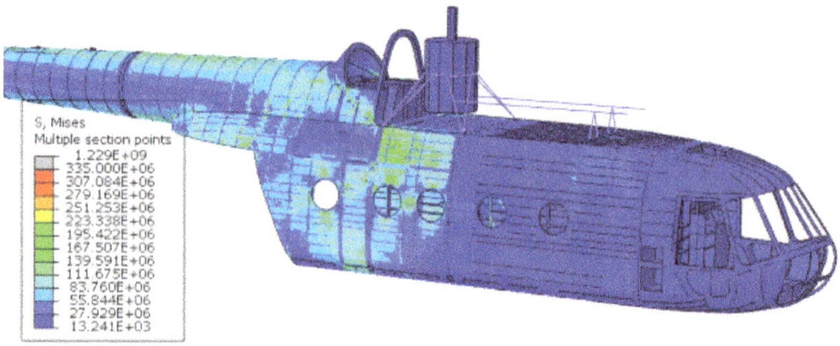

Figure 17. Equivalent Von Mises stress field for the 0.78 m drop.

As already performed for the displacement control strategy, also the inertia relief method outcomes were compared with the experimental test findings, i.e., the fuselage skin optical scans and the FBG strain measurements. First, a qualitative comparison was carried out by observing the optical scanning procedure outcomes, which are presented in Figure 15; the inertia relief analysis provided consistent results (the 0.78 m drop was considered). Indeed, it correctly detected the most stressed areas and their maximum stress magnitude was relatively adherent with the experimental findings (Figure 17). Then, to quantitatively assess the analysis, the strains derived from the inertia relief method analysis were compared to the experimental FBG strain measurements; the comparison was carried out for the 0.78 m drop, considering the time instant in which the main landing gear strut forces display their maximum value, i.e., 0.67 s after the release instant.

Frames 7 and 10, and the tail boom, respectively, were compared (Figure 18). Unfortunately, not all of the experimental sensor measurements were available, since some signals were corrupted entirely by noise, and their corresponding strain measurements were discarded (Figure 18b). The comparison showed that the finite element model was entirely accurate in describing the strains which occur in frame 7, as can be observed for the static comparison (Figure 12). However, some difficulties were encountered in the characterization of frame 10. Indeed, the same kind of mismatch encountered for the static strain comparison occurred, i.e., the experimental strain values corresponding to the central section of the frame (from sensor 6 to sensor 11, Figure 18) showed a different structural behavior with respect to the finite element analysis findings. More precisely, the strain values relative to the experimental measurements were located in the negative strain half-plane, and behaved somewhat irregularly. In contrast, the finite element analysis strain values were positive and approximately around the same strain level. However, this attitude was expected, since the sensitivity of frame 10 to the external loading—the main landing gear strut was directly connected to frame 10—was already identified during the static comparison. Concerning the tail boom strain comparison, it has to be recalled that the inertia relief method is lacking the structural dynamic contribution of the tail boom structure, which behaves quite differently from the fuselage, due to its geometry and boundary conditions. Indeed, as also experimental measurements (i.e., FBG strain measurements) revealed, the tail boom undergoes oscillations as soon as the helicopter is released, since the elastic return is not negligible for the tail boom structure, which can be thought as a cantilever beam with a distributed load (its weight), plus a concentrated load on its free extremity, representing the tail rotor. Thus, when the helicopter is released from the hung position, the loads related to the gravitational field are canceled, and the elastic return of the tail boom structure causes the free vibration of the same according to the initial conditions enforced by the gravitation field. Then, when the impact on the ground takes place, the system gets excited once more, and the structure starts vibrating according to the new conditions imposed by the forces which rule the dynamic landing phase. Thus, the inertia relief method is unsuitable for assessing the tail boom strain field, and it is not generally legitimate for the simulation of systems that, according to the load spectrum, are undergoing non-negligible vibrations. In the present scenario, since the investigated instant occurred immediately after the ground impact, and the major dynamic structural contribution corresponding to the large displacements-low frequency vibration modes (i.e., large strains) was still unsettled, the inertia relief assumptions were acceptable, depending on the investigated area.

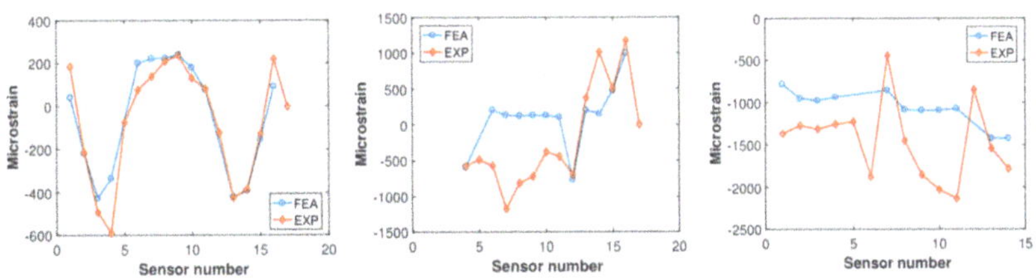

(a) (b) (c)

Figure 18. Experimental (EXP) vs. FEA strain comparison for (**a**) frame 7, (**b**) frame 10 and (**c**) tail boom.

7. Conclusions

In the present work, a method for rotorcraft landing simulation was presented. This method fills a gap in the rotorcraft landing related open literature, providing an effective way to overcome a complex and challenging problem. The technique adopted herein allows to simulate the landing event by means of a sequentially coupled analysis strategy, which initially relies on a multibody analysis of the rotorcraft landing, and then provides the necessary input parameters to a finite element model of the rotorcraft, to assess the landing from a structural point of view.

The investigated analysis method was critically evaluated, and its legitimacy was corroborated by means of experimental data resulting from a full-scale experimental activity conducted on the examined rotorcraft. Experimental data were fundamental in the modeling phase since they enabled the numerical models of the helicopter—and the whole method also—to be objectively verified. The experimental tests also allowed the identification or adjustment of some unknown parameters, based on the experimental measurements (e.g., shock absorbers characteristics). The structural domain was investigated according to two different approaches, namely the displacement control strategy and the inertia relief method, which rely on different intermediate parameters, respectively. It was proven that the displacement control method introduced additional stiffness to the airframe (except for the tail boom), biasing the related structural response. On the other hand, this method provided a time-dependent response that accounts for the structural dynamics, and depending on the investigated area, it can provide consistent results, for instance concerning the tail boom. The inertia relief method, instead, provided consistent data with regards to the airframe, but since it does not account for the dynamic structural response, it did not accurately show the tail boom structural response, which is more prone to dynamic excitation.

The disclosed method is not relegated to rotorcrafts only, and its validity can be extended to other structures. It is essential to notice that this method, apart from being a tool for landing simulation, offers the opportunity to implement on-board monitoring strategies, allowing the creation of a comprehensive collection of structural response at a relatively low cost. This, which can be useful for assessing the structural integrity under different landing conditions or to interpret real-time signals from a network of sensors, would not be possible using experimental testing activities only, especially when considering complicated and large structures.

The present study is thus forming a solid basis for the development and implementation of on-board monitoring systems, providing a means to identify the most suitable areas for the sensor location and the most appropriate sensor technology. Future works might explore the relationship elapsing between the structural response following the landing and the quantities measured by the sensors. This would allow the monitoring system to link the measured data to a complete map of the structural response and, consequently, to the corresponding landing parameters. Vice versa, if the real-time landing parameters are known, it is possible to detect potentially harmful landings conditions, allowing the adjustment of the landing configuration.

Author Contributions: Conceptualization, C.S., M.D. and M.G.; data curation, D.C., L.C. and C.S.; formal analysis, D.C., L.C. and C.S.; funding acquisition, M.G.; investigation, D.C., L.C., W.Z., C.S., F.C. and M.D.; methodology, C.S. and M.D.; project administration, C.S.; resources, M.G.; supervision, M.G.; validation, D.C.; visualization, D.C.; writing—original draft, D.C. and L.C.; writing—review and editing, F.C. All authors have read and agreed to the published version of the manuscript.

Funding: This research was funded by the European Defense Agency (EDA) grant number B 1288 ESM2 GP.

Acknowledgments: This work has been developed based on the results from ASTYANAX project (Aircraft fuSelage crack monitoring sYstem And progNosis through on-boArd eXpert sensor network), a Cat.-B project coordinated by the European Defense Agency (EDA) and involving three nations: Italy (Politecnico di Milano, AleniaAermacchi, AgustaWestland), Poland (Instytut Techniczny Wojsk Lotniczych - AFIT, Military Aviation Works No. 1, AGH University of Science and Technology) and Spain (Instituto Nacional de Técnica Aeroespacial - INTA). Authors thank M. Frovel from Instituto Nacional de Técnica Aeroespacial, Spain, for the installation and signal acquisition of the optical Fiber Bragg Grating sensors used in the experimental tests.

Conflicts of Interest: The authors declare no conflict of interest.

References

1. *Certification Specifications for Large Aeroplanes, CS25*; An European Authority in Aviation Safety (EASA) Publication; European Authority in Aviation Safety: Cologne, Germany, 2008; Available online: https://www.easa.europa.eu/document-library/certification-specifications (accessed on 28 April 2020).
2. *Airworthiness Standards: Transport Category Airplanes, FAR 25*; U.S. Government Publishing Office: Washington, DC, USA, 2008.
3. Zimmerman, R.E.; Warrick, J.C.; Lane, A.D.; Merritt, N.A.; Bolukbasi, A.O. *Aircraft Crash Survival Design Guide. Volume 3. Aircraft Structural Crash Resistance*; Tech. Rep.; Simula Inc.: Phoenix, AZ, USA, 1989.
4. Rozelle, R.; Lacagnina, M.; Rosenkrans, W. Stabilized approach and flare are keys to avoiding hard landing. *Flight Saf. Dig.* **2004**, *23*, 1–16.
5. Airbus. Avoiding hard landings. In Proceedings of the 15th Performance & Operations Conference, Puerto Vallarta, Mexico, 23–27 April 2007.
6. Farrar, C.R.; Worden, K. An introduction to structural health monitoring. *Philos. Trans. R. Soc. A Math. Phys. Eng. Sci.* **2006**, *365*, 303–315. [CrossRef]
7. Boller, C.; Buderath, M. Fatigue in aerostructures—Where structural health monitoring can contribute to a complex subject. *Philos. Trans. R. Soc. A Math. Phys. Eng. Sci.* **2006**, *365*, 561–587. [CrossRef] [PubMed]
8. Dehaven, H. Causes of Injury in Lightplane Accidents. *Aero Dig.* **1944**, *44*, 51–55.
9. Thomson, R.G.; Carden, H.D.; Hayduk, R.J. *Survey of NASA Research on Crash Dynamics*; NASA Technical Paper; NASA: Hampton, VA, USA, 1984.
10. Thomson, R.G.; Goetz, R.C.; Thornson, R.G. NASA/FAA General Aviation Crash Dynamics Program—A Status Report. *J. Aircr.* **1980**, *17*, 584–590. [CrossRef]

11. Jackson, K.E.; Fasanella, E.L.; Kellas, S. Development of a Scale Model Composite Fuselage Concept for Improved Crashworthiness. *J. Aircr.* **2001**, *38*, 95–103. [CrossRef]
12. Thomson, R.G.; Caiafa, C. Designing for Aircraft Structural Crashworthiness. *J. Aircr.* **1982**, *19*, 868–874. [CrossRef]
13. Winter, R.; Pifko, A.; Cronkhite, J.; Winter, A.P.R. Crash Simulation of Composite and Aluminum Helicopter Fuselages Using a Finite Element Program. *J. Aircr.* **1980**, *17*, 591–597. [CrossRef]
14. Annett, M.; Horta, L. Comparison of test and finite element analysis for two full-scale helicopter crash tests. Collection of Technical Papers. In Proceedings of the 52nd AIAA/ASME/ASCE/AHS/ASC Structures, Structural Dynamics and Materials Conference, Denver, CO, USA, 4–7 April 2011. [CrossRef]
15. Melosh, R.J.; Kamat, M.P. Computer simulation of light aircraft crash. *J. Aircr.* **1977**, *14*, 1009–1014. [CrossRef]
16. Lyle, K.H.; Stockwell, A.E.; Hardy, R.C. Application of Probability Methods to Assess Airframe Crash Modeling Uncertainty. *J. Aircr.* **2007**, *44*, 1568–1573. [CrossRef]
17. Kohlgrueber, D.; Kamoulakos, A. Validation of numerical simulation of composite helicopter sub-floor structures under crash loading. In *Annual Forum Proceedings*; American Helicopter Society: Fairfax, VA, USA, 1998; Volume 1, pp. 340–349.
18. Jackson, K.E.; Boitnott, R.L.; Fasanella, E.L.; Jones, L.E.; Lyle, K.H. *A History of Full-Scale Aircraft and Rotorcraft Crash Testing and Simulation*; Technical Report; NASA: Hampton, VA, USA, 2004.
19. Fasanella, E.L.; Boitnott, R.L.; Lyle, K.H.; Jackson, K. Full-scale crash test and simulation of a composite helicopter. *Int. J. Crashworthiness* **2001**, *6*, 485–498. [CrossRef]
20. Kellas, S.; Jackson, K.; Littell, J. Full-scale crash test of an md-500 helicopter with deployable energy absorbers. In *Annual Forum Proceedings*; AHS International: Fairfax, VA, USA, 2010; Volume 2, pp. 1292–1303.
21. McCarthy, M.; Wiggenraad, J. Numerical investigation of a crash test of a composite helicopter subfloor structure. *Compos. Struct.* **2001**, *51*, 345–359. [CrossRef]
22. Lyle, K.H.; Jackson, K.E.; Fasanella, E.L. Simulation of Aircraft Landing Gears with a Nonlinear Dynamic Finite Element Code. *J. Aircr.* **2002**, *39*, 142–147. [CrossRef]
23. Fuchs, Y.T.; Jackson, K.E. Vertical Drop Testing and Analysis of the WASP Helicopter Skid Gear. *J. Am. Helicopter Soc.* **2011**, *56*, 12005. [CrossRef]
24. Tho, C.-H.; Sparks, C.E.; Sareen, A.K.; Smith, M.R.; Johnson, C. Efficient Helicopter Skid Landing Gear Dynamic Drop Simulation Using LS-DYNA. *J. Am. Helicopter Soc.* **2004**, *49*, 483–492. [CrossRef]
25. Caprile, C.; Airoldi, A.; Biaggi, A.; Mandelli, P. Multibody Simulation of a Helicopter Landing with Skid Landing Gear in Various Attitude and Soil Conditions. In Proceedings of the Twentyfifth European Rotorcraft Forum, Rome, Italy, 14–16 September 1999.
26. Zhu, Z.H.; LaRosa, M.; Ma, J. Fatigue Life Estimation of Helicopter Landing Probe Based on Dynamic Simulation. *J. Aircr.* **2009**, *46*, 1533–1543. [CrossRef]
27. Airoldi, A.; Lanzi, L. Design of Skid Landing Gears by Means of Multibody Optimization. *J. Aircr.* **2006**, *43*, 555–563. [CrossRef]
28. Crist, D.; Symes, L.H. *Helicopter Landing Gear Design and Test Criteria Investigation*; Tech. Rep.; Bell Helicopter Textron: Fort Worth, TX, USA, 1981.
29. Milwitzky, B.; Cook, F.E. *Analysis of Landing-Gear Behavior*; Tech. Rep.; National Advisory Committee for Aeronautics: Langley Field, VA, USA, 1953.
30. Kiefer, J.; Ward, M.; Costello, M. Rotorcraft Hard Landing Mitigation Using Robotic Landing Gear. *J. Dyn. Syst. Meas. Control.* **2016**, *138*, 031003. [CrossRef]
31. Boix, D.M.; Goh, K.; McWhinnie, J. Modeling and control of helicopter robotic landing gear for uneven ground conditions. In Proceedings of the 2017 Workshop on Research, Education and Development of Unmanned Aerial Systems (RED-UAS), Linköping, Sweden, 3–5 October 2017; pp. 60–65.
32. Gualdi, S.; Masarati, P.; Morandini, M.; Ghiringhelli, G. A Multibody Approach to the Analysis of Helicopter-Terrain Interaction. In Proceedings of the 28th European Rotorcraft Forum, Bristol, UK, 17–20 September 2002.
33. Vallone, G.; Manes, A.; Sbarufatti, C.; Giglio, M. Helicopter Harsh Landing Events: A Computational Hybrid Methodology to Estimate Fuselage Damage. *J. Aircr.* **2013**, *50*, 1896–1907. [CrossRef]
34. Sbarufatti, C.; Vallone, G.; Giglio, M.; Stefaniuk, M.; Leski, A.; Zielinski, W. Experimental Validation of a Computational Hybrid Methodology to Estimate Fuselage Damage Due to Harsh Landing. *J. Am. Helicopter Soc.* **2016**, *61*, 4. [CrossRef]

35. Giglio, M.; Klimaszewski, S.; Kurdelski, M.; Leski, A.; Manes, A.; Sbarufatti, C.; Stefaniuk, M.; Vallone, G.; Zielinski, W. Model-based structural integrity assessment of helicopter fuselage during harsh landing. In *Annual Forum Proceedings*; AHS International: Fairfax, VA, USA, 2015; Volume 2, pp. 1023–1032.
36. LeBlanc, B.; Niezrecki, C.; Avitabile, P. Structural health monitoring of helicopter hard landing using 3D digital image correlation. *Proc. SPIE Health Monitor. Struct. Biolog. Syst.* **2010**, *7650*. [CrossRef]
37. Sagi, O.; Maynard, D.; Enikov, E. Capacitive transducer for condition based maintenance after harsh landing events. In Proceedings of the AUTOTESTCON (Proceedings), Baltimore, MD, USA, 12–15 September 2011; pp. 286–291. [CrossRef]
38. Sartor, P.; Schmidt, R.; Becker, W.; Worden, K.; Bond, D.; Staszewski, W. Conceptual design of a hard landing indication system using a flight parameter sensor simulation model. In Proceedings of the 27th Congress of the International Council of the Aeronautical Sciences 2010, ICAS 2010, Nice, France, 19–24 September 2010; Volume 5, pp. 3515–3526.
39. Sartor, P.; Schmidt, R.; Menezes, R.; Bond, D.; Staszewski, W. Validation and verification of a hard landing indication system for aircraft landing gear. In Proceedings of the Structural Health Monitoring 2009: From System Integration to Autonomous Systems—Proceedings of the 7th International Workshop on Structural Health Monitoring, Stanford, CA, USA, 9–11 September 2009; Volume 1, pp. 190–200.
40. Liao, L. A study of inertia relief analysis/Collection of Technical Papers. In Proceedings of the AIAA/ASME/ASCE/AHS/ASC Structures, Structural Dynamics and Materials Conference, Denver, CO, USA, 4–7 April 2011. [CrossRef]
41. McPherson, A.E.; Evans, J.J.; Levy, S. *Influence of Wing Flexibility on Force-time Relation in Shock Strut Following Vertical Landing Impact*; Technical Report; National Advisory Committee for Aeronautics: Washington, DC, USA, 1949.

sensors

MDPI

Article

High Speed, Localized Multi-Point Strain Measurements on a Containment Vessel at 1.7 MHz Using Swept-Wavelength Laser-Interrogated Fiber Bragg Gratings

Steve Gilbertson [1,*], Mark Pickrell [1], Dario Castano [2], Gary Salazar [1], Tom Beery [1], Samuel Stone [3] and Joshem Gibson [4]

1 Los Alamos National Laboratory, DARHT Experiments and Diagnostics, MS P940, Los Alamos, NM 87545, USA; mpickrell@lanl.gov (M.P.); gps@lanl.gov (G.S.); tbeery@lanl.gov (T.B.)

2 Los Alamos National Laboratory, Advanced Engineering Analysis, MS A142, Los Alamos, NM 87545, USA; castano@lanl.gov

3 Lawrence Livermore National Laboratory, Environmental Test Group, L-125, Livermore, CA 94550, USA; stone58@llnl.gov

4 Los Alamos National Laboratory, Dynamic Structure Design and Engineering, MS P942, Los Alamos, NM 87545, USA; gibson@lanl.gov

* Correspondence: steveg@lanl.gov

Received: 28 September 2020; Accepted: 15 October 2020; Published: 20 October 2020

check for updates

Abstract: Dynamic elastic strain in ~1.8 and 1.0 m diameter containment vessels containing a high explosive detonation was measured using an array of fiber Bragg gratings. The all-optical method, called real-time localized strain measurement, recorded the strain for 10 ms after detonation with additional measurements being sequentially made at a rate of 1.7 MHz. A swept wavelength laser source provided the repetition rate necessary for such high-speed measurements while also providing enough signal strength and bandwidth to simultaneously measure 8 or more unique points on the vessel's surface. The data presented here arethen compared with additional diagnostics consisting of a fast spectral interferometer and an optical backscatter reflectometer to show a comparison between the local and global changes in the vessel strain, both dynamically and statically to further characterize the performance of the localized strain measurement. The results are also compared with electrical resistive strain gauges and finite element analysis simulations.

Keywords: fiber Bragg grating; fiber sensing; high-speed interrogation; dynamic strain

---

1. Introduction

Fiber Bragg gratings (FBG) are routinely used for making static and dynamic strain and temperature measurements up to the kHz range [1–3]. Significant improvements have also been made to FBGs to improve strength and versatility [4–6] while also improving sensitivity [7,8]. Such measurements have been used in industry for many years, for example, in testing strain in airplanes [9] and bridges [10] or automotive traffic [11]. These relatively slow measurements are able to be made with off-the-shelf spectrometers and software packages to achieve readouts of up to a few kHz [12]. A light source illuminates a grating and the reflected light is recorded by a simple spectrometer.New measurements are made as fast as the spectrometer can read out the data, which is sufficient for measuring, for example, strain induced in bridges as cars drive over.

On the other end of the scale, destructive testing of high explosives (HE) can also be characterized using chirped FBGs. These have been demonstrated at 100 MHz repetition rates for measuring the detonation wavefront and even some strain in the detonation process [13,14]. These experiments are often very fast (few 10s of microseconds) while also having such high pressures that the gratings are sometimes cleanly destroyed in the experiment with no measurable strain behavior.While the technology for this work is well developed, it is cost prohibitive to many labs interested in non-destructive testing.Indeed, the high bandwidth scopes alone cost well over $100K while only providing recording times of up to 2 ms. Additionally, such high repetition rates in the light source limit the number of samples that can be made in a single measurement.For example, a 50 GS/s scope measuring data from a grating interrogated by a 100 MHz laser will only have 500 points in each measurement [13]. This is a sacrifice of fine resolution for fast recording speeds in order to achieve these state-of-the-art measurements. Figure 1 shows a schematic emphasizing the differences in complexity between these 2 methods.

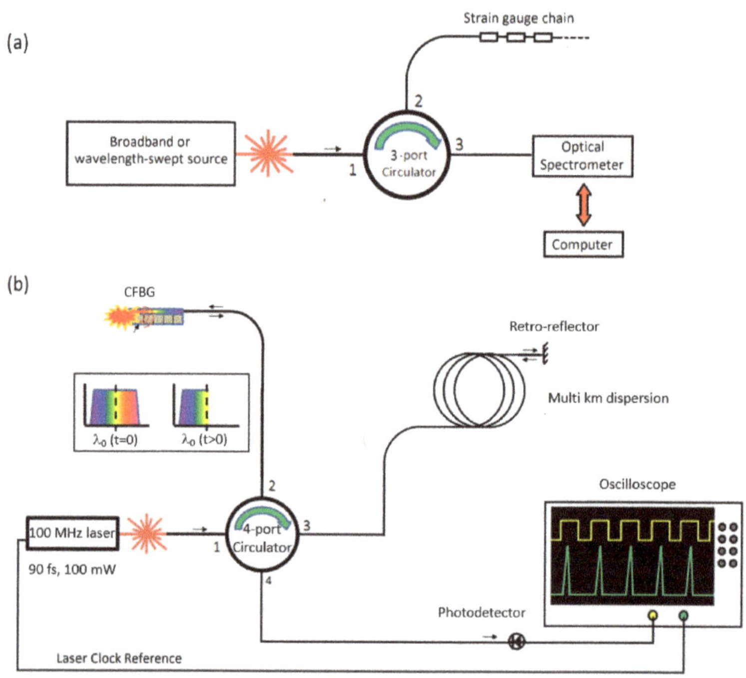

Figure 1. Examples of strain measurements at: (**a**) kHz repetition rates with Bragg gratings and (**b**) MHz repetition rates with chirped Bragg gratings.

Between these two extremes is a range of measurement rates of concern to safety qualifications.Rates from 1 kHz up to a few MHz are necessary for characterizing the peak and average strain in a high explosive containment vessel.These vessels are used at Department of Energy and Department of Defense national labs as well as many other areas and act as a first line of defense to the co-located workers and the general public [15]. Figure 2 shows a typical vessel used for containing an HE experiment at Los Alamos National Lab. The diagnostic suite described in this manuscript is also included in the figure.

The vessels are usually 0.9 to 1.8 m in diameter and made with high-strength low alloy (HSLA-100) steel nominally 2.54 cm thick for the 0.9 vessels and 6.4 cm thick for the 1.8 m vessels. The vessels also have access ports in order to insert experiments and other diagnostics or to act as windows for radiographic measurements to be made during the experiment. These ports increase the complexity of the strain behavior where the ports intersect with the spherical shape of the vessel.

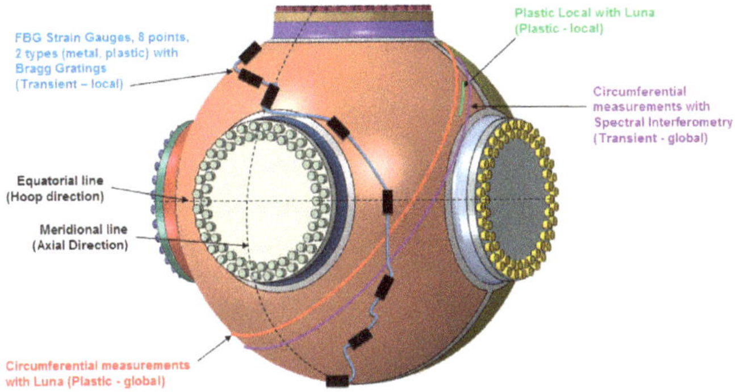

Figure 2. A typical explosive containment vessel used at Los Alamos National Lab. An example suite of diagnostics as discussed in this paper are indicated on the vessel surface.

The vessels are specifically designed to contain the pressures expected during the explosive test with calculations of the strain performed using a finite element analysis (FEA) package [16,17]. With such rigorous engineering qualifications in place, the vessels are quite expensive, exceeding $1 million each for the 0.9 m vessels and $2 million for the 1.8 m vessels. The vessels are pulled from service after several tests in order to ensure that they are not compromised during a dynamic experiment. Due to this balance between safety and economics, a suite of diagnostics capable of characterizing the vessel performance will ensure the vessels are not pulled from service too late that they are a safety risk, but also not too early that they become cost prohibitive to replace.Besides the cost of the vessel, the experiments themselves represent a multi-million dollar test of a high explosive system with specific diagnostics chosen many months in advance.

Historically, containment vessel strain has been measured on a case-by-case basis using resistive strain gauges placed in locations that can be directly compared with FEA results [18,19]. While the results have often shown agreement between experiment and calculation, other times, the results are not clear. The data areoften noisy due to additional electromagnetic or radiofrequency signals being picked up by the wires, which can act as antennas readily receiving signals between the locations where the data arerecorded and where the gauges are fielded. This noise can be several kHz, which is well above the frequencies predicted. Unfortunately, spurious strain signals cannot be fully ruled out, especially in a fast dynamic test with a high impulse from the detonation. Additionally, the gauges can break through the epoxy adhering them to the vessel surface if the impulse on the vessel is high enough. Finally, the strain measurements are localized to a few select points on the vessel, meaning the global elastic or plastic strain in the vessel is not fully characterized.

These limitations can be overcome through the use of optical methods for strain measurements. An all-optical path between the diagnostic and the recording location is impervious to electrical noise. The light weight nature of FBGs and fiber optics in general means the diagnostics are far more likely to stay adhered to the surface during the experiment. Because the gauges are light-weight, they are a bit more fragile than electrical gauges, although damage to the gauges has thus far been successfully mitigated through increased care in handling the vessels. Figure 3 shows an example of the optical gauges fielded in these experiments.

In this paper, the results from FBG gauges fielded on a recent 1.8 m diameter vessel during a high explosive test (hydrotest 3685) at Los Alamos National Lab will be shown. The 1.8 m diameter vessels are always used for the full scale hydrotests as they are capable of withstanding the HE pressures and holding blast mitigation inside the vessel. The diagnostics fielded were a set of robust light weight plastic gauges (HBM OL). Another option was a set of rubberized robust weldable gauges mounted onto steel plates (HBM OL-WA). For safety reasons, we were unable to weld these to the surface and so they were instead epoxied into place as a test. These gauges were much heavier than the plastic type and it was predicted they would break through the epoxy layer holding them onto the surface, which was generally found to be true. The measurement technique uses a swept wavelength laser to sequentially interrogate a set of 8 gauges placed on the vessel and at a repetition rate of ~1.7 MHz. This is sufficiently fast to oversample the strain data for accurate mapping of sound wave ringing in the vessel walls. This method can rule out or confirm fast strain spikes often seen in resistive strain gauge data as electrical noise since the spurious noise signals are typically well below the sample rate of the swept wavelength light source. As these are strictly localized measurements, two additional optical diagnostics, consisting of an optical backscatter reflectometer for static global plastic measurements and a spectral displacement interferometer for dynamic global elastic strain measurements will be compared to the localized data for a more complete picture of the vessel performance. No additional diagnostics were fielded on the vessel for hydrotest 3685 so the diagnostics had full reign over all available space on the vessel surface. This allowed us to add global strain measurements coving the full circumference of the vessel to compare with local strain measurements. The FBG data will also be compared directly to a set of resistive strain gauges from a recent overpressure test experiment on a 0.9 m diameter vessel as a comparison with the legacy diagnostics. The smaller vessel is used for testing diagnostic feedthroughs with a small HE charge. As long as the feedthroughs do not fail under an HE load, the experiment is considered a success. This smaller diameter vessel was planned to have electrical diagnostics to compare the strain experienced by the vessel with legacy data. The localized FBG method was fielded near the electrical gauges wherever possible. Due to the nature of the gauges, the circumferential diagnostics for measuring global strain were not fielded on the 0.9 m vessel. Finally, the results will also be compared with FEA calculations throughout the paper.

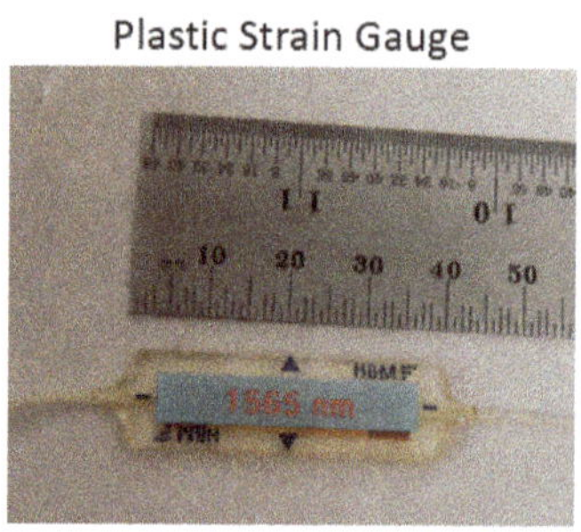

Figure 3. Optical light-weight plasticstrain gauges tested in our first experiments.

2. Materials and Methods

2.1. Real Time Localized Strain (RTLS) Measurements

FBGs are sensitive to strain and temperature and have been used for many years [1–3]. The process for manufacturing the gauge is to write a periodic index change into a single mode fiber which can be done uniformly or non-uniformly [20]. The index change periodicity determines the reflected wavelength, λ, through the relation $\lambda = 2n\Lambda$, where Λ is the periodicity, and n is the refractive index of the fiber. By stretching or compressing the fiber, the grating periodicity increases or decreases proportionally, thereby changing the reflected wavelength linearly with the change in length. Figure 4 shows a schematic of this.

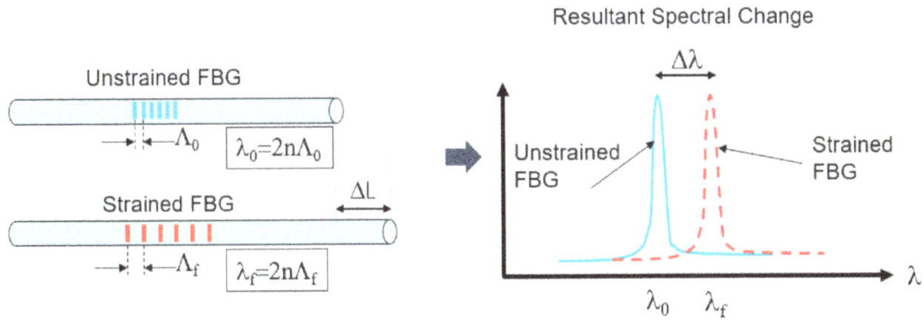

Figure 4. (**a**) Representation of a Bragg grating periodicity change when being strained. (**b**) Reflected spectrum of a strained and unstrained grating.

Once the wavelength changes, the strain and temperature change can be calculated from the equation:

$$\frac{\Delta\lambda}{\lambda} = k\varepsilon + (\alpha_\Lambda + \alpha_n)\Delta T. \tag{1}$$

In Equation (1), $\Delta\lambda$ is the wavelength change, λ is the center wavelength of the gauge, k is the so called "k-factor" which is 1-p, where p is the strain-optic coefficient for the fiber and is typically ~0.21 for the optical fiber used in our experiments, α_Λ is the thermal expansion coefficient of the fiber, and α_n is the thermo-optic coefficient. For our fast dynamic experiments, ΔT changes very slowly in comparison and is basically zero. This means the majority of our wavelength change comes from the force induced strain in the vessel. The strain, ε, in Equation (1) is the standard value of ΔL/L.

The method for measuring the strain on a vessel is the real-time localized strain (RTLS) system. A block diagram of the technique is shown in Figure 5a. The RTLS technique is presented as an alternative for resistive strain gauges. The remainder of the paper following this section will be alternative methods for verifying the response of the optical method.

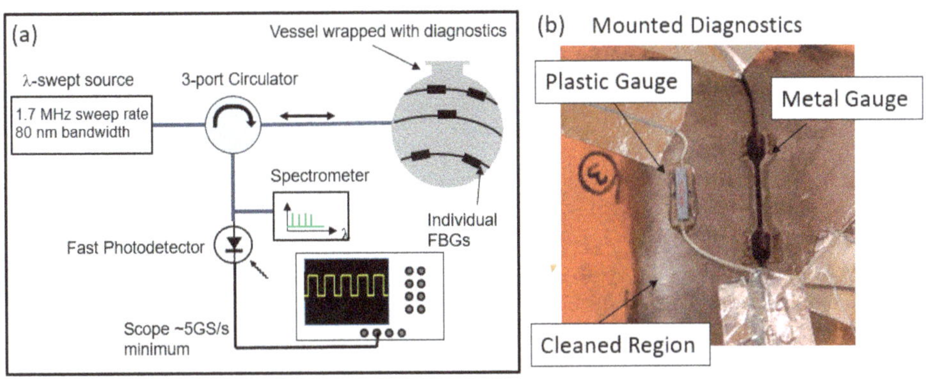

Figure 5. (**a**) Schematic representation of the real time localized strain(RTLS) method for measuring localized dynamic strain on a containment vessel. (**b**) Plastic and metal optical gauges mounted to a cleaned surface of the vessel.

The light source used was an Optores Fourier domain mode-locked (FDML) laser [21,22]. The model used in our experiments had a sweep rate of ~1.7 MHz and a bandwidth of 80 nm centered at 1550 nm. The linewidth of the source is less than 30 picometers and the average power is ~100 mW. The laser also provides a clock signal synched to the start of sweep. The model we use creates 3 identical copies of the sweep, which are consecutively delayed and superimposed to give 4 replicas of each sweep with each start-of-sweep of the laser. This means the clock signal frequency, while synched to the start of sweep, is actually a quarter of the frequency of sweeps that the laser provides. This has to be taken into consideration for the data analysis. Finally, the stability of the laser was measured to be less than 180 ps after 10,000 sweeps, which is much less than the 2.36 microsecond sweep period of the laser. This stability result is also dominated by the trigger jitter of the detector so the laser stability may be even better than this.

The output from the laser is sent to a 3 port circulator and directed to a chain of FBG strain sensors placed on the device under test (DUT), which in this case is a confinement vessel. Each sensor reflects one particular wavelength which is then sent back to the circulator. Due to the sweeping nature of the laser, each gauge is interrogated at a unique time with respect to the start of sweep that does not change in a static measurement. The sequential series of reflections is then sent to a 99:1 splitter with 1% of the light sent to a spectrometer for calibration purposes. Each peak represents one gauge placed on the DUT. The remaining 99% of the light is sent to a fast photodetector, in this case, an Optilab PR-23-M 23 GHz optical detector. The output of the detector is recorded on a digital oscilloscope. In order to record fast temporal changes in the strain on the vessel, a 4 GHz scope capable of measuring up to 25 GS/s was used. The extended memory option allowing up to 125 M points per channel was also required and even with this, the sample rate had to be reduced to 12.5 GS/s in order to record for 10 ms. The high sample rate allows measurements to be made with a frequency limited by the sweep rate of the laser source. In this case, 1.6 MHz which is over 300 times more measurements per unit time than standard off-the-shelf spectrometer-based measurement techniques.

The chain of gauges used for the localized measurements is a custom configured design manufactured by HBM. The chain was chosen to include 8 wavelengths that fit within the bandwidth of our laser source. For our proof-of-concept experiments, we focused on one type of FBG chains. The gauge is a light weight plastic-encased FBG. The gauge is particularly flexible, which aids in the installation of the gauges on a curved surface. The gauges can be prepared with any amount of optical fiber between them, thereby allowing the gauges to be placed at any location and in any orientation on the contour of the vessel. To attach the gauges, we first had the vessel surface prepared by removing the paint with a grinder. The bare metal was then scuffed with 400 grit sandpaper. Next, the surface was cleaned with phosphoric acid followed by a neutralizer. Finally, the gauges were attached to the surface with a thin layer of two part 5 min epoxy. The gauges were held in place with a Teflon cushion until the epoxy was set. An image of the gauges attached to a sample vessel is shown in Figure 5b. The gauges employ a single mode fiber etched with a grating. The maximum strain measurable for both types of gauge is +/−10,000 με and the peak reflectivity at the specified wavelength is 15%. The linewidth of a single FBG is specified at 0.13 nm and the FBG inside the gauge is 5 mm long. A sample spectrum for a series of 8 gratings is shown in Figure 6a.

A second type of weldable FBG with a rugged rubberized coating on the fiber optic was also tested. Due to safety considerations, we were not able to weld the gauges to the containment vessel, so they were attached using the same epoxy method of the plastic gauges. Due to the increased mass of the weldable type, it was expected that they would have a larger chance of breaking through the epoxy that was adhering them to the surface at any time during the dynamic response of the vessel walls following the HE detonation. These gauges generally failed for these reasons.

The peak wavelengths were chosen to coincide with our laser source bandwidth. We also designed the chain with 5 nm separation between each peak wavelength, corresponding to over 4000 με, which exceeds the peak strain often expected in the vessels. This helps ensure the peaks do not cross during the experiment for ease in data analysis.

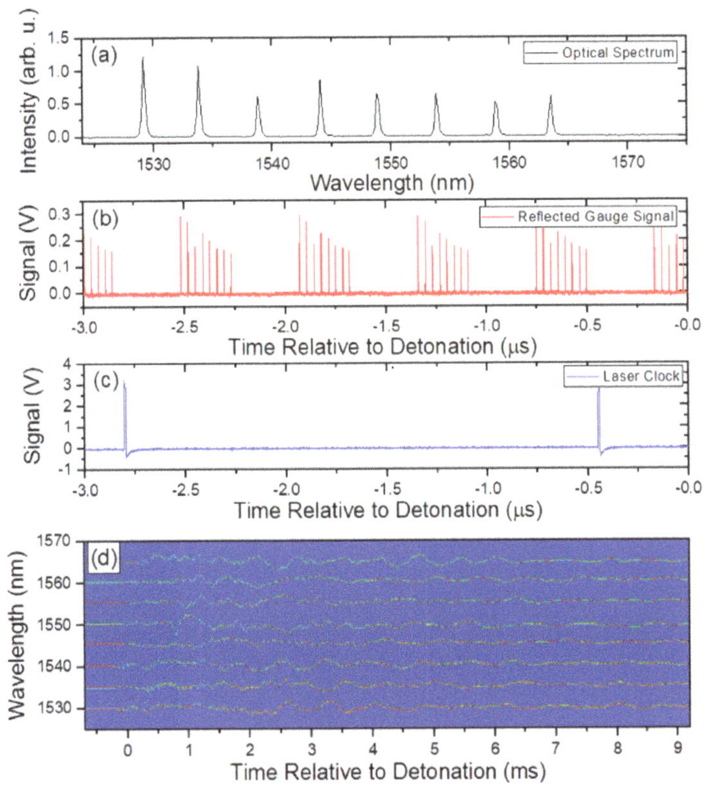

Figure 6. Representative data accumulated with the RTLS method. (**a**) Typical reflected spectrum from a chain of optical strain gauges. (**b**) Reflected response of the grating as measured as a function of time. Each set of 8 peaks represents a single measurement of all the gratings in a chain made by one sweep of the laser. (**c**) Recorded clock signal of the laser. Since the laser makes 3 replicas of the laser sweep, each peak is synched with the start of sweep of the laser but occurs every 4th pulse. (**d**) Example of the 2D array of strains measured with each chain of optical strain gauges. Each horizontal signal represents the temporal history of the strain on the vessel at one particular location during the first 10 ms after detonation.

To form an accurate measurement of the strain, the data must be first converted from the time axis as it is recorded on the scope to the spectral domain. The first step is to parse the data according to the start of each sweep. This is done by recording both the data and the clock signal during the experiment. The data shown in Figure 6b,c show the raw data and the clock signal, respectively. The clock period is then extracted and divided by 4 since the laser source we use creates 4 sweeps for each clock pulse. Each raw data sweep is then extracted using the quarter clock period with the time of each start of sweep recorded. The time axis for each sweep is relabeled from t0 to N, where N is the total number of samples multiplied by the time step in a single sweep. In the case of our light source, we record ~7800 samples for each sweep. With a 12.5 GS/s sample rate, our time step is 80 ps, giving a total time window for each sweep of nearly 625 ns. Next, the new 0 to N time axis must be converted to wavelength. A sample calibration spectrum as seen in Figure 6a is recorded shortly before each experiment. This is done as close to the experiment as possible to ensure the temperature of the vessel does not vary much from before to after the experiment. A set of thermocouples is also fielded on the vessel, which allow for independent verification of the temperature. From the sample spectrum, the peak of each gauge is found. A polynomial fit of the peak wavelengths versus the temporal point as recorded on the time axis of the scope is produced. This gives a fitting function that allows every temporal sweep of the laser to be converted into wavelength. Once this has been accomplished, a 2D array of each temporal sweep converted into wavelength and plotted against the start time of each sweep can be made. It should be noted that the spectrum shown in Figure 6a has a very similar profile to the temporal representations shown in Figure 6b, demonstrating that the spectral information is preserved in the measurement. Figure 6d shows a sample 2D array plotted as an image with the intensity of each peak plotted as a color. Each separate line is the wavelength change for each gauge. From this point, extracting the wavelength change for each gauge and converting into the strain using Equation (1) is straightforward.

2.2. Global Plastic and Elastic Deformation of Vessels

The RTLS measurements can give an overview of the localized strain at various points on the vessel. The individual measurements only record for 10 ms however, and so plastic deformation in the vessel cannot be fully deduced from the RTLS measurements alone. To further characterize the vessel and complement the local transient measurements, two additional techniques are typically fielded. These additional diagnostics give insight into the global change, both statically and dynamically, by integrating the total strain along the circumference of the vessel.

The first is a simple technique using an optical backscatter reflectometer (OBR) with high resolution. The device we chose is a Luna OBR4600 [23] which has an ultimate resolution of 10 μm. The Luna device is slow compared to the detonation event of the experiment with a single scan taking a few seconds, so we treat the measurements statically. Figure 7a shows how the Luna device is employed for global measurements. First, the vessel is prepared by removing the paint and creating a surface in a manner similar to what was described for the RTLS method. Next, a fiber is attached with epoxy onto the vessel in a great circle around the vessel. A slight amount of tension is applied to the fiber during the epoxy application to ensure the fiber is tight around the vessel. A gold reflector is sometimes attached to the end of the fiber to increase the visibility of the location of the end of the fiber, although this is not really necessary due to the large dynamic range of an off-the-shelf OBR4600. From here, static measurements can be made of the fiber length that has been stretched around the vessel. Figure 7b shows a sample Luna OBR4600 dataset.

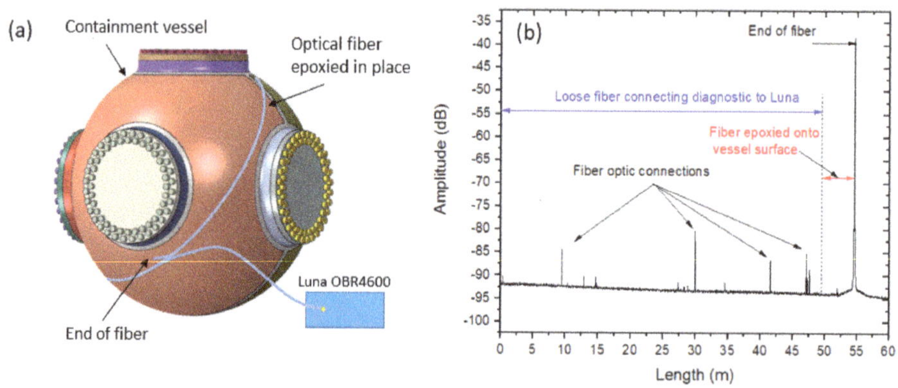

Figure 7. (**a**) Location of a single optical fiber placed around the circumference of a vessel for Luna measurements. The fiber has to avoid the ports of the vessel and is unable to follow a true great-circle of the vessel. (**b**) Luna optical backscatter reflectometer (OBR) trace for a fiber stretched around the vessel circumference.

The individual peaks represent locations of fiber connections with the final peak at 55 m being the end of the fiber. As the measurement is slow, several scans can be made to show any length changes in the vessel due to temperature changes all the way up to the experiment. Typically, we scan every 15 min for an hour before and after the experiment. The global change to the circumference of the vessel includes any plastic deformation as well as temperature changes due to the detonation event that occurred during the experiment. Length changes from temperature induced expansion must be considered in the analysis of the data using complementary thermocouples fielded on the outside and inside of the vessel. Further discussion of the temperature effect will be discussed in the results section. Additionally, a shorter piece of fiber optic was attached to the vessel. Since it is not a full circumference of the vessel, a scaled measurement of the length change can be measured and compared with the full length fiber around the total vessel circumference.

The second technique for global measurements makes a transient measurement in real time during the first 10 ms after the detonation event. This is an elastic measurement that can be more closely compared with the RTLS measurements. The technique is spectral interferometry for transient strain (SITS) measurements and a schematic is shown in Figure 8.

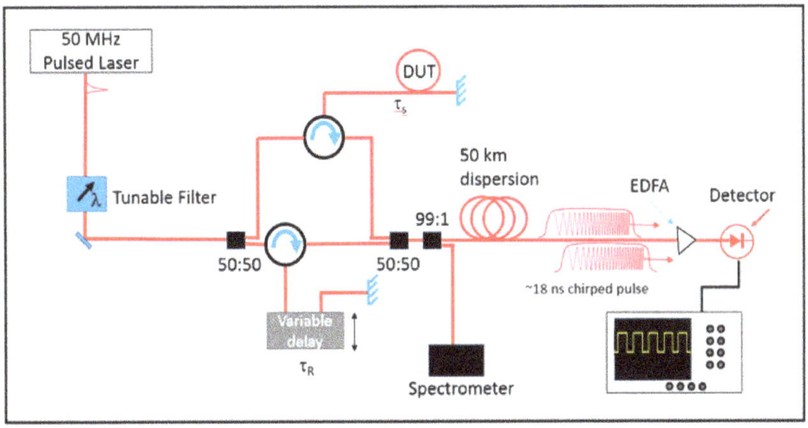

Figure 8. Schematic representation of the experimental setup for spectral interferometry for transient strains (SITS). DUT is device under test, EDFA is erbium doped fiber amplifier.

First, a laser operating at 50 MHZ with an average power of 50 mW and pulse duration of ~100 fs was sent through a tunable filter. This allowed the spectral content of the laser to be modified for either increased range in the measurement or increased resolution. Using more bandwidth reduces the range over which length measurements can be made but also increases the resolution. The laser next entered the interferometer portion where it was split into two equal replicas with a 50:50 splitter. Each half of the light was sent to a 3 port circulator. One leg of the interferometer had the light sent to the device under test (DUT). In this case, the DUT was a vessel with a single optical fiber wrapped around it. A gold reflector was attached to the end of the fiber in order to capture as much of the light as possible. The light then reflected from the gold reflector and traveled back to the circulator. The other leg of the interferometer acted as the reference and had the light traveling through a variable delay and reflecting from another gold reflector. The two legs then recombined at a second 50:50 splitter. From here, 1% of the light was split off and sent to a spectrometer for comparison purposes. The remaining 99% of the light was sent through 50 km of dispersion which, along with the optical detector, acted as a dispersive Fourier transform [24]. This allowed us to measure a representation of the spectral content of the combined signal on a fast oscilloscope. An erbium-doped fiber amplifier (EDFA) was also included for post amplification if the signal was not strong enough. The scope chosen was a 25 GHz Tektronix scope with optional high memory. This allowed record lengths of 10 ms at 50 GS/s.

As this is a spectral measurement, interference fringes can be seen on the scope. Figure 9 shows a sample set of fringes for two different delays between the two legs of the interferometer. When the delay between the two pulses is small, the fringes are large. However, when the DUT is put under strain, the fiber is stretched and the delay between the two pulses increases, causing the fringes to become much finer.

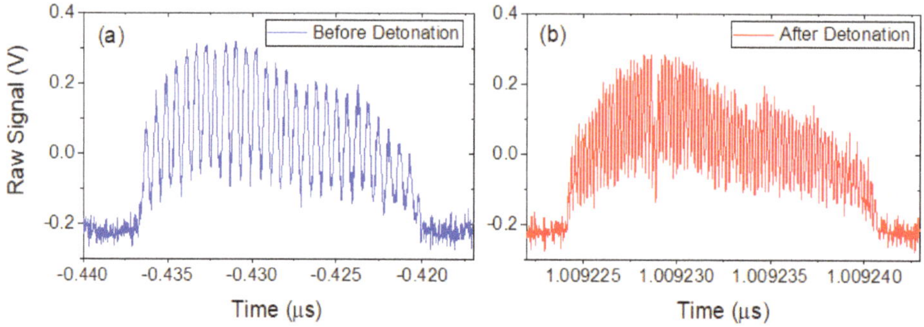

Figure 9. Sample spectral interferograms as recorded on an oscilloscope as a function of time for (**a**) closely spaced and (**b**) widely spaced optical pulses. The fringe visibility is beginning to be limited by the bandwidth of the scope in (**b**).

The longitudinal resolution of the measurement is dependent on the bandwidth of the source, while the total length change that can be measured is dependent on the total bandwidth, Δf, of the recording scope and the repetition rate, f_{rep}, of the light source. The resolution is given by:

$$\Delta z = \frac{{\lambda_0}^2}{2(\Delta\lambda)}, \tag{2}$$

where λ_0 is the center wavelength of the source and $\Delta\lambda$ is the laser source bandwidth. The total range of the measurement is limited by when the scope can no longer resolve spectral fringes. By reducing the repetition rate of the source, an individual pulse can be further stretched so as to resolve fringes corresponding to larger separations between the signal and reference pulses without pulse to pulse overlap, although this is at the sacrifice of temporal resolution. The temporal period, T, of the fringes in the interferograms seen in Figure 9 is given by $T = 2\pi\beta_2 L/\tau$ and the dispersion of optical fiber can be calculated from $D = 2\pi c\beta_2/{\lambda_0}^2$, where L is the length of the dispersion fiber in the setup, β_2 is the group velocity dispersion in the fiber, and τ is the delay between the two pulses in the interferometer [25]. The smallest value of T, and hence the largest value of τ that can be measured is limited by the scope bandwidth, Δ*f*, and is 1/Δ*f*. Solving for τand noting that the total range R_{max} that can be measured is cτ/n, the full range is given as:

$$R_{max} = \Delta f {\lambda_0}^2 DL/n, \tag{3}$$

where *n* is the refractive index of the fiber. The units of *DL* are (ps/nm). Equation (3) can be expressed in units of the desired resolution, Δz, using Equation (2) as $R_{max} = 2(\Delta z)(\Delta\lambda)(\Delta f)DL/n$. However, multiplying the total dispersion inherent in the setup, DL, by the bandwidth of the laser pulse Δλ, yields the duration of the laser pulse after propagation through the dispersive medium. The longest value this product can be without pulse-to-pulse overlap is the laser periodicity $T_L = 1/f_{rep}$.

For a desired resolution as calculated from Equation (3), the total maximum range which can be measured is therefore given by:

$$R_{max} = \frac{2(\Delta z)(\Delta f)}{n f_{rep}}. \tag{4}$$

With a 50 MHz repetition rate and 25 GHz bandwidth scope, ranges of 3 cm from resolutions of 30 μm would be possible. Additionally, a new measurement would be made every 20 ns which allows for high precision in temporal changes of the total circumference of the vessel.

To analyze data in SITS measurements, the time axis on the scope must be changed back into the spectral domain. A method similar to how the data is parsed and re-labeled with a wavelength axis in the RTLS measurements is first conducted on the raw SITS data recorded from the scope. The wavelength range was determined by comparison with the spectrometer measurements made before the experiment. The DC component of the signal is subtracted and a Fast Fourier Transform (FFT) is applied to every temporal measurement. The FFT yields two sidebands located at +/− the delay between the two pulses and is dependent on the fringe separation in the spectral domain. As the strain measurements are made by the length change in the vessel, the initial difference in delay from before the detonation event must be known. To accomplish this, static spectra were recorded a few minutes before the experiment began with known delays between the two legs of the interferometer. These acted as calibration spectra. Again, the DC signal was subtracted and an FFT was applied to each spectrum. Figure 10 shows a sample result of this analysis. The FFT determined the initial offset between the DUT leg and the reference leg to be 2 mm.

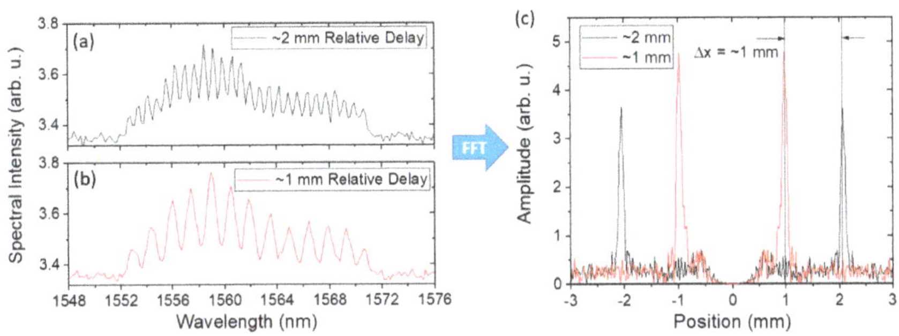

Figure 10. Sample spectral interferograms as recorded on a spectrometer for (**a**) 2 mm spaced and (**b**) 1 mm spaced optical pulses. (**c**) The fast Fourier transform of the results after subtracting the DC component. There are two sets of sidebands for each delay, symmetric about 0 delay.

When a 1 mm delay was removed from the reference leg, the sidebands showed a corresponding shift of 1 mm. A 2D image showing the location of the sideband as a function of time was made to visualize the total length change during the experiment.

For the actual experiment, care was made in the initial tuning. The zero delay point represents no fringes in the spectral domain and we chose to avoid it so that the sidebands would not cross. As it was expected that the fiber would stretch during the experiment leading to increased delay, the initial offset was adjusted such that the positive sideband would start at some pre-defined value and only move to a larger value rather than moving towards the 0 delay.

This particular diagnostic acted as a bridge between the RTLS measurements and the Luna measurements to measure the global circumferential change in the vessel during the dynamic experiment. The SITS result could be compared with the individual strain gauges to show how the total circumference relaxes dynamically or compared with the Luna measurements to show how the vessel plastically deforms.

The optical techniques listed in this section can be summarized in the following table (see Table 1). RTLS and SITS both measure dynamically the transient response of the vessel as it elastically returns to the unstrained state, but only for a limited amount of time after the start of the experiment. The short timescales of these diagnostics limit the amount of information that can be gathered due to plastic or permanent deformation in the vessel. Luna measurements are employed for static or plastic deformation in the vessels due to the slow nature of the recording method.

Table 1. The techniques described in this paper with the strain types each is capable of measuring.

Diagnostic Name	Strain Type	Local vs. Global
RTLS	Elastic	Local
SITS	Elastic	Global
Luna-long fiber	Plastic	Global
Luna-short fiber	Plastic	Local

2.3. Electrical Strain Gauges

All strain gauges were installed at a vessel preparation facility prior to the movement of the vessel to the firing site. A total of 9 strain gauge rosettes were installed on the vessel exterior. One of the 9 strain gauge rosettes was used as a dummy gauge and all three gauges on that rosette were wired. This gauge acted as a dummy gauge because it was not provided with excitation power during the test, so a response seen on this gauge is electrical noise. Vishay micro-measurements CEA-06-250UR-350 strain gauges were selected based upon the vessel material, expected strain, mounting configuration, and sensitivity. This part number corresponds to a constantan grid, completely encapsulated in a polyimide rosette that has a 0.25 inch gauge length and a nominal resistance of 350 ohms. The gauges selected are also temperature compensated for steel, though this is not a necessary requirement for the dynamic strain measurement being made during this test. The gauges have a nominal gauge factor of 2.1 for grid 1 (horizontal), 2.125 for grid 2 (diagonal), and 2.1 for grid 3 (vertical).

The resistance change of a strain gauge is very small and often poses a challenge to accurately measure. While many different methods exist to measure change in resistance, a Wheatstone bridge powered by a precision voltage source is commonly used as it provides the capability to measure both static and dynamic signals. The purpose of the Wheatstone bridge is to provide a means of measuring a voltage centered at 0 V and correlating that voltage to a resistance. A balanced bridge provides a reference voltage centered at 0 V. The sense voltage can be amplified via an operational amplifier. This method is an improvement over directly measuring the gauge resistance as resolution error is reduced. Adding such complexity into a measurement system brings additional challenges regarding the uncertainty and increased introduction of noise.Accuracy can be degraded by factors such as voltage losses in the lead lines attached to the gauge. Three wire gauges used in the quarter bridge configuration minimize this gauge desensitization effect [26] as opposed to a two wire configuration. While the three wire strain gauge configuration possesses other attributes such as temperature compensation, the thermal capacitance of the vessel negates thermal dynamics.

Nearly all the components necessary in a measurement including the operational amplifier, low pass filter, 5 MHz analog to digital converter, and precision bridge excitation are packaged into a commercially available data acquisition system. Bridge completion is external to the data acquisition system by using rack mounted precision 350 Ohm resistors. Care is taken to minimize the length of wire from the gauge to bridge completion reducing the desensitization effect. It is also desirable to minimize the length of wire before amplification. This has the potential to reduce the amount of noise that is amplified. The overpressure test required about 15 feet of wire to bridge completion and another 10 feet of wire before amplification. Both the vessel and data acquisition chassis are tied to the building electrical ground. This method introduces a potential ground loop and potential 60Hz noise but ensures components are electrically safe to touch.

During the shot, vessel acceleration is violent enough to rip wires off solder tabs due to inertial cable loading. Through practice, Lawrence Livermore National Lab has found this effect best mitigated by using small gauge wire (30 AWG) and strain relief wire loops between the gauge and solder tabs shown in Figure 11.

Figure 11. Foil strain gauge with adjacent solder tabs.

Proper gauge application is also necessary to ensure intimate contact of the gauge with the surface. Vishay surface preparation recommendations are followed including paint removal with an angle grinder, directional sanding, and pre-bond surface preparations applied. Vishay M-Bond AE-10 is applied to gauges and solder tabs to bond to the vessel surface. A vacuum pad is placed over the gauge to apply pressure to the gauge and the vessel surface for a day while the bond cures. A layer of Vishay M-Coat-A polyurethane is applied to the strain gauge and the solder tabs to protect gauges and provide electrical insulation.

2.4. FEA Simulations

Prior to an explosive test in a vessel, a hydrodynamic computation (for example, with CTH from Sandia National Laboratories) is conducted to simulate the HE blast and to calculate pressure history profiles at many tracer points located next to the internal wall of the vessel. The development of optical diagnostics capable of delivering low noise and high accuracy measurements of the strain are critical to improving the modeled vessel performance. As such, the experimental data is being used to verify or improve the FEA code rather than the code being used to explain the model. Normally we use 2D, axisymmetric models for the hydrodynamic computation, but depending on the HE setup inside the vessel, we also compute pressure profiles using 3D models. Next, with an explicit, dynamic Abaqus 3D FEA, we verify the vessel's structural integrity by subjecting it to the internal pressures computed with the hydrodynamic software. The pressure profiles are applied to areas assembled with element faces that are close to the location of their corresponding tracer point. Figure 12 depicts the FE model of the 0.9 m vessel with a shell thickness equal to 2.82 cm that was subjected to an overpressure test.

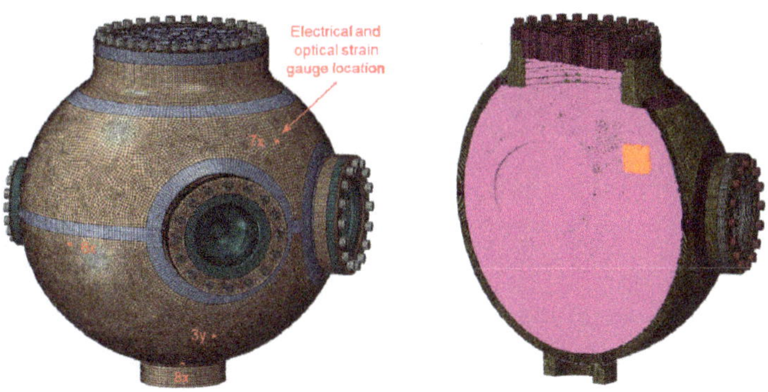

Figure 12. FE model of the 0.9 m vessel. Shown on the left are the elements used to compare the FE strains to the test measurements. The orange area highlighted on the cut-out on the right are element faces in the neighborhood of a tracer; the inside wall of the vessel is subjected to about 770 different pressure profiles.

Highlighted on the left figure are the finite elements that were used to correlate the computed strains to the test measurements, and on the right we show an small area to which we apply a pressure profile next to a tracer. To secure the covers to the nozzles, we preload the FE model for the bolts to the actual preload for the test; for example, in the 1.8 m vessel with a 6.35 cm thick shell shown in Figure 13, the preload for each bolt (of 64 per cover) is about 245,000 N (55,000lbf). It should be noted that the bolt preload affects the FE computation. Based on our experience with the tests and FE analyses, we run the FE computation to 10.0 ms, which is when the vibration of the vessel is diminished because of the natural material damping and the friction between covers and nozzles.

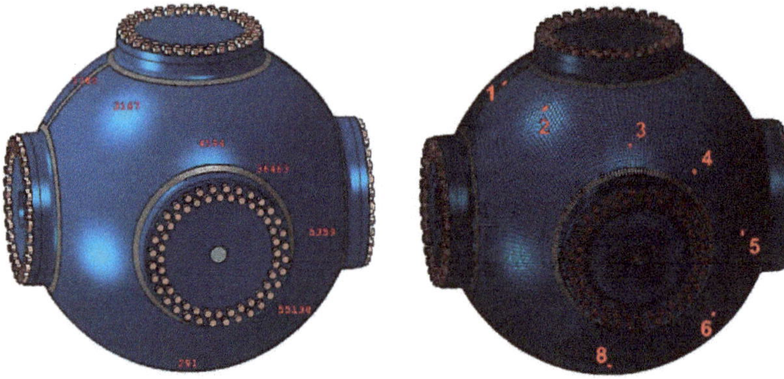

Figure 13. FE model of the 1.8 m vessel. The elements numbers for the FE correlation are shown on the left figure and the optical gauge number is on the right next to its corresponding element.

Briefly, these are the sequence of events during the transient FE results of an 1.8 m confinement vessel (Figure 13) when subjected to in internal HE blast: Following the first pressure surge at the tracers on the internal wall, which might last for about 0.5 ms, the vessel's shell expands and contracts in 1.0 ms. In the 1.8 m vessel, the heavier ports are slower to react to the initial pressure and the internal stresses from the expanding shell, but once they catch up with the dynamics of the shell, the whole vessel attains at near 2.5 ms a natural vibration. In general, when vibrating at its natural frequency of about 1300 Hz, there are locations on the vessel that reach their highest strain/stress levels. At 7.0 ms, the elastic stresses that sustained the natural vibration begin to dissipate as the results of damping and friction. Figure 13 shows the element numbers chosen for the finite element correlation in the left figure. The right figure shows the optical gauge number in the experimental FBG chain. For example, number 1 is the first gauge corresponding to the 1530 nm central wavelength and number 8 is the final gauge corresponding to the 1565 nm central wavelength.

To assess the structural soundness of the vessels subjected to impulsive loading, the FE results are compared to the limits stated in Code Case 2564 of the ASME Boiler and Pressure Vessel Code, Section VIII, Division 3. A Mises stress field and the equivalent plastic strain (named PEEQ in Abaqus) are shown in Figure 14. Note that the south pole of the vessel, located on the y-axis, accumulates large plastic strains. This is because the south pole is active in most modes of vibration of the vessel, and furthermore, it is locally blasted with a high pressure. Additionally, we compute transient loads in the bolts, which should not exceed the proof load of a bolt, and the relative displacements between covers and nozzles to ensure that there will not be any outgassing. Therefore, we recognize that to comply with the structural requirements of the vessels to the large amount of HE in the tests and the inherent difficulty when computing their internal pressures and dynamic response, we must validate our integrated hydrodynamic and FEA against strain measurements obtained on the shell of the vessels. Because this is our first effort, we have not adjusted parameters for the hydrodynamic and FE computations in order to obtain better correlations. Currently, the metrics for the comparisons simply include: visuals of the FE strain vs. the optical strain measurement; the short time Fourier transform

(STFT) to assess the frequency vs. time strain responses; and a FFT to assess the natural frequency of the vessels when the strains become large. Comparisons between these metrics will be shown in later sections.

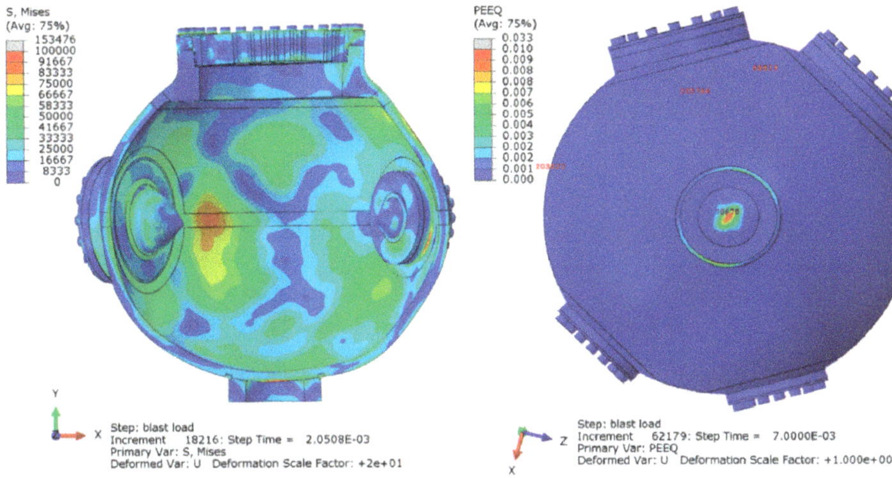

Figure 14. Mises stress at 2 ms, shown on a deformed cut-out, which has been scaled by a factor of 20. As shown the equivalent plastic strain (PEEQ) value is 1% plastic strain at the south pole of the vessel.

3. Results

3.1. Hydrotest 3685

3.1.1. RTLS Results

The first experiment where the vessel strain diagnostics were fielded was a proof-of-concept test of the strain on a containment vessel surface for Hydrotest 3685. A detonation event was staged inside a 1.8 m containment vessel at the Dual Axis Radiographic Hydrodynamic Test (DARHT) facility at Los Alamos National Lab. The experiment consisted of 8 FBG based strain gauges for use with the RTLS technique and 2 single fibers wrapped around the vessel for the Luna measurements and the SITS measurement. The combination of these three strain measurements allowed the overall vessel performance to be recorded and characterized both locally and globally.

The RTLS measurements are the primary way to characterize the vessel performance as they can be compared with legacy experiments that had electrical gauges fielded on them. For the 3685 hydrotest, the FBG gauges were laid out as seen in Figure 15. Figure 5b shows a sample of two gauges fielded on the vessel for hydrotest 3685.

For this first experiment, the gauges were positioned so that they would not interfere with other diagnostics. The mounting procedure was described in the Materials and Methods section. The FE analysis was performed with the gauge locations and orientations as measured to show the overall agreement. The gauges consisted of an FBG chain of the small light weight plastic type. Before the initiation of the detonation, sample reference spectra were saved for the chain of gauges. The results can be seen in Figure 16 for the plastic gauges.

In the static case, a single gauge can undergo 4200 $\mu\varepsilon$ of strain before the wavelength shift would cause the gauge data of two neighboring gauges in the chain to overlap. If the gauges moved in opposite directions, the amount of strain before overlap could be much less. Along with the measured spectra shown in Figure 16, a snapshot of the static result as measured on the oscilloscope is also shown. It is clear that the relative peak heights and spacing are preserved in the purely spectral measurement and with the swept wavelength laser reflecting from each grating in the chain. This is important for calibration purposes during the data processing portion of the experiment.

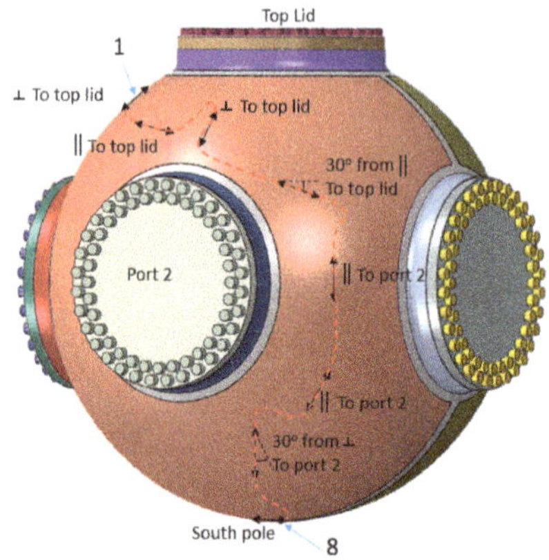

Gauge	Length along vessel from fiducial to center of gauge
1	36 cm to top lid
2	37 cm to top lid
3	58 cm to top lid
4	76 cm to top lid
5	42 cm from port 2
6	45 cm from port 2
7	59 cm to port 2
8	South pole

Figure 15. Locations and orientations of the gauges on the vessel for hydrotest 3685. The distances to a known fiducial are also shown. These maps are provided to the modeling team in order to compare the measured strain with the pre-shot predictions at these particular locations.

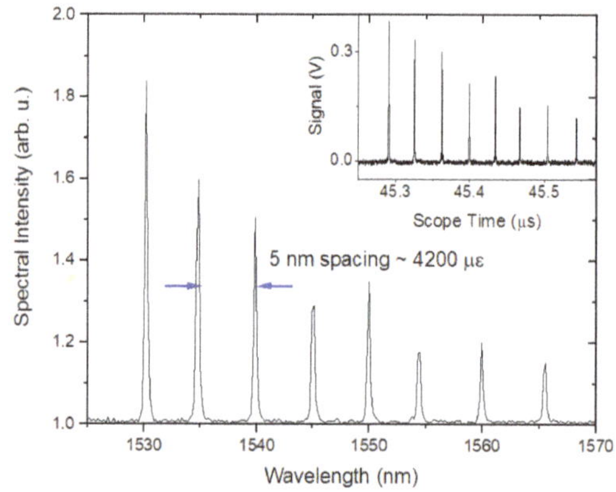

Figure 16. Reflected calibration spectra recorded with a spectrometer for the plastic-backed optical strain gauges. The insets show the reflected spectrum as recorded with the RTLS technique before the detonation event. In both cases, the RTLS representation qualitatively matches the spectral response very well.

After analyzing the data, 2D plots of the signal recorded in each gauge were made as shown in Figure 17. Each horizontal line is a different gauge plotted as a function of time. For the plastic gauges shown in Figure 17, 8 clean, individual signals with no crossings were recorded.

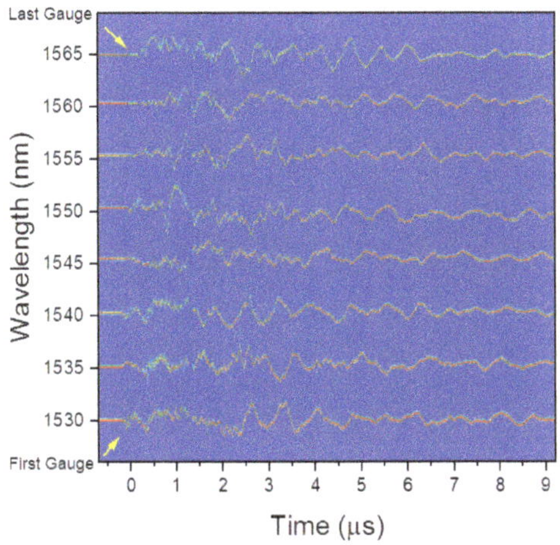

Figure 17. Experimental RTLS measurements for an 8 optical strain gauge chain on hydrotest 3685 for plastic-backed gauges. All 8 of the plastic-backed gauges remained on the surface. The center wavelength of each gauge is listed on the vertical axis.

The plastic gauges yielded strain measurements at all 8 points for the 3685 hydrotest. Figure 18 shows the results for all 8 gauges. A ringing periodicity of ~0.7 ms is seen in all 8 gauges, consistent with the expected frequencies calculated in the FEA. The source of the ringing is the natural breathing of the vessel after HE detonation when the masses of the port covers are taken into consideration. Peak strains of nearly 2000 με were seen after the initial transient from t0 to ~2.5 ms. This was well below the yield strength of the vessel and indicated that the strain at the 8 points tested did not exceed the safety requirements for the vessel.

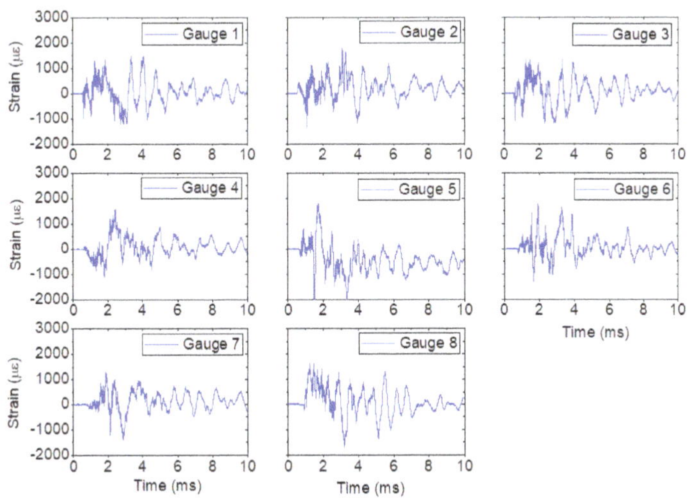

Figure 18. Optical strain measurements for plastic-backed strain gauges for hydrotest 3685. All results show consistent peak strains and ringing. Gauge 1 is shown at the bottom of Figure 17a and corresponds to center wavelength 1530 nm. The other gauges follow in 5 nm increment center wavelengths.

3.1.2. Luna and Spectral Interferometry for Transient Strain (SITS)

The local transient measurements of the elastic strain on the vessel can be combined with the global transient and static measurements for characterization of the elastic and plastic deformation of the vessel, respectively. The vessel used for hydrotest 3685 had a diameter of 1.8 m. The circumference was 5.7 m but, due to the locations of the ports on the vessel, a true great circle around the vessel was not available. The fiber for this shot was routed around the vessel as close to a great circle as possible. For the global plastic measurements, Luna OBR4600 data wereused. For several hours prior to the start of the experiment, Luna scans were saved in 15 min intervals. Figure 19a shows the stability of the circumference of the vessel before the detonation event. A spread in the data of 490 um was observed over the course of 4 h which can easily be attributed to thermal expansion of the vessel as the temperature rises from the morning until shot time. After detonation, scans were again saved in 15 min intervals. As the vessel cooled, small changes due to thermal expansion could be recorded as well.

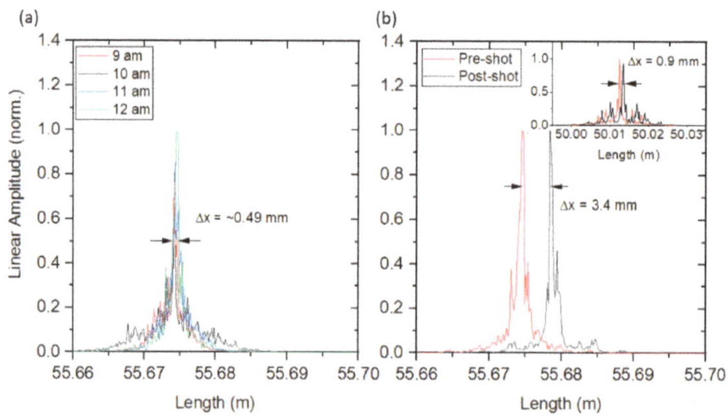

Figure 19. Luna OBR results for hydrotest 3685. (**a**) OBR length measurements prior to detonation in 1 h increments. The total length of the fiber varied by 0.49 mm over this time and can be attributed to the vessel thermally expanding as the temperature increase during the day. (**b**) OBR length measurement immediately after the detonation event. A total length change over the full circumference of the vessel was measured as 3.4 mm. The inset in (**b**) shows the result of a fiber 25% the length of the circumference epoxied onto the surface. The total length change was found to be 0.9 mm, which was consistent with the full circumference fiber.

Only small changes were observed after the detonation event, implying the vessel either stayed at an elevated temperature for an extended period of time, was elastically deformed due to the increased pressure from the phase change of the HE products from solids to gas, or was plastically deformed. As the RTLS data did not imply plastic deformation at any of the 8 measured points, the increased circumference was probably due to pressure and temperature changes in the vessel. Figure 19b shows the total change recorded by the Luna OBR4600 from 4 min before the shot until 1 min after the shot. A total length change of 3.4 mm was recorded around the full vessel, which did not change over the course of an hour. Additionally, the inset of Figure 19b shows the result of a 1.4 m fiber also attached to the vessel. This fiber was ~25% of the length of the total circumference fiber and so a corresponding reduced length change was expected. The results were consistent with a 0.9 mm length change as measured in the shorter fiber.

The dynamic change in the total circumference was also recorded using the SITS technique. The total length change of the vessel was calculated for each pulse of the laser, resulting in 50,0000 individual length measurements for the 10 ms time window that data were recorded. Figure 20 shows the 2D image of the sideband plotted as a function of time.

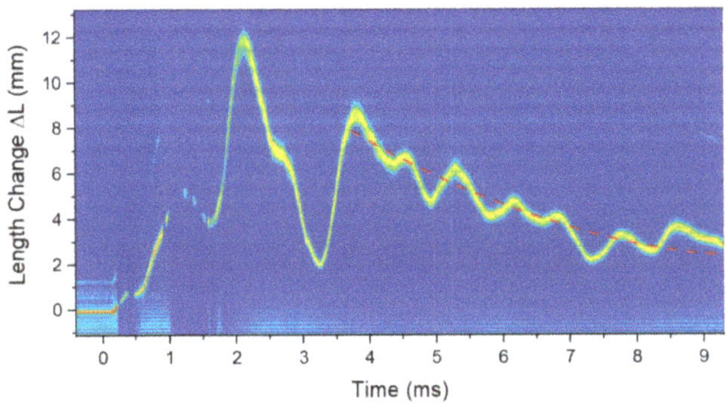

Figure 20. SITS result for the total circumferential change of the vessel during the first 10 ms after the detonation event for hydrotest 3685. Peak values of over 1 cm were seen and the overall result decayed back to an increased circumference by the end of the 10 ms recording time.

The vertical axis has been calibrated as the change in length of the total fiber and the axis has been shifted so that 0 change occurs prior to t0. This is the original offset in delay the two legs of the interferometer were set to. The overall effect is consistent with the RTLS measurements. A peak value is found around 2 ms which was seen in many of the individual gauges fielded with the RTLS technique. Additionally, the minimum in the data was found around 3 ms when the vessel quickly relaxes after the initial shock from the detonation. Since this is a total length change, the results in Figure 20 show that the vessel circumference increased by over 1.2 cm at the peak. Dividing this by the total vessel circumference yields a total value of $\Delta L/L$ of 0.002, or 2000 $\mu\varepsilon$. The breathing periodicity is ~1.4 ms, in agreement with the RTLS method. Finally, it can be seen that the total length change asymptotically relaxes to a value above the state before t0. Fitting a curve to guide the eye to this portion shows a final value of ~3 mm. While this is only a 10 ms snapshot of the total vessel circumference change, the result shows agreement with the OBR4600 results. The final state of the vessel circumference can again be attributed to either plastic deformation, or increased temperature, or some combination of these two variables.

3.1.3. Comparison with Simulations

In the case of the 1.8 m vessel used for hydrotest 3685, as compared to the optical gauge measurements, the FE model accurately predicts the magnitude of the strains as seen in Figure 21. The figures show the raw experimental data (blue lines) overlaid with the FE prediction (red lines). The locations of the gauges were determined from pre-shot measurements of the gauges on the vessel surface. Figure 21a shows the results for gauge location 1 and Figure 21b shows the locations for gauge position 3 (see Figure 13 for reference).

The central goal of the FEA predictions is to show that the peak strain values and the primary breathing oscillation are predicted. Currently, improvements to the model are being made using the low noise and repeatable results from the optical diagnostics. The model itself is complex with the results heavily influenced by the shape of the vessel with the entrance and exit ports, the mass of all covers, and particularly the influence of the HE charge and any shielding contained within the vessel. Differences seen between the experimental results and the prediction can be attributed to all of these effects, but the results are showing a close match in the behaviors of most importance to qualifying the safety performance of the containment vessel-mainly the peak strain values and oscillation frequency.

In addition to the strain amplitudes, an analysis of the vibration of the vessel was performed. Figure 22 shows the experimental data ((a) upper figure) and the FE prediction ((b) upper figure) for gauge location 3.An STFT was performed on the data using 8192 data points for the experimental data and 128 points for the FEA and can be seen in the middle plots of Figure 22.The results show similar spectral responses across the full temporal signal with a main peak around 1300 Hz. This is the calculated breathing mode for the vessel and appears once the vibrations including the mass of the lids and covers reaches a steady-state. An additional lower amplitude and frequency peak is seen under 500 Hz in the experimental results. It is unknown what the source of this is as it was not predicted in the FE. This is the subject for ongoing research. The STFT results also agree with the FFT seen in the lower plots in Figure 22. The FFT is applied to the time window from 2.6 to 7.6 ms. In both cases, the main frequency peak is found at 1300 Hz.

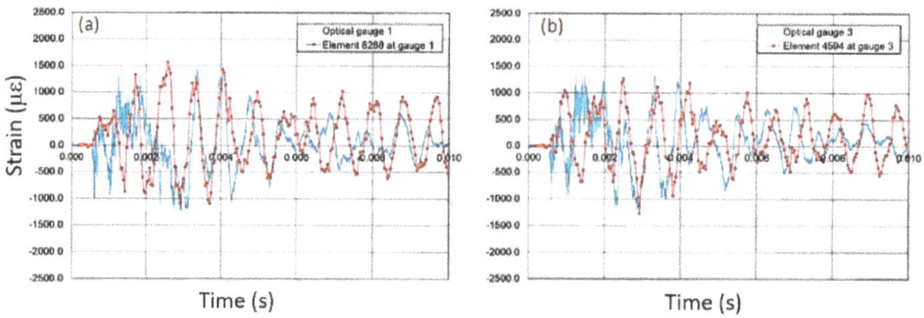

Figure 21. Comparisons between optical gauge and FE strains at (**a**) location 1 and (**b**) location 3 (see Figures 13 and 15) for the 1.8 m vessel used in hydrotest 3685.

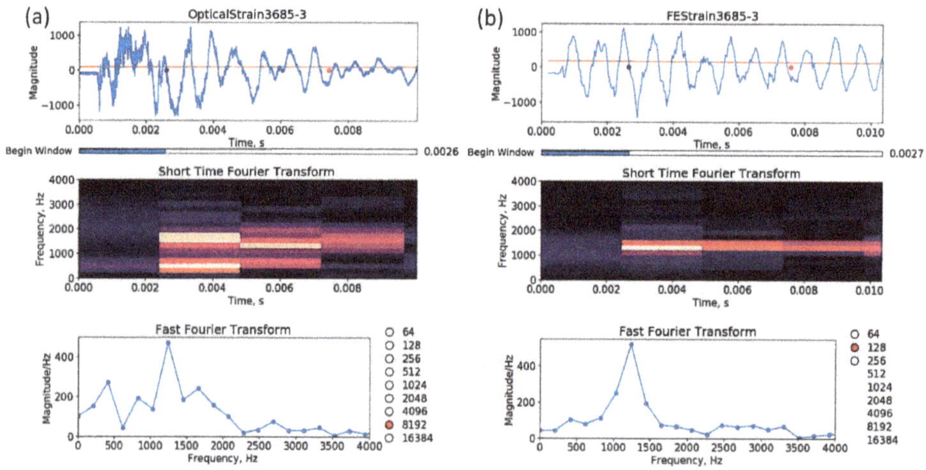

Figure 22. Comparisons between an optical strain measurement (**a**) and its corresponding FE computation (**b**) for gauge 3 on the 1.8 m vessel (see Figure 13). The top plots are the strain magnitude in micro-strains versus the time in seconds (note that between the plots, the x- and y-axis are scaled differently); STFTs of frequency in Hz versus time are plotted in the middle; and at the bottom are the FFTs between 2.6 and about 7.6 ms. From the optical strain data, the STFT and the FFT are computed using $2^{13} = 8192$ data points; and from the FEA, we used $2^7 = 128$ points.

3.2. *Overpressure Test (OPT)*

RTLS Results

A second experiment, an overpressure test or OPT was carried out on a 0.9 m vessel with only the RTLS techniques. This vessel test was different than hydrotest 3685 in that the vessel was a 0.9 m diameter vessel rather than the larger 1.8 m diameter vessel. The point of an OPT is to ensure that the vessel and any diagnostic feedthroughs in the vessel ports are able to withstand the pressures generated during the dynamic experiment. The 0.9 m vessels have a thinner wall of 2.82 cm nominal. The OPT consisted of 4 hand-packed cylinders of HE, in this case, composition 4 (C4) with a total mass of 0.277 kg per cylinder. However, the OPT also fielded a set of resistive electrical gauges on the surface. As the electrical gauges were a legacy diagnostic, they were primary and were given preference on the placement locations. The optical gauges were fielded nearby whenever possible so that direct comparisons could be made. Two identical chains of optical gauges of the plastic variety were fielded on the vessel for redundancy. Ensuring the optical gauges had the same orientation as well as being placed in the same locations was not always possible however so only a small subset of the optical and electrical gauges can be compared. Figure 23 shows an image of the optical gauges and electrical gauges as fielded on the vessel.

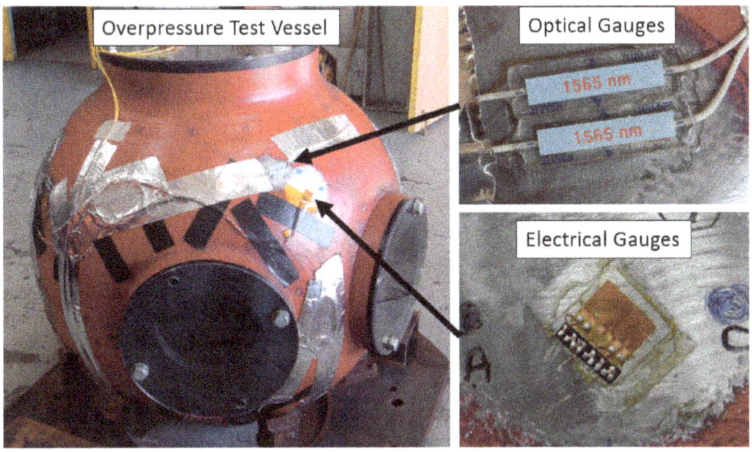

Figure 23. Optical gauges as placed on the overpressure test (OPT). Only plastic gauges were used for the optical measurements. A set of electrical gauges was also fielded which can be compared directly to the optical gauges.

The gauge locations were chosen by the modelers as points of particular interest for measuring the strain on the vessel. The electrical gauges were applied to the surface first as they were the primary diagnostic for strain. The mounting technique used double bubble epoxy and the same mounting technique as described before. The optical gauges were then attached to the surface as close to the electrical gauges as possible. The optical gauges chosen for this experiment measured strain only along the FBG length while the electrical gauges chosen could measure in 3 directions. FBG gauges 3, 6, and 8 were fielded with locations and orientations very near the corresponding electrical gauges.

Figure 24 shows a spectrogram for an optical gauge chain on the OPT vessel. In the lowest trace, the strain measurements show crossings, indicating that the strain exceeded the amount expected for the design of our FBG chains. This gauge was partially peeled from the surface during movement of the vessel so it did not record real vessel strain data.

The peak strain values as measured for all of the FBG gauges in a single chain (see Figure 25) are very close to those seen on hydrotest 3685 shown in Figure 18.

While the charge was smaller for the OPT, the 0.9 m vessel walls were also thinner, resulting in a comparable strain observed. The 8 independent measurements of the strain were obtained around the vessel with ringing periodicity of ~390 μs, again consistent with the FEA predicted vessel breathing after including the mass of the lids and covers. Two gauge chains were fielded and Figure 26 shows the agreement between two of the gauges. Many of the gauges on the vessel were not fully adhered to the surface but the agreement was still quite obvious. Gauges 6 and 7 as seen in Figure 26 showed the best agreement.

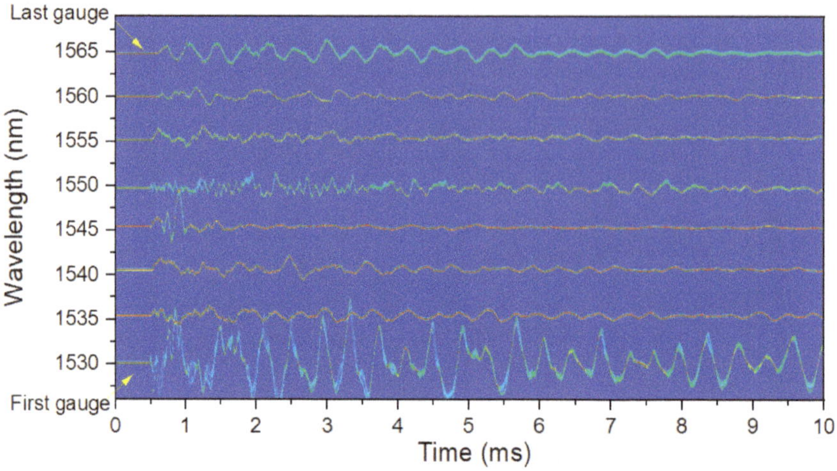

Figure 24. RTLS measurements for 8 plastic-backed gauges on the OPT vessel. The lowest trace corresponding to center wavelength 1530 nm was not fully adhered to the surface.

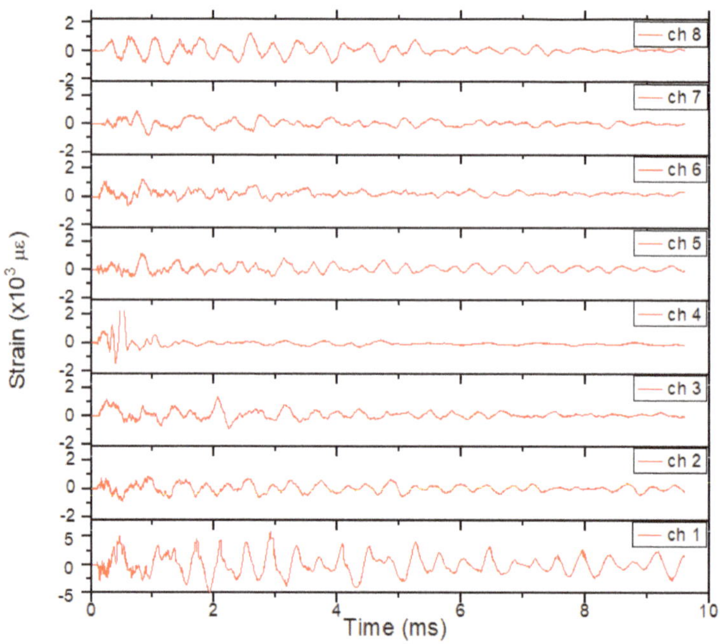

Figure 25. Extracted strain plots for each gauge fielded on the OPT. These correspond directly with the array seen in Figure 23.

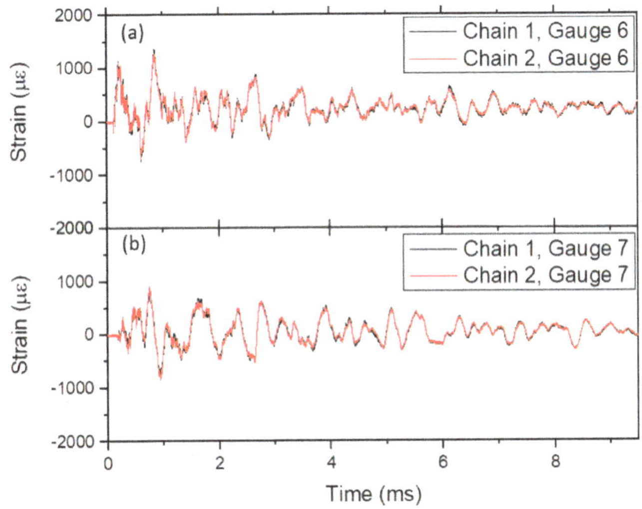

Figure 26. Strain plots for closely located optical gauges on the vessel. The gauges come from two distinct diagnostic chains (red and black lines) and were mounted as close as possible to each other. (**a**) Gauge 6 with wavelength 1555 nm and (**b**) gauge 7 with wavelength 1560 nm.

Maximum differences of ~7% of the strain are seen at certain times and can possibly be attributed to how the two gauges were bonded to the surface of the vessel but more likely to the small offset in position between the two gauges. The peak values were well below the maximum strain allowed for this particular vessel.

3.3. Comparison of Optical and Electrical Diagnostic with FEA Simulations

Due to vessel surface area and time constraints, the electrical and optical strain gauges were placed at various locations that were not always in close proximity or at the same orientation to each other. Since the purpose of this section is to show the differences between the optical and electrical gauges, three locations were chosen where the electrical and optical gauges were fielded under nearly identical conditions. Figure 12 shows the locations of these gauges as 3y, 6x, and 7x. Figure 27a shows the data for electrical gauge 6x after a 250 kHz low pass filter was implemented in software using a 53rd order finite impulse response filter. As the electrical gauges have long coaxial cables running between the diagnostic room and the firing point near the vessel, they are able to pick up a significant amount of electrical noise during the explosive test. This is seen in the figure as a high frequency signal remaining after filtering around 250 kHz which is much higher than any predicted strain (black curve). Additionally, spurious peaks of strain that exceed predictions by several times are also seen. To overcome even more noise in the data, the signal was further smoothed with 20 points adjacent averaging (red curve). This was done to see the overall strain amplitudes and slower periodic structure in the signals for a closer approximation to the optical signals. Figure 27b shows the optical data (blue curve) overlaid with the smoothed electrical signal (red curve). It should be emphasized that the optical data is in its raw form-no smoothing has been done to the temporal signal.

As an optical diagnostic, the result shows none of the high frequency noise seen with the electrical gauge even though the optical diagnostic is sampled at nearly 2 MHz which is several times higher frequency than the noise seen with the electrical signals. To see the overall agreement between the legacy electrical gauges and the new optical gauges, the electrical signals will be smoothed with a 20 point adjacent average throughout this section. Figure 28 shows the results of the smoothed electrical gauges (red lines) plotted with the raw optical gauges (blue lines) for locations 3y and 7x. In all figures, the data for only the first 5 msareshown in order to demonstrate the agreement in the short timescale features of the strain. As can be seen in Figures 27b and 28, the optical and electrical gauges show overall good agreement in terms of strain values, periodicity, and fine features in the temporal profiles. For a direct comparison between the electrical and optical data, the electrical data can only be justified to be smoothed by a ~3 point adjacent average since the data was sampled at 5 MHz as compared to the optical data at 1.7 MHz. However, a 20 pointtapboxcarfilter was used to most closely match the noise in the optical data. This demonstrates clearly the electrical gauges are highly sensitive to electrical noise. This was in addition to a 53rd order finite impulse response filter with an upper limit of 250 kHz applied to the raw signals. The largest source of this noise comes from the broad spectrum x-ray source used for radiography of the dynamic experiments. The electrical diagnostic picks up this noise while the optical diagnostic is not sensitive to it. While there is no physical basis for this degree of filtering, the optical gauges alleviate the issue of such significant post-processing of the raw signals. Differences in the measured strains seen in each plot can be attributed to small offsets in gauge positions on the surface of the vessel.

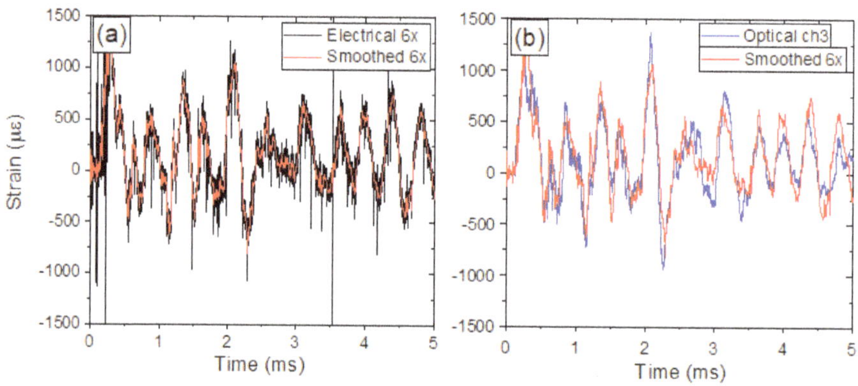

Figure 27. (**a**) Electrical gauge 6x raw (black line) and smoothed (red line). (**b**) Raw optical gauge (blue line) and smoothed electrical gauge 6x.

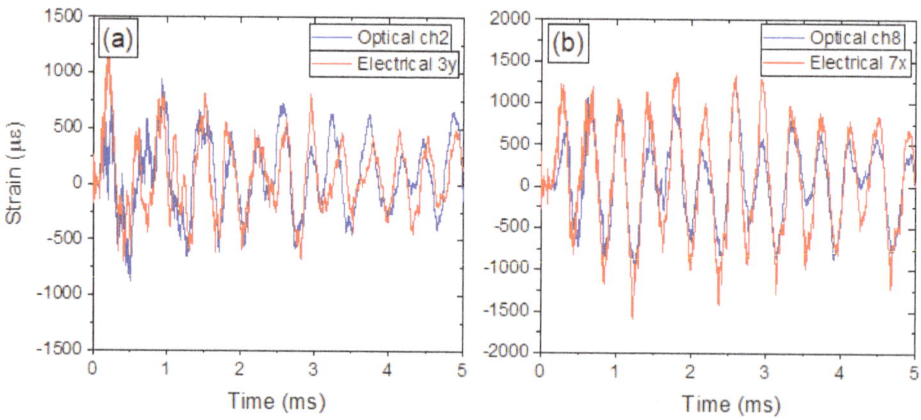

Figure 28. (**a**) Smoothed electrical gauge 3y raw (red line) and corresponding raw optical gauge (blue line). (**b**) Smoothed electrical gauge 7x raw (red line) and corresponding raw optical gauge (blue line).

Figure 29 shows the FE results for the 0.9 m vessel used in the OPT for a single optical gauge for comparison.

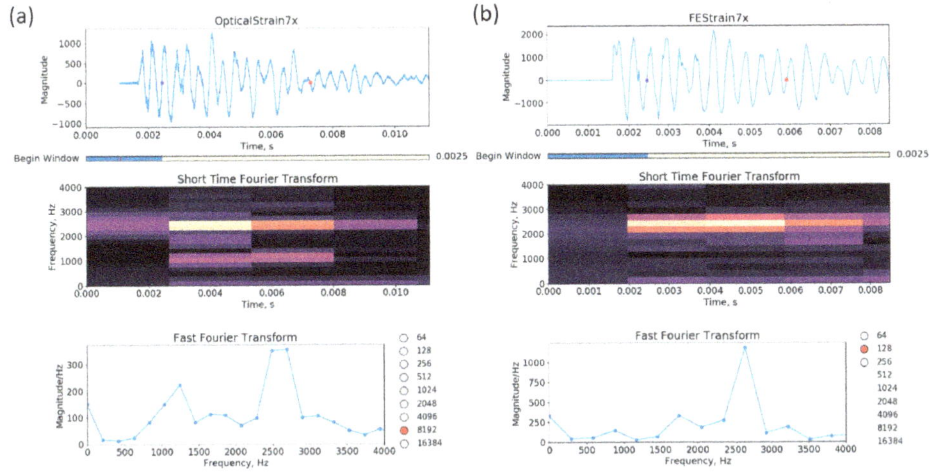

Figure 29. Optical strain measurement (**a**) versus its corresponding FE computation (**b**) for gauge 7x. On the plots, the x- and y-axis are scaled differently. The blue and red dots mark the beginning and end of the windows for the FFTs; with the optical gauge, the window contains 2^{13} = 8192 data points, and with the FE, 2^7 = 128 data points.

In the case of the 0.9 m vessel, the strain magnitude plots show that the FEA over-predicts the strain by a factor of two at the location of optical gauge 7x by nearly a factor of 2 (Figure 29 upper plots). We can see that we should add more damping to the FE model to reduce the vibration after about 7 ms. The STFTs (middle plots) are reasonably similar, and the FFT (lower plots), which is applied to the strain response between 2.5 and 6 ms, depicts a natural frequency of vibration at about 2600 Hz. Although the response is expected to be transient, it is interesting that for a few milliseconds the vessel is at a natural frequency, which produces the larger strains in the FEA and the test.

4. Discussions

Sections 3.1.3 and 3.3 show the agreement between the experimental optical gauges and the FEA for a 1.8 m vessel and a 0.9 m vessel, respectively. Overall, the model predicts the strain amplitude well, particularly for the 1.8 m vessel. The frequency response for both vessels also agrees very well with the FEA. As was discussed previously, the RTLS diagnostics showed excellent agreement with the SITS and Luna diagnostics as fielded on hydrotest 3686. Peak values of localized strain, as recorded by the optical strain gauges, were recorded to be nearly 2000 με. The total integrated strain around a single great circle of the vessel was also found to be a peak value of ~1.2 cm/5.7 m, or ~2100 με. Additionally, the total vessel circumference over the course of 10 ms asymptotically approached a final increased length of ~3mm, consistent with the Luna measurements made at the same time. A length change of 3 mm/5.7 m corresponds to a strain of ~500 με. The RTLS measurements also show a final positive value for the local strain in the vessel surface of ~50–100 με. Further work needs to be conducted to calibrate the gauges for use with such small values of strain.

The electrical gauges fielded on the OPT showed overall very good agreement with the optical gauges. As the optical gauges are not sensitive to electrical or RF noise, the signals were significantly cleaner that that observed with the electrical gauges. After significant smoothing of the electrical data, many features were shown to be consistent in the data, for example the ringing periodicity and the amplitudes of the strain. Many of the finer features were also represented in both diagnostics, although often with slightly different strain amplitudes.

The inclusion of optical strain diagnostics on a containment vessel is a new diagnostic option for the Los Alamos National Lab hydrotest program. Due to the developmental nature of this work, a series of optical diagnostics was simultaneously fielded on the 3685 hydrotest as described in the preceding sections. It was found with the RTLS diagnostics that the lightweight plastic versions of the FBG strain gauges were significantly more reliable than the heavier metal-backed gauges. Properly welding the metal gauges may improve reliability but this is not an option as welding components onto the vessel is not allowed from a safety qualification standpoint. However, if the impulse on the vessel is expected to be small or the gauges can be ensured to remain adhered to the vessel, it is expected that the metal-backed gauges would respond equally to the plastic gauges. This would be valuable as the metal gauges are specifically designed to be far more robust than the plastic gauges and would survive in harsher environments.

The RTLS results were further verified through comparisons with two additional diagnostics. The first diagnostic fielded for verification of the RTLS results was the SITS technique. This diagnostic made a precise measurement of the total vessel circumference during the dynamic experiment. Peak strain values of ~2000 $\mu\varepsilon$ were observed as determined from the total peak vessel circumference change of 1.2 cm out of 5.7 m. This was consistent with the localized peak strain measurements made with the RTLS technique. The ringing periodicity was found to be consistent between the RTLS and SITS techniques as 1.4 ms in agreement with the natural vessel breathing as seen in the FEA predictions. Additionally, the SITS technique showed that as the peak strain values integrated around the total circumference decayed, a final value above the initial state was measured. Further analysis of this final value is required however since SITS measurements ended 10 ms after the initiation of the experiment. Results indicate similar behaviors as compared with previous hydrotest experiments [15–19] when comparing the electrical diagnostic results. Hydrotest 3685 and the OPT are the first experiments showcasing the behavior of the optical diagnostics, and show excellent agreement with the electrical diagnostics although without the noise associated with the hydrotest experiments. The reduction in noise means the results can be compared directly to modeled predictions without the ambiguity associated with filtering the noisy data.

The second verification diagnostic was a series of nearly static vessel circumference length measurements made with a Luna OBR4600. The measurements showed a total length increase of 3.4 mm from before the experiment, consistent with the final result of the SITS technique. Over the course of an hour, this length did not change. However, an analysis of the thermal expansion of the vessel caused by the temperature increase from the dynamic experiment also showed an upper limit on the expected circumference change of 3.6 mm, consistent with the measured results. This showed that the increase could be fully attributed to thermal effects and the vessel was expected to relax back to its original circumference after the vessel fully cooled.

Finally, the results were also compared with legacy electrical strain gauges as fielded on an additional experiment. When the orientation and placement of the FBG and electrical gauges was consistent, the results showed excellent agreement. The large sources of noise inherent to hydrotests pose challenges to accurately predicting peak strains in the vessel. We have shown that the optical techniques show such a significant reduction in the noise, that no further smoothing is required to match the signals expected in the simulations.

All of the combined results showed good agreement with the FEA modeling. The overall peak values of strain and periodic oscillations were correctly reflected in the modeled predictions. The FEA modeling is currently being further improved using the results from the optical RTLS method on these two experiments. Further testing with simplified experimental parameters is being developed to improve the modeled predictions. The results implied that the vessels did not exceed the maximum strain that might compromise the vessel wall strength. With this evidence, the RTLS technique was demonstrated to be accurate for fast dynamic strain measurements even when high values of strain are expected. The reduced noise present in optical strain gauges versus electrical shows improved accuracy of the instantaneous strain measurement.

Author Contributions: Conceptualization, S.G. and M.P.; Formal analysis, D.C., G.S. and T.B.; Investigation, S.G. and S.S.; Project administration, J.G.; Software, M.P. and G.S.; Writing—original draft, S.G. All authors have read and agreed to the published version of the manuscript.

Funding: This research received no external funding.

Conflicts of Interest: The authors declare no conflict of interest. The funders had no role in the design of the study; in the collection, analyses, or interpretation of data; in the writing of the manuscript, or in the decision to publish the results.

References

1. Sahota, J.; Gupta, N.; Dhawan, D. Fiber Bragg grating sensors for monitoring of physical parameters: A comprehensive review. *Opt. Eng.* **2020**, *59*, 60901. [CrossRef]
2. Campanella, C.; Cuccovillo, A.; Campanella, C.; Yurt, A.; Passaro, V. Fibre Bragg grating based strain sensors: Review of technology and applications. *Sensors* **2018**, *18*, 3115. [CrossRef]
3. Jung, J.; Nam, H.; Lee, B.; Byun, J.; Kim, N. Fiber Bragg grating temperature sensor with controllable sensitivity. *Appl. Opt.* **1999**, *38*, 13. [CrossRef] [PubMed]
4. Broadway, C.; Kinet, D.; Theodosiou, A.; Kalli, K.; Gusarov, A.; Caucheteur, C.; Megret, P. CYTOP fibre Bragg grating sensors for harsh radiation enviornments. *Sensors* **2019**, *19*, 2853. [CrossRef]
5. Marques, C.A.F.; Pospori, A.; Demirci, G.; Cetinkaya, O.; Gawdzik, B.; Antunes, P.; Bang, O.; Mergo, P.; Andre, P.; Webb, D.J. Fast Bragg grating inscription in PMMA polymer optical fibres: Impact of thermal pre-treatment of preforms. *Sensors* **2017**, *17*, 891. [CrossRef]
6. Leal-Junior, A.; Theodosiou, A.; Diaz, C.; Marques, C.; Pontes, M.J.; Kalli, K.; Frizera-Neto, A. Polymer optical fiber Bragg gratings in CYTOP fibers for angle measurement with dynamic compensation. *Polymers* **2018**, *10*, 674. [CrossRef]
7. Leal-Junior, A.G.; Rocha, H.R.O.; Theodosiou, A.; Frizera, A.; Marques, C.; Kalli, K.; Ribeiro, M.R.N. Optimizing linearity and sensitivity of #D-printed diaphragms with chirped FBGs in CYTOP fibers. *IEEE Access* **2020**, *8*, 31983.
8. Min, R.; Ortega, B.; Broadway, C.; Hu, X.; Caucheteur, C.; Bang, O.; Antunes, P.; Marques, C. Microstructured PMMA POF chirped Bragg gratings for strain sensing. *Opt. Fiber Technol.* **2018**, *45*, 330. [CrossRef]
9. Ma, Z.; Chen, X. Fiber Bragg gratings sensors for aircraft wing shape measurement: Recent applications and technical analysis. *Sensors* **2019**, *19*, 55. [CrossRef] [PubMed]
10. Tiwari, U.; Garg, R.; Vishwakarma, M.; Das, B.; Bhatnagar, R. Strain monitoring of concrete bridge using fiber Bragg grating sensors. In Proceedings of the 12th International Conference on Fiber Optics and Photonics OSA Technical Digest (online), Kharagpur, India, 13–16 December 2014.
11. Yuksel, K.; Kinet, D.; Chah, K.; Caucheteur, C. Implementation of a mobile platform based on fiber Bragg grating sensors for automotive traffic monitoring. *Sensors* **2020**, *20*, 1567. [CrossRef]
12. Kim, H.; Song, M. Fiber laser FBG sensor by using a spectrometer demodulation. *Proc. SPIE* **2011**, *8073*, 80730B.
13. Rodriguez, G.; Gilbertson, S. Ultrafast fiber Bragg grating interrogation for sensing in detonation and shock wave experiments. *Sensors* **2017**, *17*, 248. [CrossRef] [PubMed]
14. Rodriguez, G.; Jaime, M.; Balakirev, F.; Mielke, C.H.; Azad, A.; Marshall, B.; La Lone, B.M.; Henson, B.; Smilowitz, L. Coherent pulse interrogation system for fiber Bragg grating sensing of strain and pressure in dynamic extremes of materials. *Opt. Express.* **2015**, *23*, 14219. [CrossRef]
15. Duffey, T.; Rodriguez, E. *Overview of Pressure Vessel Design Criteria for Internal Detonation (Blast) Loading*; Los Alamos National Laboratory: Los Alamos, NM, USA, 2001; pp. 1–13, Technical Report: LA-UR-01-2621.
16. Rodriguez, E.; Nickell, R.; Pepin, J. Design considerations for blast loads in pressure vessels. In Proceedings of the 19th International Conference on Structural Mechanics in Reactor Technology (SMiRT-19), IASMiRT, Toronto, ON, Canada, 12–17 August 2007.
17. Giglio, M. Spherical vessel subjected to explosive detonation loading. *Int. J. Pres. Ves. Piping* **1997**, *74*, 83. [CrossRef]
18. Lewis, B. *Confinement Vessel Analysis Final Report*; APTEK, Inc.: Colorado Springs, CO, USA, 1992.
19. Karpp, R.; Duffey, T.; Neal, T. Response of containement vessels to explosive blast loading. *J. Pres. Ves. Tech.* **1983**, *105*, 23. [CrossRef]

20. Raja, S.; Govindarajan, A.; Mahalingam, A.; Lakshmanan, M. Tailoring inhomogeneous PT-symmetric fiber Bragg grating spectra. *Phys. Rev. A* **2020**, *101*, 33814. [CrossRef]
21. Wieser, W.; Klein, T.; Draxinger, W.; Huber, R. Fully automated 1.5 MHz FDML laser with 100 mW output power at 1310 nm. *Proc. SPIE* **2015**, *9541*, 954116.
22. Wieser, W.; Klein, T.; Adler, D.; Trepanier, F.; Eigenwillig, C.; Karpf, S.; Schmitt, J.; Huber, R. Extended coherence length megahertz FDML and its application for anterior segment imaging. *Biomed. Opt. Express* **2012**, *3*, 2647. [CrossRef]
23. Soller, B.; Gifford, D.; Wolfe, M.; Froggatt, M. High resolution optical frequency domain reflectometry for characterization of components and assemblies. *Opt. Express* **2005**, *13*, 2. [CrossRef]
24. Goda, K.; Jalali, B. Dispersive Fourier transformation for fast continuous single shot measurements. *Nat. Photonics* **2013**, *7*, 102. [CrossRef]
25. Xia, H.; Zhang, C. Ultrafast ranging lidar based on real-time Fourier transformation. *Opt. Lett.* **2009**, *34*, 14. [CrossRef] [PubMed]
26. Micro-Measurements. *The Three Wire Quarter Bridge Circuit*; Micro-Measurements: Raleigh, NC, USA, 2015; Tech Tip-612; p. 11092.

Publisher's Note: MDPI stays neutral with regard to jurisdictional claims in published maps and institutional affiliations.

sensors

MDPI

Article

Experimental Study on Glaze Icing Detection of 110 kV Composite Insulators Using Fiber Bragg Gratings

Jie Wei [1], Yanpeng Hao [1,*], Yuan Fu [1], Lin Yang [1], Jiulin Gan [2] and Han Li [1]

[1] School of Electric Power, South China University of Technology, Guangzhou 510640, China; epjie1123@mail.scut.edu.cn (J.W.); epyuanfu@mail.scut.edu.cn (Y.F.); eplyang@scut.edu.cn (L.Y.); 201630214199@mail.scut.edu.cn (H.L.)

[2] State Key Laboratory of Luminescent Materials and Devices, South China University of Technology, Guangzhou 510640, China; msgan@scut.edu.cn

* Correspondence: yphao@scut.edu.cn; Tel.: +86-134-5043-7306

Received: 23 February 2020; Accepted: 24 March 2020; Published: 26 March 2020

Abstract: Icing detection of composite insulators is essential for the security and stability of power grids. As conventional methods have met difficulties in harsh weather, a 110 kV composite insulator with embedded Fiber Bragg Gratings (FBGs) was proposed for detecting glaze icing in this paper. FBG temperature compensation sensors in ceramic tubes were adopted for simultaneous measurement of icicle loads and temperature. Then, temperature calibration experiments and simulated icicle load experiments were carried out to obtain temperature and icicle load characteristics of FBGs. The results showed that temperature sensitivities of FBG strain sensors and FBG temperature compensation sensors were 18.16 pm/°C, and 13.18 pm/°C, respectively. Besides, wavelength shifts were linearly related to icicle loads within the polar angle range of −60° to 60°, and the load coefficient of FBG facing the icicle was -34.6 pm/N. In addition, the wavelength shift generated by several icicles was equal to the sum of wavelength shifts generated by each icicle within the polar angle range of −15° to 15°. Finally, icicles can cause wavelength shifts of FBGs within a big shed spacing. The paper provides a novel icing detection technology for composite insulators in transmission lines.

Keywords: Fiber Bragg Grating (FBG); composite insulator with embedded FBGs; glaze icing; icing detection

---

1. Introduction

Since the 21st century, the global climate has changed dramatically, and the shortage of conventional energy has become increasingly severe. It is significant to construct smart grids for the electric power industry [1]. Currently, the sensing technologies of electrical equipment are an essential foundation to ensure safe operation, promote controllability, and achieve intelligence of smart grid [2].

In March 2019, the Internet Department, State Grid Corporation of China, published an outline for the construction of the Ubiquitous Power Internet of Things (UPIoT). UPIoT includes four layers: sensing layer, network layer, platform layer, and application layer. In particular, the sensing layer is comprised of various sensors, edge computing equipment, and local communication networks to achieve acquisition, aggregation, and processing of data from conditions of power equipment. In summary, intelligent sensing technologies are one of the core foundations of the sensing layer in UPIoT [3].

Insulators in transmission lines are easily covered by ice in cold and high humidity regions, resulting in flashover accidents and even severe disasters of power grids. To date, there have been several methods (e.g., ultraviolet imaging, unmanned aerial vehicle (UAV), and video camera) to detect icing conditions on insulators. Wu and Liao [4,5] processed aerial insulators images from UAV based on a texture segmentation algorithm and a robust detection algorithm. On the other hand, some researchers monitored icing conditions of insulators and conductors using video cameras installed on power towers. Huang proposed an online technology for measuring icing shape on transmission lines based on vision and force sensors. The technology can accurately represent the icing shape on transmission lines [6]. Hao proposed a GrabCut segmentation algorithm and processed iced insulator images captured under artificial icing simulations and natural conditions. The results demonstrated that the method can recognize the icing conditions of in-service glass insulators [7]. The above technologies were helpful to improve performance of detecting iced insulators. However, in harsh environments, unmanned aerial vehicles are not able to fly long distances. In addition, cameras on power towers are easy to lose power supply and covered by ice causing blurry images. Consequently, the technologies are not suitable for long-term monitoring iced insulators in all weather.

Currently, optical fiber sensors (OFS) have been widely used in various electrical equipment, such as transmission lines [8,9], power transformers [10,11], and insulator strings [12]. OFS is small in size, immune to electromagnetic interferences, chemically resistant, and has no power supply requirements. Besides, the composite insulator is long and stick-shaped. Therefore, OFS can be embedded into composite insulators, which establish distributed optical fiber sensing networks of the power grid.

Fiber Bragg grating (FBG) is one of the typical OFS, and it has been successfully developed in monitoring conditions of composite insulators. In 1992, Seike et al. [13] first proposed an optical fiber composite insulator for detecting fault points in electric power transmission networks. An optical fiber was embedded in a central through-hole of the composite insulator. However, the insulation properties of the method were not satisfactory. Trouillet et al. [14] investigated strain distribution and internal temperature of a core rod using an optical fiber with six FBGs. The method provided the basis for detecting operating conditions of core rod in the composite insulator based on FBGs. Central China Power Group [15] and China Electric Power Research Institute [16–22] implanted FBGs into the core rod of composite insulators, and performed calibration tests to research the relationship between the center wavelength of FBG and the temperature/stress of composite insulators. From 2010 to 2019, the composite insulators with FBG embedded have been operated on 110 kV [21], 220 kV [17], 500 kV [17], and UHV [22] transmission lines, which validated the feasibility of the proposed method in the field. However, the methods embedded optical fibers into the core rod, which were not suitable for detecting glaze icing on the composite insulator sheds.

For theoretical analysis, Kumosa [23,24] and Portnov [25] established numerical models that can be used to evaluate the relationship between Bragg wavelength shift and internal stress of the core rod in composite insulators. Kerrouche et al. [26] calculated strain/shear stress from a substrate to FBGs. The results showed a relatively uniform stress distribution along the substrate. Chen et al. [27,28] simulated reflection spectrum of FBGs based on analysis of the stress distribution of a core rod in a composite insulator.

The above studies have validated the feasibility of embedding FBG sensors into the core rod of composite insulators. In addition to those studies, few studies have investigated glaze icing detection of composite insulator based on FBG, although glaze icing detection is one of the significant characteristics for fault detection in insulators.

In this study, a 110 kV composite insulator with embedded FBGs was designed to detect glaze icing. Firstly, the FBG sensing principle and the design of FBG temperature compensation sensors were described. Secondly, the FBG arrangement in the composite insulator was presented in detail. Then, temperature calibration experiments were performed, and FBG temperature characteristics errors were analyzed as well. The results indicate that the compensation method can measure temperature and icicle loads of FBGs simultaneously. Moreover, a polar coordinate system was proposed for describing positions of simulated loads and FBGs on insulator sheds. Finally, simulated icicle load experiments were performed to find the relationship between an icicle, several icicles, icicle positions, and wavelength shifts. The proposed method introduces a novel way for glaze icing detection of composite insulators in transmission lines.

2. Detection Principle

2.1. FBG Sensing Principle

An FBG is a spatial phase grating inscribing into the photosensitive fiber core. The FBG has a significant property of serving as a narrow band filter or reflector. More specifically, it can reflect a narrow band of light with a specific wavelength from an incident broad source, and the rest of the light with other wavelengths passes through the FBG. The FBG reflection wavelength is defined as λ_B [29]:

$$\lambda_{\mathrm{B}} = 2n_{\mathrm{eff}} \cdot \Lambda \tag{1}$$

where n_{eff} is the effective refractive index of the optical fiber, and Λ denotes the periodicity of the grating.

Temperature, stain, or other environmental parameters would change the refractive index and periodicity of the FBG resulting in a wavelength shift. In this case, the FBG wavelength shift caused by temperature change (ΔT) and axial strain (ε) can be expressed as [30]:

$$\begin{aligned} \Delta\lambda_{\mathrm{B}} &= \lambda_{\mathrm{B}}(\alpha_f + \xi)\Delta T + \lambda_{\mathrm{B}}(1 - P_e)\varepsilon \\ &= K_{\mathrm{B},T}\Delta T + K_{\mathrm{B},\varepsilon}\varepsilon \end{aligned} \tag{2}$$

where α_f denotes the thermal expansion coefficient, ξ is the thermo-optic coefficient, P_e is the elastic-optic coefficient, and $K_{\mathrm{B},T}$ and $K_{\mathrm{B},\varepsilon}$ are both constant which denote the temperature sensitivity (pm/°C) and the axial strain sensitivity (pm/με), respectively.

2.2. Simultaneous Measurement of FBG Temperature and Strain

Glaze icing can change the temperature and strain of insulator sheds, which both cause the wavelength shifts of FBG at the same time. Therefore, it is necessary to propose a method for measuring FBG temperature and strain simultaneously. Specifically, two FBGs (i.e., FBG-1 and FBG-2) are inscribed in an optical fiber. FBG-1 is affected by temperature and strain, whose wavelength shift can be described as Equation (3). In contrast, FBG-2 is packaged by a designed housing that is not affected by strain. Thus, the wavelength shift of FBG-2 can be described as Equation (4).

$$\Delta\lambda_{\mathrm{B1}} = (K_{\mathrm{B1},T}\Delta T + K_{\mathrm{B1},\varepsilon}\varepsilon_1) \tag{3}$$

$$\Delta\lambda_{\mathrm{B2}} = K_{\mathrm{B2},T}\Delta T \tag{4}$$

Therefore, the wavelength shift of FBG-1 caused by strain can be described as:

$$\Delta\lambda_{compensation} = \Delta\lambda_{\mathrm{B1}} - \frac{K_{\mathrm{B1},T}}{K_{\mathrm{B2},T}}\Delta\lambda_{\mathrm{B2}} \tag{5}$$

3. Experiments

3.1. Fiber Bragg Grating

In this paper, FBGs with different reflection wavelengths have been inscribed in single mode optical fibers with polyimide coating using phase mask technology. The diameter, effective length and reflectivity of each FBG are about 250 μm, 10 mm and 90%, respectively. The initial wavelengths of FBGs in room temperature are within a range of 1510 to 1590 nm. The strain measurement of FBG ranges from −2500 to 2500 με, and the accuracy rate is 1‰ F.S. Moreover, the operating temperature ranges from −40 to 120 °C. More specifically, three optical fibers were embedded into a composite insulator and labeled as 1#, 2#, and 3#, respectively.

In order to measure icicle loads and temperature simultaneously, the FBG temperature compensation sensor was packed in a ceramic tube, which is cylindrical, with a length of 30 mm and a diameter of about 2 mm (Figure 1). Besides, epoxy resin adhesive was injected into the ceramic tube for insulation. The FBG temperature compensation sensors are not affected by icicle loads due to the high elasticity modulus of the ceramic tube.

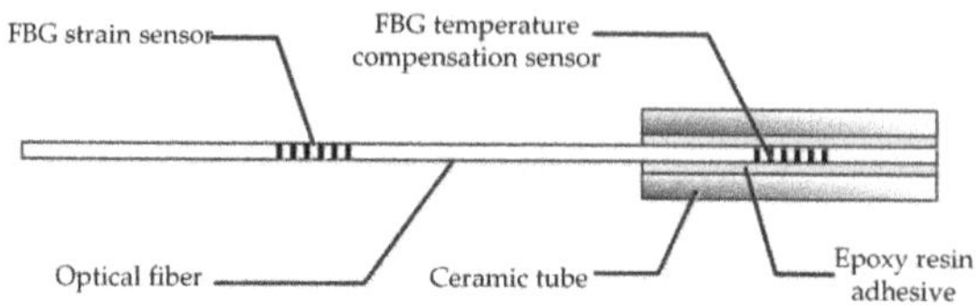

Figure 1. Fiber Bragg Grating (FBG) strain sensor and FBG temperature compensation sensor in an optical fiber.

3.2. A 110 kV Composite Insulator with Embedded FBGs

The type of 110 kV composite insulator is FXBW-110/100, whose parameters are listed in Table 1. The insulator has 13 big sheds, which are projected from the insulator trunk, and is intended to increase the creepage distance [31], as shown in Figure 2a.

Table 1. Parameters of a composite insulator FXBW-110/100.

Description		Values
Structural height		1240 mm
Insulation height		1036 mm
Leakage distance		4953 mm
Diameter of the core rod		18 mm
The big insulator shed	Shed overhang	61 mm
	Shed spacing	75 mm
	Number	13
The small insulator shed	Shed overhang	37 mm
	Shed spacing	25 mm
	Number	26

Three optical fibers, 1#, 2#, and 3#, were embedded vertically into the composite insulator during silicone rubber injection process. Besides, the optical fibers were at a 120° interval around the central axis of the composite insulator (Figure 2). Fourteen FBGs were inscribed in 1# and 2# optical fiber respectively, and thirteen FBGs were inscribed in 3# optical fiber. The FBGs were labeled as FBG-mn, where m was the label of the optical fibers (i.e., 1, 2, and 3), and n was the label of the FBGs (i.e., 1, 2 ... and 14). For example, the fourteen FBGs in the 1# optical fiber were separately labeled as FBG101 - FBG114. Specifically, FBG101 - FBG113 were all FBG strain sensors, which were designed above each big shed root in silicone rubber housing (Figure 2b), and FBG114 was an FBG temperature compensation sensor. Similarly, FBG201 - FBG213 in 2# optical fiber and FBG301 - FBG313 in 3# optical fiber were all FBG strain sensors. In addition, FBG214 was also an FBG temperature compensation sensor.

After proper preparation of the core rod surface, three optical fibers with FBGs were put on the core rod surface according to the arrangement in Figure 2c. Then, two ends of optical fibers were fixed, and midpoints between two FBGs were adhered on the core rod using room temperature vulcanized silicone rubber (RTV) until curing. The core rod and end fittings were joined by a pressing process. Finally, the core rod and optical fibers were placed in a vacuum injection machine, and the pressure remained constant during the process of injecting silicone rubber and curing.

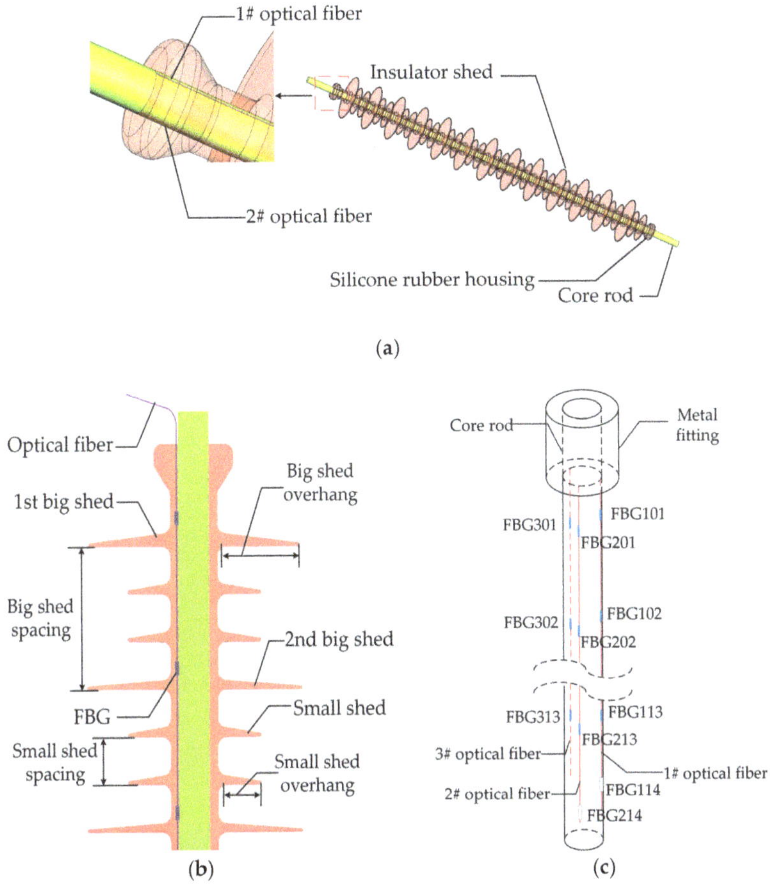

Figure 2. Diagram of a 110 kV composite insulator with embedded FBGs. (**a**) The parts of a composite insulator with embedded FBGs; (**b**) Relative positions between FBGs and insulator sheds; (**c**) Distribution diagram of three optical fibers with FBGs.

3.3. The Sensing System

The sensing system was made up of three parts, i.e., the 110 kV composite insulator with embedded FBGs, an FBG interrogation system, and an artificial climate chamber, as depicted in Figure 3. The type of interrogation system is JEME-iFBG produced by Shenzhen JEMETECH Company in China. Specifically, the interrogation system can acquire signals of 6 channels simultaneously based on a high-power wavelength scanning laser, and each channel can detect 48 FBGs at most. With a wavelength range of 1510 to 1590 nm and a wavelength accuracy of 1 pm, the interrogation system can satisfy the requirement for detecting glaze icing on composite insulators.

The artificial climate chamber was used to simulate the low-temperature environment during glaze ice accretion on composite insulators. Specifically, the artificial climate chamber offered a temperature change from −15 °C to room temperature (20 °C) with an accuracy of 1 °C. In experiments, three optical fibers were pulled out from the chamber, and the interrogation system was linked to the computer using the USB port for data recording, analysis, and transfer.

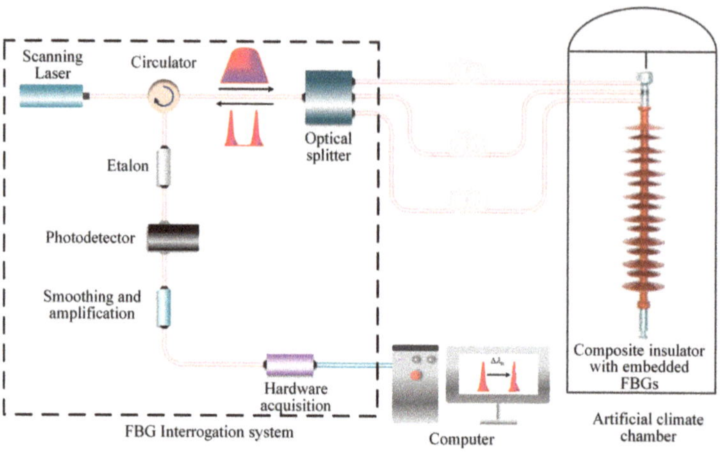

Figure 3. Sensing system of the composite insulator with embedded FBGs.

3.4. Temperature Calibration Experiments

Previous studies have demonstrated that glaze ice is easy to accrete on insulator sheds when the temperature is between −8 and 0 °C in the field [32]. Besides, according to IEEE Standard 1783-2009, the air temperature should remain between −15 and −5 °C for depositing artificial ice on insulator sheds in a climate chamber [33]. Thus, in the temperature calibration experiments, the air temperature in the climate chamber was set between −15 and +15 °C. The process was as follows:

(1) The 110 kV composite insulator with embedded FBGs was put into the artificial climate chamber. Three optical fibers were pulled out from the chamber, and connected with the demodulator and computer through outlet fibers. The wavelengths of all FBGs were recorded as initial wavelengths when the temperature was 0 °C.

(2) The temperature range of the artificial climate chamber was set from −15 to +15 °C, and the wavelength shifts were recorded at intervals of 3 °C. A previous research has indicated that the temperature of FBG in the composite insulator was equal to the temperature in the chamber after 8 hours [18]. Therefore, after being placed into the chamber for 8 h, the temperature of the composite insulator could be considered as a constant and recorded if a wavelength rate was less than 3 pm/min [18]. After averaging 10 sets of data, the results of temperature calibration experiments were created.

3.5. Simulated Icicle Load Experiments

Many studies have indicated that the common glaze ice accretion on insulators causes the highest probability of ice flashover occurrence [34,35]. Therefore, in order to find the relationship between glaze ice and wavelength shifts, weights were hung on the edge of a big shed for simulating glaze ice loads generated by icicles [36] (Figure 4).

In order to describe icing conditions of composite insulators quantitatively, icicle length and icicle bridged degree were proposed, which can be described as [37–39]:

$$\eta = \frac{S-d}{S} \times 100\% = \frac{L}{S} \times 100\% \tag{6}$$

where η denotes the bridged degree of a big shed spacing, S is the big shed spacing, d is the air gap distance, and L is the icicle length.

In order to describe positions of simulated loads and FBGs on insulator sheds quantitatively, a polar coordinate system was proposed, where the midpoint of the core rod was the original point (Figure 5).

The position was considered as (r, θ, n), where r denoted the radius (mm), θ denoted the polar angle (°), and n denoted the number of big sheds in the insulator. For example, the coordinates of FBG101, FBG201, and FBG301 were (9, 0°, 1), (9, 120°, 1), and (9, −120°, 1), respectively. When the simulated load was on the edge of the 1st big shed facing the 1# optical fiber, its coordinate was (61, 0°, 1). In this paper, the r of simulated loads were all 61 mm, so the coordinates of simulated loads could be simplified to $P(\theta, n)$.

The process was as follows:

(1) The 110 kV composite insulator with embedded FBG was placed in the artificial climate chamber, and the air temperature was set as -10 °C during the experiments. The optical fibers were pulled out from the chamber, and connected with the demodulator and computer.

(2) Geometric characteristics of an icicle on insulators were proposed in [37]. In this paper, the weights were set as 0.5, 1.0, 1.5, 2.0, and 2.5 N, corresponding to icicle lengths and bridged degrees, as shown in Table 2. The icicle lengths and bridged degrees are common parameters in icicle growth experiments [38,39].

(3) Weights were hung on insulator big sheds. The wavelength shifts were continuously recorded when wavelength rates were stable (< 3 pm/min) [18]. After averaging 10 sets of data, the results were created.

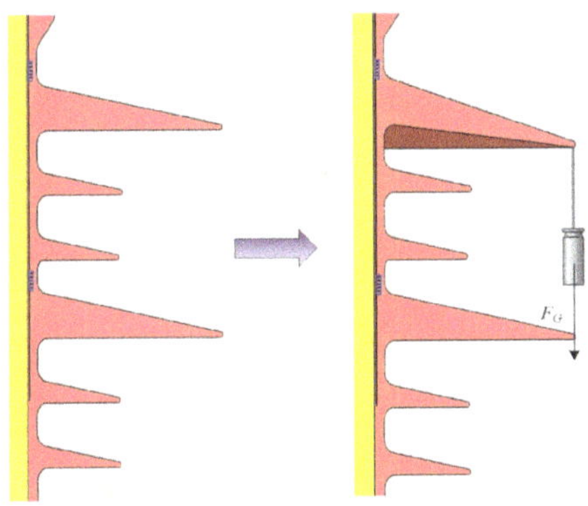

Figure 4. Hanging weights on a big shed for simulated loads of icicles [36].

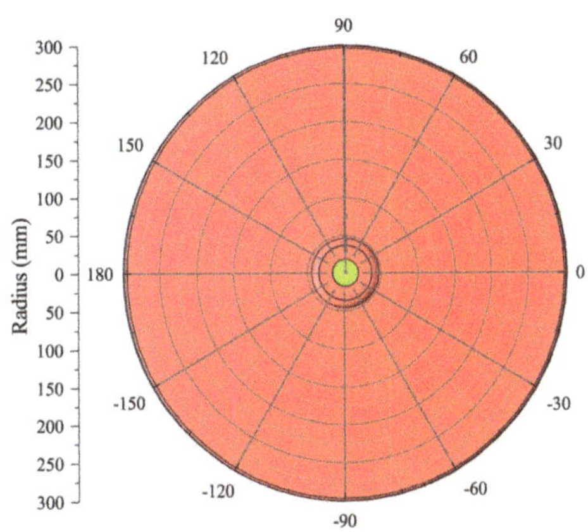

Figure 5. The polar coordinate system on insulator sheds.

Table 2. Icicle load corresponds to the icicle length and icicle bridged degree with density of 0.9 g/cm^3 [36,37].

Weights	Corresponding to the Icicle Length L	Corresponding to the Icicle Bridged Degree η
0.5 N	11 mm	15%
1.0 N	22 mm	29%
1.5 N	33 mm	44%
2.0 N	44 mm	59%
2.5 N	55 mm	73%

4. Results and Discussion

4.1. The Relationship between Wavelength Shifts of FBGs and Temperature

The wavelength shift mean deviation (WSMD) was proposed to discuss the temperature characteristics of FBG strain sensors in an optical fiber. For example, the WSMD of 13 FBG strain sensors in 1# optical fiber, i.e., FBG101 - FBG113, was expressed as f_1, which can be defined as:

$$f_1 = \frac{\sum_{n=1}^{13}\left|\lambda_n - \overline{\lambda}\right|}{13} \tag{7}$$

where n denotes the label of FBG strain sensor in 1# optical fiber, λ_n denotes the wavelength shift of FBG1n at a temperature, and $\overline{\lambda}$ denotes the average value of wavelength shifts of 13 FBG strain sensors in 1# optical fiber at the temperature.

Similarly, f_2 and f_3 represent the WSMD of FBG strain sensors in 2# and 3# optical fiber, respectively. As shown in the right Y axis of Figure 6, the mean deviations were all within 8 pm, which showed that wavelength shifts of FBG strain sensors in an optical fiber were seen as the same at a temperature.

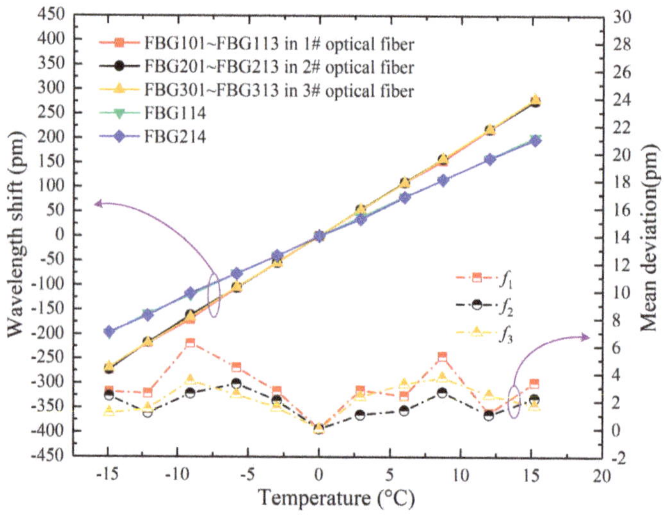

Figure 6. The relationship between wavelength shifts of FBGs and temperature.

Therefore, wavelength shifts dependent on temperatures of FBG101 - FBG113, FBG201 - FBG213, and FBG301 - FBG313 were shown in red, black, and yellow solid lines, respectively (Figure 6). The solid lines showed that wavelength shifts of all FBG strain sensors were linear with temperature, and goodness of fit R^2 were all greater than 0.997. Therefore, the temperature sensitivities of all FBG strain sensors were approximately 18.16 pm/°C.

The green and blue lines showed that wavelength shifts of FBG114 and FBG214 were linear with temperature, and goodness of fit R^2 were approximately 0.997. The temperature sensitivities of FBG temperature compensation sensors were 13.18 pm/°C. Therefore, according to Equation (5), the wavelength shift of an FBG strain sensor caused by strain was as follows:

$$\begin{aligned} \Delta\lambda_{compensation} &= \Delta\lambda_S - \tfrac{18.16}{13.18}\Delta\lambda_T \\ &\approx \Delta\lambda_S - 1.38\Delta\lambda_T \end{aligned} \tag{8}$$

where $\Delta\lambda_S$ denoted the wavelength shift of an FBG strain sensor, and $\Delta\lambda_T$ denoted the wavelength shift of an FBG temperature compensation sensor.

4.2. A Simulated Load of an Icicle on a Big Shed

The wavelength shift of FBG102 changed with the simulated load of one weight at $P_1(0°, 2)$, as the red line in Figure 7. When a simulated load on the edge of the 2nd big shed moved at a 30° interval, the wavelength shifts of the FBG102 were shown in Figure 7.

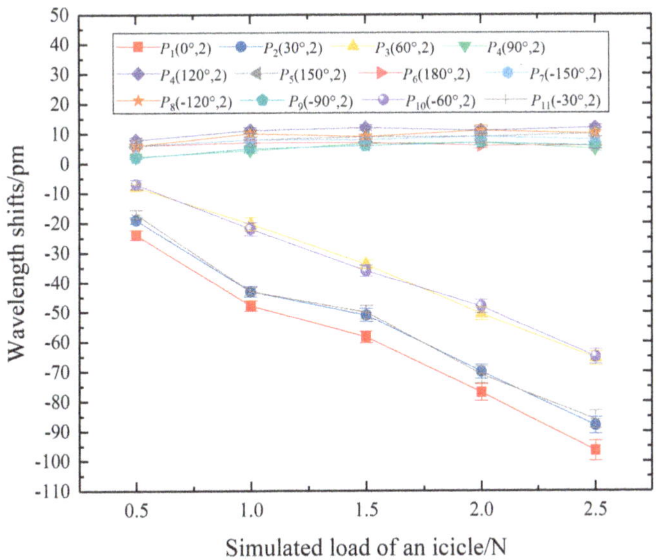

Figure 7. The relationship between the wavelength shift of FBG102 and simulated load generated on different positions.

It was found that the wavelength shifts of FBG102 were linearly related to the simulated load of one weight at $P_1(0°, 2)$, $P_2(30°, 2)$, $P_3(60°, 2)$, $P_{10}(-60°, 2)$, and $P_{11}(-30°, 2)$, respectively, corresponding to a polar angle range of -60° to 60°. The linear fitting formulas were described as:

$$\Delta\lambda_1 = -34.6F_1 - 8.9,\ R^2 = 0.990 \tag{9}$$

$$\Delta\lambda_2 = -33F_2 - 4.7,\ R^2 = 0.984 \tag{10}$$

$$\Delta\lambda_3 = -28.8F_3 + 7.8,\ R^2 = 0.997 \tag{11}$$

$$\Delta\lambda_{10} = -28.4F_{10} + 7,\ R^2 = 0.998 \tag{12}$$

$$\Delta\lambda_{11} = -33.2F_{11} - 3.6,\ R^2 = 0.978 \tag{13}$$

where $F_i|_{i=1,2,3,10,11}$ were the simulated load generated at $P_i|_{i=1,2,3,10,11}$, $\Delta\lambda_i|_{i=1,2,3,10,11}$ were the wavelength shifts of the FBG102, respectively.

According to the above formulas, the absolute value of the load coefficient in Equation (9) was the largest (i.e., 34.6 pm/N). Therefore, the load coefficient of FBG facing the load was the highest. Besides, with the increase of the polar angle of the load, the absolute value of the load coefficient gradually decreased.

Comparing the Equation (10) with Equation (13), Equation (11) with Equation (12), the load coefficients of FBG102 were approximate when the loads were symmetrical about the polar angle of 0°. It can be concluded that the wavelength shifts of FBG were equal when two loads were symmetrical about the extension line from the original point to the FBG.

In contrast, when the polar angles of icicle load were from 90° to −90°, with the increase of the icicle load, the wavelength shifts appeared to remain stable. The reason for this was that FBG102 was located 86.27 mm away from P_4 and P_9, even 140 mm away from P_6. The strain transferring to FBG102 was so small that the wavelength shifts were all less than 12 pm.

4.3. Several Simulated Loads of Several Icicles on A Big Shed

In order to simulate several icicles accretion on different positions of an insulator shed, three simulated loads were hung on a big shed in the insulator. It is clear from Section 4.2 that wavelength shifts were easily affected by the icicle load with polar angle ranges of −60° to 60°. Therefore, a simulated load F_a of 1 N was at $P_a(0°, 3)$, and the other two simulated loads F_b and F_c of 0.5 N were at $(\theta°, 3)$ where θ changed in a polar angle range of -60° to 60°. The relationship between the wavelength shifts and the changing polar angles (θ_b, θ_c) were shown in Figure 8.

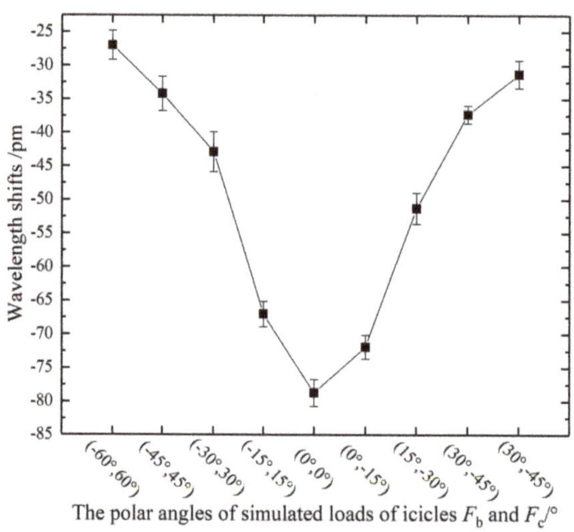

Figure 8. The relationship between the wavelength shift of FBG103 and the positions of simulated loads.

It was found that the absolute value of wavelength shift decreases with the distance from $P_a(0°, 3)$ to load point (e.g., F_b and F_c). In addition, when the polar angle of F_b and F_c were between -15° and 15°, the wavelength shifts of FBG103 were between 67 and 79 pm. In contrast, the wavelength shift generated by the sum of F_a, F_b, and F_c at 0° was 79 pm. The deviations of the wavelength shifts were within 12 pm. In view of the experimental error, it can be found that when the polar angle of F_b and F_c were between −15° to 15°, the wavelength shifts were similar to that generated by the sum of F_a, F_b, and F_c in 0°.

In conclusion, when the polar angles are between −15° and 15°, the wavelength shift generated by several icicles was equal to the sum of wavelength shifts generated by each icicle. In contrast, when the polar angles are between ±15° and ±60°, the absolute values of wavelength shifts would decrease with the distance from icicle positions to P_a.

4.4. Wavelength Shifts of All FBGs in an Optical Fiber

To analyze wavelength shifts of all FBGs in an optical fiber when an icicle was generated on a big shed, a simulated load of 2 N was generated at (0°, 1), (120°, 1), and (−120°, 1), respectively.

It was found from Figure 9 that the wavelength shifts of the FBGm01 were -80 pm, those of the FBGm02 were less than −10 pm, while those of FBGmn (3 ≤ n ≤ 11) were almost 0 pm. The reason for this is that the silicone rubber is soft material, whose elasticity modulus is 2.6 MPa. The deformation of the 1st big shed was large under high loads, meanwhile, the strain transferring to FBGm02 was large enough to detect the wavelength shifts. As for FBGmn (3 ≤ n ≤ 11), the transferred strain was too small to change the wavelength shifts. In conclusion, icicle loads can cause wavelength shifts of the FBGs within one big shed spacing.

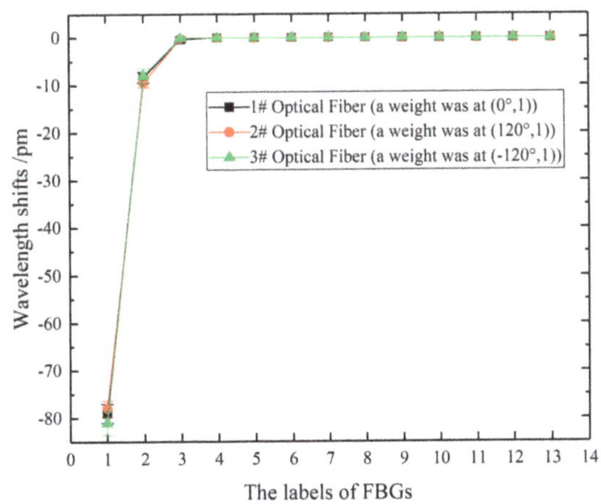

Figure 9. The wavelength shifts of FBG1n, FBG2n and FBG3n when a simulated load of 2 N was at (0°, 1), (120°, 1) and (−120°, 1), respectively.

5. Conclusions

In this paper, a detection method for glaze icing of 110 kV composite insulators has been proposed. The 110 kV composite insulator with embedded FBGs is helpful to sense glaze icing conditions in all weather, offering much better versatility than conventional approaches. Three optical fibers with FBGs were embedded symmetrically into a 110 kV composite insulator. Besides, the FBG temperature compensation sensors were packaged in ceramic tubes so as to solve the problem of icicle loads and temperature cross-sensitivity. Then, temperature calibration experiments and simulated icicle load experiments were performed.

The results showed that temperature sensitivities of FBG strain sensors and FBG temperature compensation sensors were 18.16 pm/°C, and 13.18 pm/°C, respectively. FBG wavelength shifts were linearly related to the icicle loads within the polar angle range of −60° to 60°. In particular, the load coefficient of FBG facing the icicle was −34.6 pm/N. Moreover, within the polar angle range of −15° to 15°, the wavelength shift generated by several icicles was equal to the sum of wavelength shifts generated by each icicle. Finally, icicle loads can cause wavelength shifts of the FBGs within one big-shed spacing. Therefore, the icicle loads of composite insulators can be detected using FBGs in the laboratory.

In further research, a mathematical model and a location algorithm will be proposed for monitoring multiple icicle loads and positions simultaneously on the composite insulator. In addition, uncontrollable outdoor factors (e.g., sunlight, wind, rain, etc.) should be taken into considerations as well.

Author Contributions: Y.H., J.W., and L.Y. conceived and designed the experiments; J.W., Y.F., and H.L. performed the experiments; Y.H., J.W., L.Y., and J.G. analyzed the data; J.G. contributed reagents/materials/analysis tools; J.W., Y.H., and Y.F. wrote the paper. All authors have read and agreed to the published version of the manuscript.

Funding: This work was supported by Smart Grid Joint Fund Key Project between National Natural Science Foundation of China and State Grid Corporation U1766220.

Conflicts of Interest: The authors declare no conflict of interest.

References

1. Farhangi, H. The path of the smart grid. *IEEE Power Energy Mag.* **2010**, *8*, 18–28. [CrossRef]
2. Marris, E. Energy: Upgrading the grid. *Nature* **2008**, *454*, 570–573. [CrossRef] [PubMed]
3. Jiang, X.; Liu, Y.; Fu, X.; Xu, P.; Wang, S.; Sheng, G. Construction ideas and development trends of transmission and distribution equipment of the ubiquitous power internet of things. *High Volt. Eng.* **2019**, *45*, 1345–1351.
4. Wu, Q.; An, J.; Lin, B. A texture segmentation algorithm based on PCA and global minimization active contour model for aerial insulator images. *IEEE J. Sel. Top. Appl. Earth Obs. Remote Sens.* **2012**, *5*, 1509–1518.
5. Liao, S.; An, J. A robust insulator detection algorithm based on local features and spatial orders for aerial images. *IEEE Geosci. Remote Sens. Lett.* **2015**, *12*, 963–967. [CrossRef]
6. Huang, X.; Zhang, F.; Li, H.; Liu, X. An online technology for measuring icing shape on conductor based on vision and force sensors. *IEEE Trans. Instrum. Meas.* **2017**, *66*, 3180–3189. [CrossRef]
7. Hao, Y.; Wei, J.; Jiang, X.; Yang, L.; Li, L.; Wang, J.; Li, H.; Li, R. Icing condition assessment of in-service glass insulators based on graphical shed spacing and graphical shed overhang. *Energies* **2018**, *11*, 318. [CrossRef]
8. Bjerkan, L. Application of fiber-optic bragg grating sensors in monitoring environmental loads of overhead power transmission lines. *Appl. Opt.* **2000**, *39*, 554–560. [CrossRef] [PubMed]
9. Wydra, M.; Kisala, P.; Harasim, D.; Kacejko, P. Overhead transmission line sag estimation using a simple optomechanical system with chirped fiber bragg gratings. Part 1: Preliminary measurements. *Sensors* **2018**, *18*, 309. [CrossRef] [PubMed]
10. Ma, G.; Zhou, H.; Cheng, S.; Li, Y.; Zhang, Q.; Li, C.; Zheng, Q. Distributed partial discharge detection in a power transformer based on phase-shifted FBG. *IEEE Sens. J.* **2018**, *18*, 2788–2795. [CrossRef]
11. Ghorat, M.; Gharehpetian, G.; Latifi, H.; Hejazi, M.; Layeghi, A. Partial discharge acoustic emission detector using mandrel-connected fiber Bragg grating sensor. *Opt. Eng.* **2018**, *57*, 074107. [CrossRef]
12. Yan, Z.; Zhang, Z.; Li, L.; Zhai, T.; Guo, R.; Jiang, C. Insulator string leakage current-monitoring sensor based on fiber Bragg grating. *Opt. Express* **2018**, *26*, 24940–24952. [CrossRef] [PubMed]
13. Seike, S.; Mima, T.; Nozaki, M.; Tani, N.; Ikeda, M. Optical Fiber Composite Insulator. U.S. Patent 5,090,793, 25 February 1992.
14. Trouillet, A.; Lepley, D.; Mureravaud, A.; Marin, E. Integration of fibre Bragg grating strain sensors into composite electrical insulators. In Proceedings of the SPIE Second European Workshop on Optical Fibre Sensors, Santander, Spain, 9 June 2004; pp. 92–95.
15. Chen, J.; Wang, J.; Sheng, G.; Yang, F.; Xiong, P. Practical researches on achieving status monitoring with the use of optic fiber composite insulator. *Insul. Surge Arresters* **2010**, *1*, 6–9.
16. Cai, W.; Luo, B.; Deng, H.; Fan, X.; Xiong, P. Stress distribution analyses of composite insulator with fiber bragg grating embedded. *High Volt. Eng.* **2011**, *37*, 1106–1114.
17. Cai, W.; Luo, B.; Meng, G.; Fan, X.; Zhu, X. Deng, H. Discuss on the intelligent monitoring technology of composite insulator with the optic fiber sensors. *Insul. Surge Arresters* **2011**, *3*, 5–9.
18. Du, S.; Cai, W.; Deng, H.; Wen, S.; Deng, W.; Xiong, P. Production and calibration test of the composite insulator with fiber bragg grating embedded. *High Volt. Eng.* **2012**, *38*, 2774–2780.
19. Wen, S.; Cai, W.; Deng, H.; Du, S.; Xiong, P. Fault simulation experiment on composite insulators with fiber bragg grating implanted. *High Volt. Eng.* **2013**, *39*, 81–87.

20. Deng, H.; Cai, W.; Liu, C.; He, J. The feasibility of the composite insulator with fiber Bragg grating embedded in the rod. In Proceedings of the IEEE PES Innovative Smart Grid Technologies, Istanbul, Turkey, 12–15 October 2014; pp. 1–4.
21. Cao, M.; Liang, S.; Qin, Z.; Ke, L.; Wang, L. Study on the application of optical sensor in composite insulator. *Insul. Surge Arresters* **2014**, *2*, 32–35.
22. Wan, X.; Nan, J.; Xu, T.; Li, Y. On-line monitoring technology of UHV composite insulator based on FBG sensor. In Proceedings of the 2019 IEEE Sustainable Power and Energy Conference, Beijing, China, 21–23 November 2019; pp. 2901–2904.
23. Kumosa, M.; Han, Y.; Kumosa, L. Analyses of composite insulators with crimped end-fittings: Part I—Nonlinear finite element computations. *Compos. Sci. Technol.* **2002**, *62*, 1191–1207. [CrossRef]
24. Kumosa, M.; Armentrout, D.; Kumosa, L.; Han, L.; Carpenter, S.H. Analyses of composite insulators with crimped end-fittings: Part II—Suitable crimping conditions. *Compos. Sci. Technol.* **2002**, *62*, 1209–1221. [CrossRef]
25. Portnov, G.; Bakis, C.E. Analysis of stress concentration during tension of round pultruded composite rods. *Compos. Struct.* **2008**, *83*, 100–109. [CrossRef]
26. Kerrouche, A.; Boyle, W.; Sun, T.; Grattan, K.; Taljsten, B. Strain measurement using embedded fiber bragg grating sensors inside an anchored carbon fiber polymer reinforcement prestressing rod for structural monitoring. *IEEE Sens. J.* **2009**, *9*, 1456–1461. [CrossRef]
27. Chen, W.; Dong, X.; Yang, F.; Zeng, X.; Ma, J.; Wang, X.; Ma, X. Theoretical analysis of temperature response of the fiber Bragg grating embedded in composite insulator. In Proceedings of the SPIE Advanced Laser Manufacturing Technology, Beijing, China, 19 October 2016; p. 101530X.
28. Chen, W.; Tang, M. Monitoring on internal temperature of composite insulator with embedding fiber Bragg grating for early diagnosis. In Proceedings of the 25th International Conference on Optical Fiber Sensors, Jeju, Korea, 24–28 April 2017; pp. 1–4.
29. Hill, K.O.; Meltz, G. Fiber bragg grating technology fundamentals and overview. *J. Lightwave Technol.* **1997**, *15*, 1263–1276. [CrossRef]
30. Ma, G.; Li, Y.; Mao, N.; Shi, C.; Zhang, B.; Li, C. A fiber bragg grating-based dynamic tension detection system for overhead transmission line galloping. *Sensors* **2018**, *18*, 365. [CrossRef]
31. *Composite Suspension and Tension Insulators for AC Overhead Lines with a Nominal Voltage Greater Than 1 000 V - Definitions, Test Methods and Acceptance Criteria*; IEC 61109, Ed.; International Electrotechnical Commision: Geneva, Switzerland, 2008; p. 9.
32. Zhang, Z.; Jiang, X.; Hu, J.; Shen, Q.; Chen, L.; Liu, G. Influence of environment parameters on the icing accretion on the surface of insulator. *High Volt. Eng.* **2010**, *36*, 2418–2423.
33. *IEEE Guide for Test Methods and Procedures to Evaluate the Electrical Performance of Insulators in Freezing Conditions*; IEEE Standard 1783–2009; IEEE: Piscataway, NJ, USA, 2009.
34. Volat, C.; Farzaneh, M. Three-dimensional modeling of potential and electric-field distributions along an EHV ceramic post insulator covered with ice—Part I: Simulations of a melting period. *IEEE Trans. Power Deliv.* **2005**, *20*, 2006–2013. [CrossRef]
35. Volat, C.; Farzaneh, M. Three-dimensional modeling of potential and electric-field distributions along an EHV ceramic post insulator covered with ice—Part II: Effect of air gaps and partial arcs. *IEEE Trans. Power Deliv.* **2005**, *20*, 2014–2021. [CrossRef]
36. Wei, J.; Hao, Y.; Fu, Y.; Yang, L.; Gan, J.; Yang, Z. Detection of glaze icing load and temperature of composite insulators using fiber bragg grating. *Sensors* **2019**, *19*, 1321. [CrossRef]
37. Fofana, I.; Farzaneh, M.; Hemmatjou, H.; Volat, V. Study of discharge in air from the tip of an icicle. *IEEE Trans. Power Deliv.* **2008**, *15*, 730–740. [CrossRef]
38. Deng, Y.; Jia, Z.; Jiang, H.; Guan, Z.; Zhou, J. Analysis of icicle growth process of composite insulator under energized condition and its impact factors. *IEEE Trans. Dielectr. Electr. Insul.* **2015**, *22*, 1613–1622. [CrossRef]
39. Deng, Y.; Jia, Z.; Zhou, J.; Sun, L. Ice flashover performance and its characterization parameter of composite insulator with booster sheds. *IEEE Trans. Dielectr. Electr. Insul.* **2016**, *3*, 1021–1029. [CrossRef]

sensors

MDPI

Article

Wearable Hand Module and Real-Time Tracking Algorithms for Measuring Finger Joint Angles of Different Hand Sizes with High Accuracy Using FBG Strain Sensor

Jun Sik Kim [1,2,†], Byung Kook Kim [1,3,†], Minsu Jang [1,4], Kyumin Kang [1,5], Dae Eun Kim [3,*], Byeong-Kwon Ju [2,*] and Jinseok Kim [1,*]

1 Center for Bionics, Korea Institute of Science and Technology, Seoul 02792, Korea; kjs414@kist.re.kr (J.S.K.); cfg10x6p1@kist.re.kr (B.K.K.); minsujang@kist.re.kr (M.J.); kyuminkang@kist.re.kr (K.K.)
2 Display and Nanosystem Laboratory, School of Electrical Engineering, Korea University, Seoul 02841, Korea
3 School of Mechanical Engineering, Yonsei University, Seoul 03722, Korea
4 School of Chemical Engineering, Sungkyunkwan University, Suwon 16419, Korea
5 Department of Electrical Engineering, Korea University, Seoul 02841, Korea
* Correspondence: kimde@yonsei.ac.kr (D.E.K.); bkju@korea.ac.kr (B.-K.J.); jinseok@kist.re.kr (J.K.); Tel.: +82-2-958-6745 (J.K.)
† These authors contributed equally to this work.

Received: 24 February 2020; Accepted: 27 March 2020; Published: 30 March 2020

check for updates

Abstract: This paper presents a wearable hand module which was made of five fiber Bragg grating (FBG) strain sensor and algorithms to achieve high accuracy even when worn on different hand sizes of users. For real-time calculation with high accuracy, FBG strain sensors move continuously according to the size of the hand and the bending of the joint. Representatively, four algorithms were proposed; point strain (PTS), area summation (AREA), proportional summation (PS), and PS/interference (PS/I or PS/I_α). For more accurate and efficient assessments, 3D printed hand replica with different finger sizes was adopted and quantitative evaluations were performed for index~little fingers (77 to 117 mm) and thumb (68~78 mm). For index~little fingers, the optimized algorithms were PS and PS/I_α. For thumb, the optimized algorithms were PS/I_α and AREA. The average error angle of the wearable hand module was observed to be 0.47 ± 2.51° and mean absolute error (MAE) was achieved at 1.63 ± 1.97°. These results showed that more accurate hand modules than other glove modules applied to different hand sizes can be manufactured using FBG strain sensors which move continuously and algorithms for tracking this movable FBG sensors.

Keywords: fiber Bragg grating strain sensor; algorithm; hand motion capture; real-time tracking

1. Introduction

Accurate measurement of hand motions is useful for various applications such as rehabilitation, virtual reality (VR) and augmented reality (AR) technology, and robotics [1–4]. For example, after a stroke, the damaged functions of the brain can be improved through continuous and repetitive finger rehabilitation [5,6]. The hand serves as a conduit for direct interaction between objects in a VR/AR world and users in the real world [7]. For robot-assisted surgery (RAS), accurate measurement of the finger movements is needed to ensure smooth teleoperation with the surgical robot and prevent mistakes [8]. Thus, studies have focused on using various sensors and equipment such as optical marking methods (i.e., camera recognition), inertial measurement unit (IMU) sensors, electrical resistance strain sensors, and fiber optic sensors to accurately measure and evaluate finger movements [1–3,9,10].

The optical marking method visually identifies changes in marked positions using an optical camera. This method can be used with marker recognition algorithms to check the movement of joints [2]. Although it provides excellent accuracy, the installation cost of equipment such as cameras is too high, and it is best suited to limited spaces where user motion is limited [11]. The algorithms are also very complex [12,13]. Research on overcoming the limitations of the optical marking method has focused on fabricating glove-type sensor modules where sensors (e.g., IMU, strain sensor, optical fiber sensor) are attached directly to the glove to measure the movement of finger joints. Using gloves ensures high accuracy while increasing the user's freedom of movement.

The IMU sensor has multiple degrees of freedom and measures linear acceleration and angular velocity along three axes [14–16]. For a more accurate measurement, an inertial navigation system (INS) also uses triaxial magnetic sensors and barometric pressure sensors [16]. While an IMU sensor can measure finger movements at a relatively low cost and is easy to install, the measurement data can be affected by electromagnetic fields, which can result in large errors. Errors from magnetic or angular velocity sensors accumulate over time to cause drift [17,18]. This is a chronic drawback of IMU sensors that many researchers are working on to reduce; however, it cannot be eliminated because it is an accumulation of errors over time [17]. Another disadvantage of IMU sensors is the complexity of the calculation process to locate an object [19]. The x, y, and z positions and the yaw, pitch, and roll angles are obtained by applying compensation and the Kalman filter to acceleration and angular velocity sensor data; however, the human hand has 27 degrees of freedom, and some joints depend on the movement of other joints [20]. Obtaining the position/rotation angle information for each finger phalange becomes very complex if all finger movements are acquired with multiple IMU sensors and data are fused and decoupled during the calculation process. An alternative approach is to attach a resistive strain sensor to the position of each finger joint to directly measure the strain of a bending joint [21–23]. While this approach reduces drift and has a light calculation load, the positions of the strain sensors on the glove are fixed; thus, it is difficult to produce a hand module with excellent accuracy for users with different hand sizes.

Unlike previous studies that focused on accurately measuring finger joint angles, recent studies have focused on developing a dexterous hand module with a small and repeatable measurement error for various users [6,22,24]. Wise et al. used optical fibers to fabricate finger angle tracking gloves with three different sizes so that users can select the right size themselves [25]. Borghetti et al. used a slightly different approach [24] and developed a calibration procedure to measure the joint angles based on the hand size of the new user. However, these studies did not assess the angle error when various users equip the glove. Even if different types of gloves are made, people have a wide range of hand sizes. Thus, a system needs to be developed that can use the same calibration process to measure any hand size. Li et al. developed a hand module that uses a stretchable bending sensor and achieved an absolute angle error of 4.5–8.0° for a hand size range of 17–21 cm [22]. Although they used stretchable sensors, the sensor positions in the hand module were still fixed, so reducing the measurement angle error would be difficult when applied to various hand sizes. Jha et al. used a fiber Bragg grating (FBG) sensor to measure the angles of two different hand sizes; they achieved a very small angle error of 0.13° with a mechanical setup and 0.67° with a human hand with very small maximum standard deviations of 0.30 and 0.67°, respectively [6]. However, they did not measure the joint angles of the thumb and little finger in real time. In addition, they used the angle of the distal interphalangeal (DIP) joint to approximate the angle of the proximal interphalangeal (PIP) joint. Thus, more research is needed to realize an extremely small angle error that can be used for practical applications in the industry.

As indicated above, hand motion tracking glove modules with high accuracy need to be developed for various hand sizes. Such equipment needs to be applicable to different rehabilitation patients, VR/AR users, and surgeons [4,5,8]. In particular, RAS requires high accuracy. This study focused on evaluating the performance of a wearable hand module. Unfortunately, the scope of this study was limited to adult males. A key reason for this is the length of the silicone ring structure. If the silicone ring structure is reduced to match the size of an adult female's hand, this causes the silicone to reach its maximum tension before a male can clench his fist. Meanwhile, if the silicone ring structure is enlarged to match the size of an adult male's hand, the thimble can easily escape from the end of the finger when a female opens her fist. Thus, 11 adult males aged between 26 and 48 years old were evaluated to determine the range of the hand size, and the angle error of the wearable hand module was measured.

Configuring a versatile system that is applicable to various hand sizes and has a small angle error is difficult. The first problem is the sensor attachment in the hand module [26–28]. Since the sensor positions in the hand module do not change, the relative positions of the joints and sensors differ among users with different hand sizes, which does not allow repeatable data to be obtained. The second reason is the algorithm. Even if the attachment problem is resolved, the sensor data are measured differently for different users. Algorithms need to be developed that can identify trends in data despite differences in the measurement among users with different hand sizes.

In this study, a system was developed for accurate and real-time tracking of the finger joint angles of adult males regardless of hand size. Hand replicas were 3D printed for different finger sizes (index-little fingers: 77–117 mm, thumb: 68–78 mm) because using actual hands could introduce many critical variables such as hand tremors, temperature, and limitations of the goniometer or other sensors. Unlike the traditional method of fixing the sensor position in the glove, a smaller angle error was achieved by developing an algorithm that converts data from movable sensors for real-time calculation. According to the literature, the allowable angle error of the finger joint for clinical purposes is less than 5° [29]. Therefore, the objective of this study was to realize a maximum angle error of no more than 5°. In the experimental results, the maximum angle error of the wearable hand module with the movable FBG strain sensor was 4.6°; the mean angle error and standard deviation were 0.47 ± 2.51°, and the mean absolute error (MAE) was 1.63 ± 1.97°.

2. FBG Strain Sensor and Hardware Design

2.1. Principle of the FBG Sensor

The FBG sensor is a thin optic fiber strand consisting of core, cladding, and buffer layers with lattice patterns evenly engraved on the core [30]. When the core emits a broadband light, only a certain wavelength is reflected from the engraved lattice pattern. This reflected wavelength is called the Bragg wavelength, and it expresses the relationship between the effective refractive index of the fiber core and distance from one lattice to the next lattice [31]:

$$\lambda_B = 2n_{eff}\Lambda \tag{1}$$

Here, λ_B is the Bragg wavelength reflected by the lattice pattern, n_{eff} is the effective refractive index of the fiber core, and Λ is the period of the lattice pattern. Equation (1) shows that the Bragg wavelength is linearly proportional to the effective refractive index of the fiber core and distance between lattice patterns. The effective refractive index of the fiber core is a constant value determined at the time of manufacture. The Bragg wavelength changes with the distance between grid patterns, which is affected by temperature or strain. Therefore, an FBG sensor can be used to detect strain or temperature. In this experiment, a quantitative assessment was performed with a hand replica at the same temperature as the environment, not with an actual human hand. In addition, because the data were measured after the temperature was stabilized, the effect of the temperature was not considered to be significant. The variation in the Bragg wavelength due to strain can be expressed by [32]:

$$\Delta\lambda_B = \lambda_B(1 - \rho_\alpha)\Delta\varepsilon_{axial} \tag{2}$$

where λ_B is the shift in the Bragg wavelength due to axial strain, ρ_α is the coefficient of photo-elasticity of the optical fiber, and $\Delta\varepsilon_{axial}$ is the axial strain. Thus, the change in the Bragg wavelength is only affected by the axial strain.

2.2. Design of the Embedded FBG Strain Sensor

To measure the angle of a finger joint, an FBG sensor should be able to measure not only the axial strain but also the bending strain. Researchers have studied attaching an FBG sensor to a surface [33,34] or embedding it in a polymer matrix [35,36] to measure the bending strain. An adhesive [33] or polyimide backing patch [37,38] is usually used to attach an FBG sensor to a bending surface. In this case, nonuniform hardening of the adhesive may cause nonuniform bending at the FBG node or reduce the durability of the sensor over a long bending time because of the reduced adhesion of the tape [33,34,37]. Therefore, FBG strain sensors are embedded into a polymer matrix to measure the bending strain more reliably [39,40]. When an FBG strain sensor is bent, the strain delivered to the FBG nodes increases by the offset distance between the sensor and neutral bending axis. He et al. determined the relationship between the position of the FBG sensor from the bending center and the change in the Bragg wavelength if the sensor is subjected to a bending strain [39,40]:

$$\Delta\lambda_B = \lambda_B(1-\rho_\alpha)C{\cdot}h_t \tag{3}$$

$$x = h_1 - (h_t/2) = h_1 - (h_1+h_2)/2 \tag{4}$$

where C is the bending curvature. h_1 and h_2 are the positions of the optical fiber core from the top and bottom, respectively, of the polymer matrix and are constant values determined during the manufacturing process. h_t is the summation of h_1 and h_2, and refers to the thickness of the FBG strain sensors. The change in the Bragg wavelength of the embedded FBG sensor is predominantly affected by the curvature C and varies from the neutral bending axis to the offset distance x of the FBG. As you can see, the data of FBG strain sensors can be changed by either the curvature C or the offset distance x. The curvature C is not a value we can designate, but an external factor that we should measure with sensors. So only what we can handle is the offset distance x. By (3) and (4), as offset distance x increases, peak wavelength change increases. This means that the sensitivity of the FBG strain sensor increases. Moreover, too much sensitivity of the FBG strain sensor can cause a collision between peak wavelengths of adjacent nodes, and if the sensitivity is too low, the effect of noise will increase. Based on this principle, Figure 1 illustrates the concept of measuring the angles of finger joints.

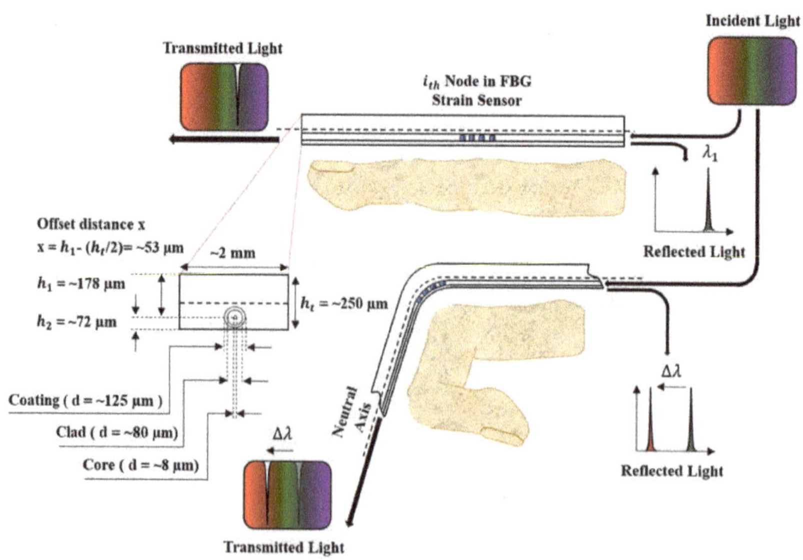

Figure 1. Principle of the fiber Bragg grating (FBG) strain sensors used to measure the finger joint angles.

In this study, considering the abovementioned issues regarding selection of appropriate offset distance, the core of the FBG strain sensor was placed at an offset distance x of approximately 53 μm from the neutral axis within a rectangular epoxy matrix with a height of approximately 250 μm (Epotek-301, Epoxy Technologies Inc., Billerica, MA, USA). The sensor was rectangular in shape with a width of 2 mm, and it measured the bending strain only in the upward and downward directions, not the leftward and rightward directions. The embedded FBG sensor (FBGS Technologies GmbH, Jena, Germany) had FBG nodes with a 3 mm length engraved on a 125 μm diameter optical fiber coated with Ormocer ®. The distance between FBG nodes was 19 mm. Due to differences in finger lengths, nine nodes of the FBG sensor were used for the index, middle, and ring fingers; seven nodes were used for the little finger; and five nodes were used for the thumb. Table 1 details the specifications of the FBG sensors.

Table 1. Specifications of the FBG sensors comprising the wearable hand module.

Finger Type	Sensor Length	Node Length (Distance)	Wavelength Range (Period)	Number of Nodes
Index Finger	155 mm	3 mm (19 mm)	1532.0~1562.4 nm (3.8 nm)	9
Middle Finger	155 mm		1532.0~1562.4 nm (3.8 nm)	9
Ring Finger	155 mm		1532.0~1562.4 nm (3.8 nm)	9
Little Finger	125 mm		1576.0~1600.0 nm (4.0 nm)	7
Thumb	85 mm		1584.0~1600.0 nm (4.0 nm)	5

2.3. *Fabrication of the Wearable Hand Module*

Figure 2 shows the detailed design and configuration for a wearable hand module manufactured with FBG strain sensors. First, to protect against ambient temperature and impact, the FBG strain sensors are fixed inside the thermoplastic polyurethane (TPU) guide (Cubicon 3DP-310F, Cubicon Inc., Gyeonggi-do, Korea). While the TPU guide is fixed at the nail part, the FBG sensor is attached to the opposite end inside of the TPU guide. If the FBG sensor is attached to the nail part where the TPU guide is also attached, an axial strain will be applied to the FBG strain sensor when the finger bends. To avoid this problem, the FBG strain sensor is attached to the opposite end of the fixing points of the TPU guide. Using this method, only the strain caused by finger bending is transferred to the FBG strain sensor. The TPU guide is designed to be movable when the hand is clenched into a fist or the palm is opened with a passage structure (flattener) made of VeroWhite Plus (Objet350 Connex, Stratasys, Eden Prairie, MN, USA) on the back of the hand. Teflon films minimize friction at the contact surface with the TPU inside the flattener. The flattener flattens the FBG nodes on the back of the hand, which helps distinguish the curvature of the metacarpophalangeal (MCP) joint more effectively. The fingertips are connected to the flattener structure with silicone rings, which hold the FBG strain sensors within a certain distance from the fingers. This mitigates the friction inside the flattener because the resilience of the silicone ring structure is transferred to the TPU when the fist is clenched. The wearable hand module was used to quantitatively evaluate a hand replica with different finger sizes. A rendering program was used to track the joint angles in real time as the module was worn by different hands.

2.4. *Evaluation of the FBG Strain Sensor*

Bending tests were performed to assess the linearity and durability of the FBG strain sensor for measuring the joint angles of the fingers. An experiment was carried out with a bending tester that could apply an angle of 90°, and the sensor was bent 10,000 times using a stepper motor. Not many nodes corresponded to the curvature of the bending tester, so 10,000 bending cycles were performed on two nodes, and the linearity was tested before and after the experiment. The Bragg wavelength during the 10,000 bending cycles was recorded to observe any significant changes in the Bragg wavelength. Figure 3a shows the linearity of the sensor data before and after the high-cycle bending test of the FBG strain sensor. The FBG strain sensor indicated no major problems with linearity before and after the

10,000 cycle test with R^2 values of 0.997–0.998. Figure 3b shows no significant change in the Bragg wavelength during the bending test. However, if the sensor data are obtained as a function of strain and the sensor is attached to the wearable hand module, the linearity of the system results may be significantly degraded. This is because the curvature acting on the node must be constant; however, the curvatures of the finger joints are very close to each other and can have different values. This problem cannot be solved solely through the sensor design and needs to be addressed by using an algorithm to process the raw sensor data. Section 3 presents the process of reading data from the FBG strain sensor and using algorithms for calculations with the data.

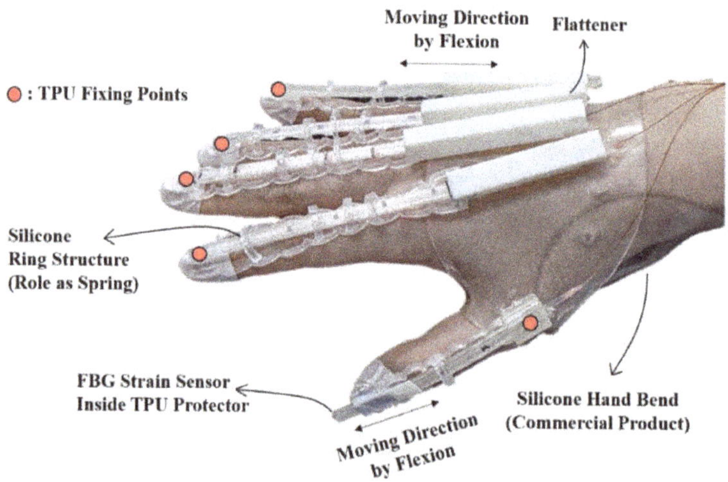

Figure 2. Wearable hand module and its components.

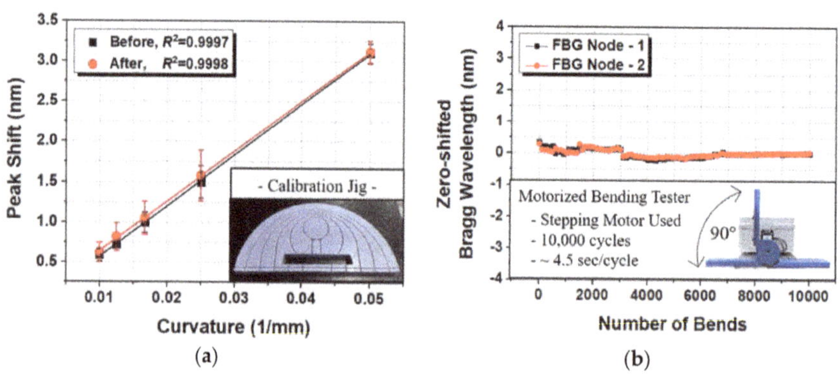

Figure 3. Evaluation of the FBG strain sensor: (**a**) Peak shift change by curvature and (**b**) zero-shifted Bragg wavelength during the high-cycle durability test.

3. FBG Interrogation System and Real-Time Tracking Algorithm

3.1. Interrogation Process

A commercial interrogator (Smart Fibres Limited, Bracknell, UK) was used to measure the changes in the Bragg wavelength of the C and L bands for the FBG strain sensor in four channels. As discussed in Section 2, bending the fingers changed the Bragg wavelength of the FBG node within the FBG strain sensor. The interrogator was used to check the shift in the Bragg wavelength of the FBG nodes and perform the algorithm calculations in real time. Figure 4 shows the process of converting the shift in the Bragg wavelength of the FBG strain sensors within the wearable hand module with the commercialized interrogator.

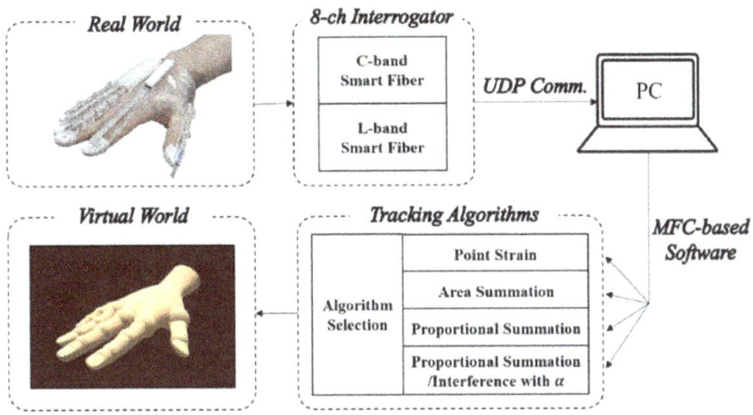

Figure 4. Schematic diagram of the FBG interrogation system for real-time tracking of the finger joint angle.

To calculate the algorithm, Bragg wavelength data are sent from the smart fiber interrogator to a PC via User Datagram Protocol (UDP) communication. A tracking algorithm in the Microsoft Foundation Class (MFC)-based program translates the Bragg wavelength data into the angle for each joint of the finger. The algorithm calculates the change in the Bragg wavelength of the FBG nodes attached to the wearable hand module every 10 ms within the thread function. The buttons in the dialog box are used to select one algorithm and track the finger joint within 20–40 ms.

Before the algorithm can be performed, the reference data for the finger joint angles when the user's fist is clenched need to be entered. The required reference data are the angles of the MCP, PIP, and DIP joints for a clenched fist and the Bragg wavelengths for a clenched fist and open palm. These five input variables are used to determine the exact finger joint angles between a clenched fist and open palm in real time. If the peak shift is outside this range, it is not used as data. For certain nodes, very small peak shifts occur between the clenched fist and open palm, and this data can lead to very large angle measurements. Thus, they are regarded as noise and excluded. If the denominator is very small, the angle will be infinite for an open palm. In this case (<1°), the angle is input into the 3D hand graphic using an alternative algorithm (e.g., PTS, AREA, PS) that does not have a denominator of zero for any given angle.

3.2. Proposed Algorithm

Various algorithms have been studied to accurately track the angles of finger joints. However, universally applicable algorithms are difficult to implement because of differences in the type of hand module, sensors measuring strain or curvature, and reference calibration method. Park et al. fabricated a hand module from Eco-flex and used changes in the resistance of the internal conductive liquid metal to convert the strain from a bending finger into the angle of a finger joint [7]. The key concept of their algorithm was decoupling to distinguish the flexion/extension motion from the abduction/adduction motion. Wang et al. used 3D printing to fabricate structures that they attached to the fingers to track the angles of finger joints during abduction/adduction, circulation, and flexion/extension motions [9]. They used the weighted average method to develop an algorithm for calculating angular data by comparing the voltage amplitudes obtained from two or more channels. However, the algorithms used in these studies were data isolation processes to obtain pure flexion or pure abduction; in contrast, the aim of the present study was to accurately measure the flexion and extension of each joint. Park et al. tried to distinguish the flexion/extension motion of a joint by using a flexible wire and a linear potentiometer [41]. The resistance of the linear potentiometer attached to the PIP joint during a clenched fist was calculated by subtracting the angle of each joint, while the resistance of the potentiometer attached to the MCP joint was added. Thus, they mainly calculated the joint angle through a sensor/spring system where the spring structure was directly attached to the sensor and the strain in the axial direction was measured [6,41]. In the present study, a spring structure was similarly used within the wearable hand module; however, the spring does not pull the sensor but instead helps the sensor easily enter the flattener structure when the motion shifted from flexion to extension. Since the FBG strain sensor moves continuously inside the flattener structure according to the finger's movement, the positions of the FBG nodes measuring each joint (especially the MCP joint) can move. Experiments showed that the FBG node positions for the DIP and PIP joints did not significantly differ with respect to the finger length and motion because they were close to the fixed point. However, the FBG node positions for the MCP and interphalangeal (IP) joints varied greatly. Thus, the key concept of the proposed algorithm is to analyze the change in sensor data due to changes in these positions. Figure 5 details the positional change process.

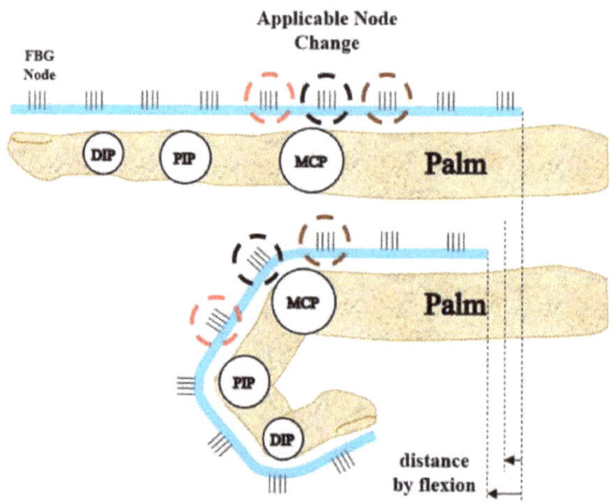

Figure 5. Positional changes of FBG nodes in the FBG strain sensor measuring the metacarpophalangeal (MCP) joint angle.

In contrast to the traditional method of directly measuring the bending strain of a joint by attaching a strain sensor, the periodically engraved FBG nodes change their position as the DIP, PIP, and MCP joints bend (especially the PIP joint). For the DIP and PIP joints, the FBG node positions do not change much between a clenched fist or open palm, but the node positions applicable to the MCP joint change significantly. Therefore, using only one FBG node for the MCP joint introduces too large an error. If an FBG node is fixed to each joint, then the joint angle can be converted to data that can easily be obtained from a single node. However, if it is not fixed and is moving, the nodes corresponding to each joint (especially the MCP joint) will also change, and an algorithm is required to convert the corresponding FBG node positions into the angle in real time. Various algorithms were considered for this purpose, but the most basic point strain methods are introduced first for comparison.

The most basic algorithm for tracking the joint angle is the point strain (PTS) method, which uses data from only one FBG node for each joint. The peak shift that occurs at one node for each joint between a clenched fist and open palm is stored; when the current peak shift (current peak shift = (current Bragg wavelength)—(Bragg wavelength during an open palm)) increases linearly, the measured joint angle increases linearly from the open palm to a clenched fist. This is expressed by:

$$\theta_{mea} = \frac{\Delta\lambda_{i\_c}}{\Delta\lambda_{i\_m}} \times \theta_{r-p} + \theta_p \tag{5}$$

where θ_{mea} is the joint angle calculated in real time by the sensor/algorithm. θ_{r-p} and θ_p are the actual angles (angle measured by the goniometer for a clenched fist that should subtract the initial angle for an open palm θ_{r-p}) for the clenched fist and open palm (initial angle θ_p). For an actual human hand, the initial angle (θ_p) with an open palm can be set to 0 because of the difficulty in measuring the exact initial joint angles. For the hand replica, however, the initial angle measured in the quantitative evaluation was entered. $\Delta\lambda_{i\_c}$ is the current peak shift of the ith node. $\Delta\lambda_{i\_m}$ is the maximum peak shift between the open palm and clenched fist of the ith node. Thus, $\Delta\lambda_{i\_c}/\Delta\lambda_{i\_m}$ has a value from 0 to 1 and is called the bending percentage (B%) of the ith node.

In the aforementioned method, it is necessary to assume that only one FBG node corresponds to each joint when moving from an open palm to clenched fist. However, because two or more FBG nodes can be matched to each joint and the FBG node positions change continuously as the fingers bend, matching only one node to each joint could be a fatal error. Therefore, setting two or more FBG nodes to measure each joint angle can help avoid this error. This is the concept behind the area summation method (AREA):

$$\theta_{mea} = (\frac{\Delta\lambda_{i\_c}}{\Delta\lambda_{i\_m}} + \ldots + \frac{\Delta\lambda_{n\_c}}{\Delta\lambda_{n\_m}}) \times \frac{1}{(n-i)+1} \times \theta_{r-p} + \theta_p \tag{6}$$

In contrast to the PTS algorithm given in Equation (5), AREA requires two or more nodes for each joint angle. Thus, the number of B% (bending percentage) terms increases ($\Delta\lambda_{i\_c}/\Delta\lambda_{i\_m}$, $\Delta\lambda_{n\_c}/\Delta\lambda_{n\_m}$). The ith node is the first node that corresponds to a joint, and the nth node is the last node to correspond to the joint. For example, nodes 1 and 2 ($i = 1$, $n = 2$) may correspond to the DIP joint, nodes 3 and 4 ($i = 3$, $n = 4$) may correspond to the PIP joint, and nodes 5, 6, 7, and 8 ($i = 5$, $n = 8$) may correspond to the MCP joint for the index-little fingers. Equation (6) separately adds B% of the i and n nodes ($\Delta\lambda_{i\_c}/\Delta\lambda_{i\_m}$, $\Delta\lambda_{n\_c}/\Delta\lambda_{n\_m}$) and distributes the number of nodes at the reference angle ($1/(n-i+1) \times \theta_{r-p}$), which does not greatly differ from simple summation. Equation (6) can greatly reduce errors that may occur with Equation (5). AREA adds the changes at each node ($i - n$) equally, so it is recommended if all FBG nodes are well-seated on the curvature of the joints.

However, depending on the size of the hand or bending, the FBG nodes may be dislodged from a joint. If the FBG nodes are not well-seated and dislodged from the joints, they should be weighted less. This is the concept behind the proportional summation (PS) algorithm:

$$\theta_{mea} = \{B\%_i \times (\frac{\Delta\lambda_{i\_c}}{\Delta\lambda_{i\_c} + \ldots + \Delta\lambda_{n\_c}}) + \ldots + \{B\%_n \times (\frac{\Delta\lambda_{n\_c}}{\Delta\lambda_{i\_c} + \ldots + \Delta\lambda_{n\_c}})\} \times \theta_{r-p} + \theta_p \tag{7}$$

If an FBG node does not receive the same strain from the joint, the summation is reduced with the strain in real-time. Equation (7) means that the proportion of θ_{r-p} is divided according to the peak shift in the present state compared to each FBG node corresponding to the joint. For example, if the MCP joint is 90° ($\theta_{r-p} = 90°$, $\theta_p = 0°$) during a closed fist, the reference angle is set as the highest peak shift among the corresponding nodes. Using this algorithm can reduce the errors implied by Equations (5) and (6).

However, differences could occur in nodes directly corresponding to the MCP joint between large and small hands. Figure 6 shows the peak shifts of nodes 5–9 corresponding to the MCP joint with large (117 mm) and small (77 mm) hands. For the small hand (77 mm), nodes 5 and 6 were directly strained by the MCP joint; for the large hand (117 mm), nodes 7 and 8 were directly strained by the MCP joint. Nodes 6 and 7 were indirectly or directly affected by the MCP joint with either hand size. In contrast, node 5 was at the center of the PIP and MCP joints for the large hand in a clenched fist, so it was not affected by any joint. In addition, if only the PIP joint moved, the peak shift may occur at node 5, which would increase the unwanted measured angle of the MCP joint.

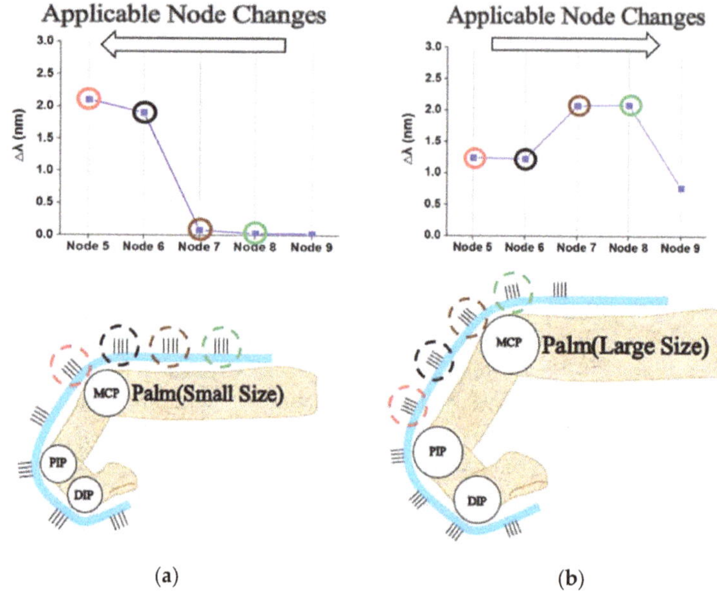

Figure 6. Peak shift changes of five nodes applicable to the MCP joint with different hand sizes: (**a**) Smaller (77 mm) and (**b**) larger (117 mm).

To solve this problem, the bending percentage of the PIP joint should be used to reduce the influence of node 5. This is proportional summing/interference (PS/I) algorithm, which prevents interference by the PIP joint during the PS method and is expressed in Equation (9). In Equation (8), C_{I_α} is the coefficient for the change in the bending percentage ($\Delta\lambda_{4\_c}/\Delta\lambda_{4\_m}$) of the PIP according to the hand size.

$$C_{I\_\alpha} = (1 - \alpha\frac{\Delta\lambda_{4\_c}}{\Delta\lambda_{4\_m}}) \quad (8)$$

$$\theta_{MCP,\,IP} = \{B\%_i \times (\frac{C_{I\_\alpha}\times\Delta\lambda_{i\_c}}{C_{I\_\alpha}\times\Delta\lambda_{i\_c}+\ldots+\Delta\lambda_{n\_c}}) + \ldots + \{B\%_n \times (\frac{\Delta\lambda_{n\_c}}{C_{I\_\alpha}\times\Delta\lambda_{i\_c}+\ldots+\Delta\lambda_{n\_c}})\} \times \theta_{r-p} + \theta_p \quad (9)$$

Nodes 5–8 ($i = 5$, $n = 8$) correspond to the MCP joint. The influence of the PIP joint angle on node 5 is calculated using Equation (9) based on Equation (7) with the coefficient from Equation (8). $\Delta\lambda_{4\_c}$ is the current peak shift of node 4, which is the bending percentage of the PIP joint. $\Delta\lambda_{4\_m}$ is the maximum peak shift of node 4 during the open palm and clenched fist in the reference process. Therefore, $\Delta\lambda_{4\_c}/\Delta\lambda_{4\_m}$ represents the bending percentage of the PIP joint. This term ranges from 0 to 1 and is subtracted from 1. Thus, a larger bending angle for the PIP joint reduces the effect of node 5.

The role of the bending percentage of the PIP joint was confirmed to depend on the hand size. For a small hand, the role of $\Delta\lambda_{4\_c}/\Delta\lambda_{4\_m}$ should be reduced because of the large influence of node 5. When the role of $\Delta\lambda_{4\_c}/\Delta\lambda_{4\_m}$ decreases, the MCP joint can be measured more accurately by using all of nodes 5–8. Conversely, for a large hand, the role of $\Delta\lambda_{4\_c}/\Delta\lambda_{4\_m}$ should be greater. Therefore, the angle of the MCP joint should be measured with only nodes 6–8, and node 5 should be excluded. By changing the value of α in Equation (8), the role of $\Delta\lambda_{4\_c}/\Delta\lambda_{4\_m}$ can be increased or decreased.

Different hand sizes between 77 and 117 mm were tested, and the results confirmed that $a = 0.00$ was optimal for small hands (77 mm), $a = 0.65$ was optimal for moderately sized hands (97 mm), and $a = 1.30$ was optimal for large hands (117 mm). In particular, when $\alpha > 1$, the angle calculated from node 5 should be subtracted from the calculation for the MCP joint for a more accurate coefficient. The a values in the experiment were optimized by comparing the maximum peak shifts of nodes 6 and 7. As the hand size increased, the peak shift of node 6 decreased; as the hand size decreased, the peak shift of node 7 decreased with a clenched fist. For different hand sizes, the optimal values of α that most closely matched the experimental results were 1.70 (117 mm), 1.30 (107 mm), 0.59 (97 mm), 0.22 (87 mm), and 0.04 (77 mm). For the thumb, the optimal α values were 0.50 (78 mm) and 0.13 (68 mm). Thus, α for nodes 3 and 4 corresponding to the MCP joint of the thumb were set to 0.57 (78 mm) and 0.30 (68 mm) based on the relational formula $\Delta\lambda_{4\_m}/(\Delta\lambda_{3\_m} + \Delta\lambda_{4\_m})$. The optimal a values obtained from the experiment led to the best accuracy for the MCP joint angle. However, they varied with hand size, so the input of α to each experiment was limited. A better approach would be to find a parameter with the most similar value to the optimal a and apply it through relational formulas for the peak shifts, such as $\Delta\lambda_{7\_m}/\Delta\lambda_{6\_m}$ for the index–little fingers and $\Delta\lambda_{4\_m}/(\Delta\lambda_{3\_m} + \Delta\lambda_{4\_m})$ for the thumb.

The PS/I_α algorithm is the PS algorithm when $\alpha = 0$, and the PS/I algorithm is when $\alpha = 1$. Depending on the hand size, the PS or PS/I algorithm may be applicable. A suitable α can be automatically applied depending on the hand size, which will significantly reduce the error in angle measurement.

Figure 7 shows a flow chart for the process of selecting the appropriate algorithm using qualitative criteria. The PTS method uses only one node per joint, so the condition "Multiple FBG nodes are used to measure a joint." has been added. The AREA method is used to calculate finger joint angles using a simple sum of multiple nodes without weighting certain nodes, so the condition "Sometimes, certain nodes are dislodged from the same curvature." has been added. Finally, for PS and PS/I_α, it is important to ensure that a particular node is completely deviated from that curvature, so the condition "In some cases, certain nodes completely deviate from the influence of curvature." has been added. This flow chart will help in selecting the appropriate algorithm for any situation when producing a wearable hand module described in this paper.

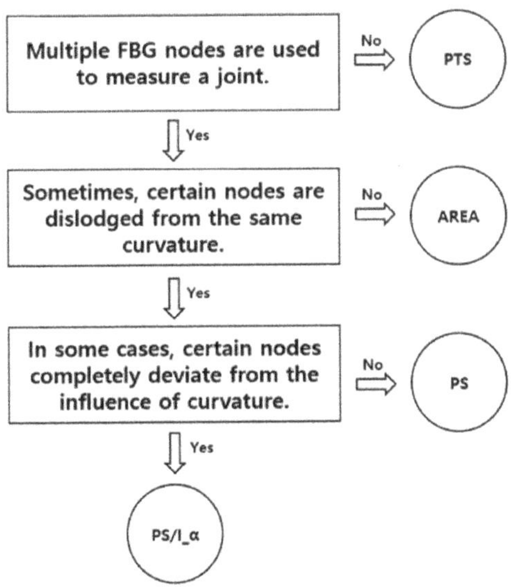

Figure 7. Flow chart diagram for selecting algorithms for measuring finger joint angles with FBG strain sensors.

4. Joint Angle Measurement

4.1. Investigation with Different Hand Sizes

Figure 2 shows that the fixed point of the FBG strain sensor was the same from the index finger to little finger but differed for the thumb. Therefore, the 3D printed hand replica was tested by printing two different finger models: The index finger and thumb with length ranges of 77–117 and 68–78 mm, respectively. The FBG strain sensors used for the index-little fingers only differs in the number of nodes according to the finger length. Eleven adult males were selected to measure the lengths of their fingers, and the lengths of the index-little fingers had a range of 105 ± 11.4 mm while the thumb had a range of 75 ± 4.5 mm as shown in Table 2. Therefore, if the quantitative evaluation considered these ranges, the results should be applicable to all normal adult male hands. Some studies also included the length of the palm, but this was excluded from the present study because the data at each joint were read slightly differently as the sensors moved [22].

Table 2. Different hand sizes of adult males investigated in this study.

Subject	Investigated Value
Number of People	11
Age Range	26~48
Index~Little Finger Length	105 ± 11.4 mm
Thumb Length	75 ± 4.5 mm

4.2. 3D Printed Hand Replica and Measurement Process

Reference angles were obtained using a goniometer for comparison with the measured angles obtained by the sensor data and algorithms. However, measuring angles of an actual human hand can introduce many errors. For example, the finger joint angle measured by goniometer deviated by 5–10° depending on whether the fist was clenched tightly or loosely. For a more accurate quantitative assessment, the wearable hand module was evaluated according to the finger joint angles of a 3D printed hand replica, which is shown in Figure 8.

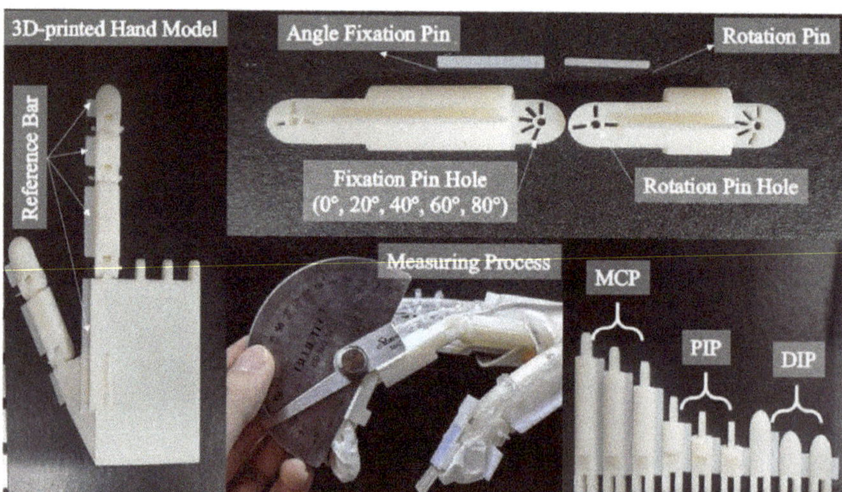

Figure 8. 3D printed hand model used for quantitative evaluation of the wearable hand module.

The 3D printed hand replica was printed using VeroWhite Plus. The MCP, PIP, and DIP joint angles were maintained using angle fixation pins. The rotation pin was inserted into the center of rotation of each joint, and two joints were rotated in a concentric circle. Then, the angle fixation pin was inserted to secure a certain joint angle so that the sensor value of the hand module would be fixed. The joint angle was fixed at 0°, 20°, 40°, 60°, 80°, and 90°. The replica had pins to hold the angle, but there was some amount of play because of the resolution (30 μm) of the 3D printer. Since the hand module contained spring components, the MCP joint angle deviated by 10–15° during the mounting and removal processes. Therefore, the quantitative assessment would be more accurate if the hand module was equipped and the sensor values were logged while the actual angle was measured with the goniometer at the same time. Therefore, reference bars were designed and printed on the side of each joint phalanx so that the sensor data and goniometer measurements could be recorded simultaneously.

4.3. *Quantitative Measurement with the Hand Replica*

A wearable hand module was mounted on a hand replica printed in different sizes. The angles of the MCP, PIP, and DIP joints were measured using different algorithms (i.e., PTS, AREA, PS, PS/I, and PS/I_α). Figure 9 shows the measured angle errors of the finger joints at 0°, 20°, 40°, 60°, 80°, and 90° for different finger sizes: 77–117 mm for the index-little fingers and 68–78 mm for the thumb. All tests were performed five times at room temperature, and a statistical analysis was performed on the angle errors of the samples under various conditions.

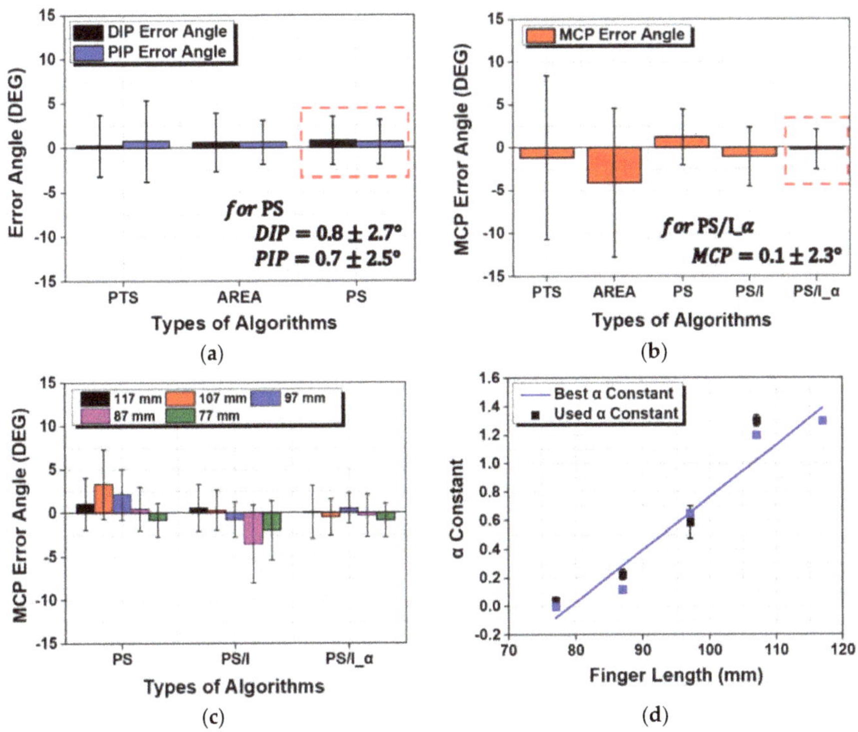

Figure 9. Joint angle errors of the wearable hand module according to different lengths of the index-little fingers: (**a**) Distal interphalangeal (DIP) and proximal interphalangeal (PIP) joint angle errors, (**b**) MCP joint angle error, and (**c**) MCP joint angle error according to the lengths of the index-little fingers, and (**d**) best and used α constant values for accurate measurement of the MCP joint.

Figure 9a shows that the DIP joint had the smallest mean angle error of 0.26° with the PTS algorithm, but the standard deviation was relatively large at ±3.43°. When an algorithm using multiple nodes such as AREA was used instead of just one node as PS, the deviation of the angle error became more stable. For the DIP joint, the optimal choice for the various hand sizes was confirmed to be the PS algorithm; the angle error was 0.82 ± 2.71° with a minimum error of 3.53° and maximum error of 1.89°. Nodes 1 and 2 corresponded to the DIP joint, even though node 1 was positioned a little bit away from the joint. For the PIP joint, the PTS algorithm had the largest error angle and standard deviation at 0.77 ± 4.57°, but a smaller angle error was observed when two or more nodes were used (such as with AREA or PS). With the PS algorithm, the angle error of the PIP joint under all finger length conditions was 0.65 ± 2.50° with a minimum angle error of 3.15° and maximum angle error of 1.85°.

For the MCP joint shown in Figure 9b, the PTS method showed a very large mean angle error and deviation of −1.17 ± 9.52°. This larger mean angle error decreased with AREA and PS but failed to meet the allowable angle error (which is larger than 5.0°, as mentioned in the introduction) for various hand sizes. With AREA, a small hand (77 mm) resulted in a large angle error of −6.23 ± 14.71°. With PS, a large hand (107 mm) resulted in a large angle error of −3.30 ± 3.10°. As discussed in Section 3.2, node 5 played a very important role; when the value from node 5 was affected by the degree of bending in the PIP joint (i.e., the PS/I_α algorithm was used), the angle error of the MCP joint was dramatically reduced. With PS/I_α, the MCP joints had an angle error of −0.21 ± 2.32° with a minimum angle error of 2.5° and maximum error of 2.1°.

For the DIP and PIP joints, the measured angles did not represent changes as significant as that of the MCP joint because they were fixed at the fingertip. For the MCP joint, the algorithm needed to be adjusted because the positions of the corresponding FBG nodes varied. Changing from PTS to PS/I_α reduced the maximum angle error of 8.19°. The behavior of the MCP joint tended to match the initial concept addressed in Figures 5 and 6. Using the appropriate relational formula for FBG nodes rather than a single FBG node helped determine the correct angle. In particular, α had a very large influence; Figure 9c,d graphs the decrease in the angle error and the best and used a values for measuring the joint angle with PS. The PS/I_α algorithm is the PS algorithm at $\alpha = 0$ and PS/I algorithm at $\alpha = 1$; Figure 9c shows that PS/I_α minimized the angle error through the application of a constant.

Figure 10 graphs the relationship between the measured and actual angles corresponding to the best algorithm for each joint (PS for the DIP and PIP joints and PS/I_α for the MCP joint) in the five experiments. The linearity showed a slight decrease in linearity with the smaller hand (77 mm, Figure 10c). However, the average linearity had $R^2 = 0.9982 \pm 0.001$ with the larger hand and $R^2 = 0.9986 \pm 0.001$ with the smaller hand. Comparing the linearity R^2 values for different algorithms produced interesting results. Figure 10d shows the R^2 value for the MCP joint between the actual and measured angles using different algorithms. For the DIP and PIP joints, the measured angle error was not significant; the linearity R^2 values did not differ significantly according to the type of algorithm with 0.9971 ± 0.001 and 0.9966 ± 0.002, respectively. For the MCP joint, the linearity R^2 was 0.9778 ± 0.026 and showed a large difference depending on the algorithm. However, similar to the angle error shown in Figure 9c, the PS/I_α algorithm showed very strong linearity with 0.9983 ± 0.001. Moreover, when the algorithm was varied from PTS to PS/I_α, linearity of angular data approached that of the raw data (0.99975 ± 0.0005, in Section 2.4) of Bare FBG strain sensor obtained from the bending test. Compared to the linearity of the Bare FBG strain sensor, this linearity degradation of PTS and AREA algorithm was similarly caused by whether or not a particular node was completely under the influence of curvature. Figure 7 allows us to select algorithms such as PTS, AREA, PS, PS/I_α by noting whether a particular node is completely under the influence of curvature or indirectly affected by curvature or completely deviated from curvature. For Bare FBG strain sensor data, when measuring linearity before and after 10,000 bends, the two nodes were completely under the influence of curvature. Therefore, considering the position of the nodes on the curvature was very important to ensure high linearity of the data. In the case of the PTS and AREA methods, which did not consider the position of the nodes on the curvature, it was thought that the degradation of linearity occurred because it did not respond to the continuous movement of nodes on curvature. This outcome is precisely in line with our algorithm concept mentioned in Section 3.2 and is considered to be one of the reasons for the low error angle. Thus, the improved linearity can be used as a good basis for judging the suitability of an algorithm, in addition to the angle error.

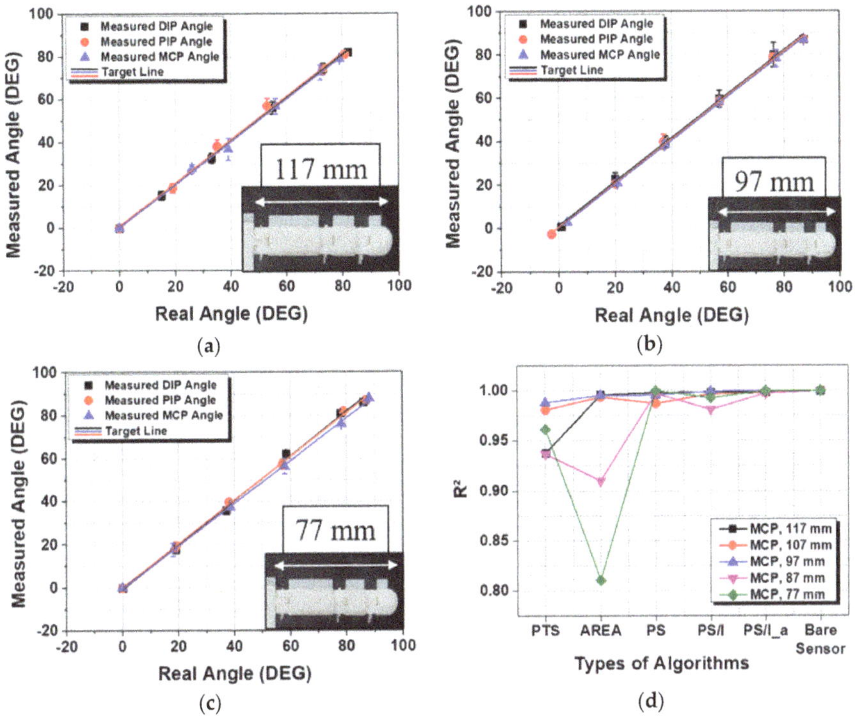

Figure 10. Measured angles versus real angles (measured by a goniometer) with respect to the length of the index finger: (**a**) 117, (**b**) 97, and (**c**) 77 mm. (**d**) Linearity of the measured MCP joint angle according to the algorithm and finger length.

This tendency was found for both the thumb and index-little fingers. Since the thumb is different, the IP joint of the thumb behaved the same as the MCP joint of the index-little fingers. Meanwhile, the MCP joint of the thumb behaved similarly to the DIP and PIP joints of the index-little fingers. Figure 11 shows the angle error and linearity of the wearable hand module attached to the thumb.

For the thumb, the PS/I_α algorithm had the smallest angle error. For the MCP joint located near the fixed point, AREA had the smallest angle error. For the IP joint, an angle error of 0.74 ± 2.84° was observed for the large thumb (78 mm), and an angle error of −0.35 ± 1.23° was observed for the small thumb (68 mm) with PS/I_α. The maximum angle error was 3.58°, and the minimum angle error was 0.88°. For the MCP joint, Figure 11b shows angle errors of 2.06 ± 2.53° and −0.05 ± 1.79° with thumb lengths of 78 and 68 mm, respectively. The maximum angle error was 4.59°, and the minimum angle error was 0.47°. Figure 11c,d shows no significant difference between the actual and measured angles. Unlike the index-little fingers, longer thumbs resulted in greater deviation from the target angle line. For the index-little fingers, a shorter finger reduced the distance between the PIP and MCP joints, and the role of node 5 constantly changed. For the thumb, however, the fixed point was on the opposite side, so the role of each joint changed, and the tendency was also reversed.

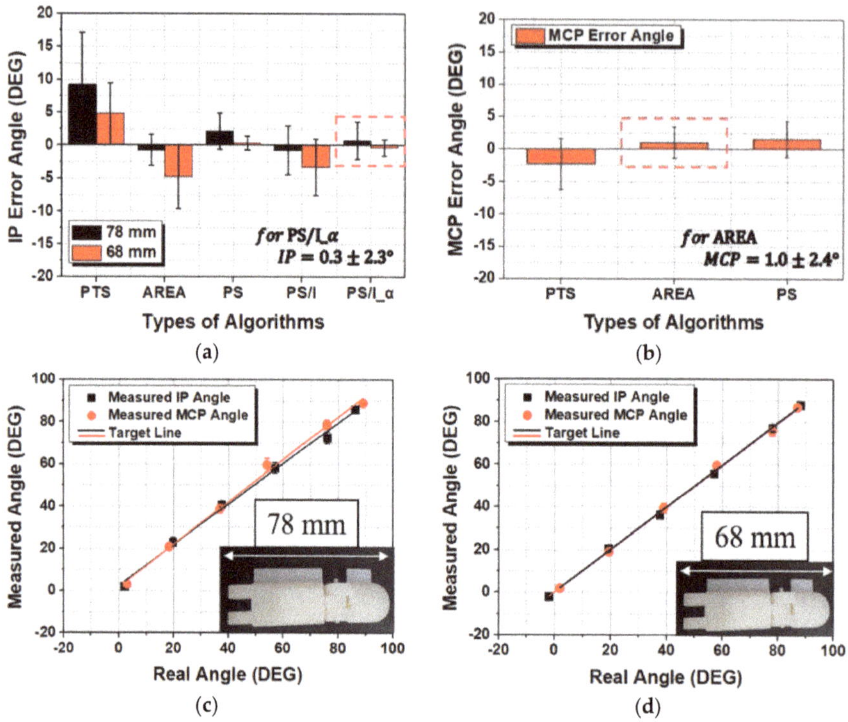

Figure 11. Measured angles versus real angles (measured by a goniometer) with respect to the size of the thumb: Angle errors of the (**a**) interphalangeal (IP) joint and (**b**) MCP joint; linearity at (**c**) 78 and (**d**) 68 mm according to the algorithm and type of joint.

Thus, the algorithm required for each joint differed depending on the fixed point. A joint closer to the fixed point meant that relatively simple algorithms such as PS and AREA were more accurate. For joints further away from the fixed point, the PS/I_α algorithm was required to accommodate the changes in the roles of intermediate nodes.

4.4. Rendering of Fingers of Different Users in Real Time

The MFC-based program was used to observe the wearable hand module when worn on different hand sizes and track the finger angles in real time. In particular, the PIP joint bending without the MCP joint or the MCP joint bending without the PIP was properly tracked when the fist was clenched. For the MCP joint, the algorithms could properly track the finger joint angle because the locations of the FBG nodes corresponding to the MCP joint changed in real time when movements other than the clenching motion was used. Figure 12 shows that users with relatively large and small hands followed the virtual hand's finger angles for various hand gestures. The results confirmed that the wearable hand module can be used to track finger joint angles in real time with high accuracy for hand gestures, not just clenching the fist.

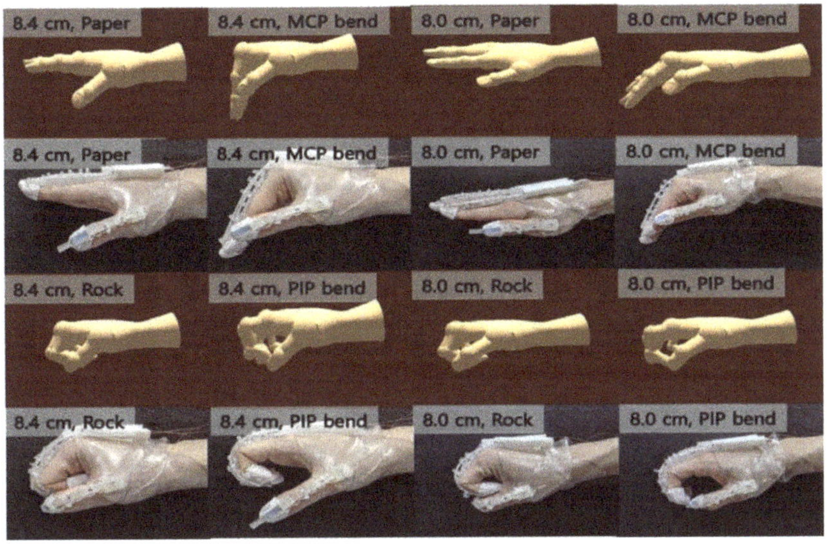

Figure 12. Comparison between real and rendered finger angle images.

5. Conclusions

Wearable hand modules using FBG strain sensors were fabricated and evaluated for their performance at accurately measuring the angles of finger joints in real time. Algorithms were developed for converting data from sensors that change position according to hand movements and sizes into joint angles, unlike conventional sensors that are fixed. PTS, AREA, PS, PS/I, and PS/I_α were used to measure the angles of the DIP, PIP, and MCP joints; PS was judged to be most suitable for the DIP and PIP joints, and PS/I_α was judged to be most suitable for the MCP joint. The maximum and minimum angle errors of the index-little fingers were 3.53 and 1.89°, respectively, for the DIP joint; 3.15 and 1.85°, respectively, for the PIP joint; and 2.53 and 2.11°, respectively, for the MCP joint. For the thumb, the maximum and minimum angle errors were 2.56 and 2.02°, respectively, for the IP joint and 3.43 and 1.41°, respectively, for the MCP joint. The MCP joint of the thumb had the largest maximum angle error of 1.01 ± 2.42°; the rest of the joints showed smaller angle errors. The average angle error of the wearable hand module for all joints, sizes, and conditions was 0.47 ± 2.51°, and MAE was 1.63 ± 1.97°. A rendering program confirmed that the wearable hand module can be fitted to users with various hand sizes and provides accurate tracking even when only the PIP or MCP joint is moved.

The results confirmed that the angles of all finger joints (MCP, PIP, DIP) could be measured with high accuracy even for different hand sizes, and the measurement accuracy differed depending on the algorithm. In addition to flexion/extension, future work should consider exact measurements of the abduction/adduction and hyperextension movements. The gap between FBG nodes should be further reduced, and corresponding algorithms should be developed to accommodate the hand sizes of adult females and adolescents.

Author Contributions: J.S.K. designed and fabricated hardware parts, such as the FBG strain sensors and wearable hand modules, and performed quantitative experiments; B.K.K. developed the software program/algorithms and wrote the manuscript; M.J. contributed to the basic concept of the FBG strain sensors; K.K. provided the principles of the interrogation system and guidance on its operation; D.E.K. provided the actual manufacturing conditions of FBG strain sensors and tribological solutions of wearable hand module; B.-K.J. provided technical advice for the software program during the initial idea and experiment of developing algorithms. J.K. provided technical and scientific guidance during this research and critiqued the manuscript. these three corresponding authors are principle investigators that contributes the technical and scientific guidance with the base of their experience. All authors have read and agreed to the published version of the manuscript.

Funding: This work was supported by a KIST intramural grant (2E30090) and the Convergence Technology Development Program for Bionic Arm through the National Research Foundation of Korea (NRF) funded by the Ministry of Science & ICT (No. 2017M3C1B2085292). This work was also supported by the National Research Foundation of Korea (NRF) grant funded by the Korea government (MSIT) (No. 2020R1A2C2004714).

Conflicts of Interest: The authors declare no conflict of interest.

References

1. Nishiyama, M.; Watanabe, K. Wearable sensing glove with embedded hetero-core fiber-optic nerves for unconstrained hand motion capture. *IEEE Trans. Instrum. Meas.* **2009**, *58*, 3995–4000. [CrossRef]
2. Cui, J.; Sun, Z. Visual hand motion capture for guiding a dexterous hand. In Proceedings of the 2004 6th IEEE International Conference on Automatic Face and Gesture Recognition, Seoul, Korea, 19 May 2004; pp. 729–734.
3. Li, C.; Cui, Y.-L.; Tian, G.-L.; Shu, Y.; Wang, X.-F.; Tian, H.; Yang, Y.; Wei, F.; Ren, T.-L. Flexible CNT-array double helices strain sensor with high stretchability for motion capture. *Sci. Rep.* **2015**, *5*, 1–8. [CrossRef] [PubMed]
4. Ma, Y.; Jia, W.; Li, C.; Yang, J.; Mao, Z.-H.; Sun, M. Magnetic hand motion tracking system for human–machine interaction. *Electron. Lett.* **2010**, *46*, 621–623. [CrossRef]
5. Kim, D.; Yoon, S.; Park, Y.; Jeon, K.; Park, S.; Jeon, J.; Seo, K. Design and Implementation of a Wearable Hand Rehabilitation Robot for spasticity patient. In Proceedings of the 2014 Korean Society of Computer Information Conference, Jeju, Korea, 17–19 July 2014; pp. 21–24.
6. Jha, C.K.; Agarwal, S.; Chakraborty, A.L.; Shirpurkar, C. An FBG-based sensing glove to measure dynamic finger flexure with an angular resolution of 0.1 up to speeds of 80/s. *J. Lightwave Technol.* **2019**, *37*, 4734–4740. [CrossRef]
7. Park, W.; Ro, K.; Kim, S.; Bae, J. A soft sensor-based three-dimensional (3-D) finger motion measurement system. *Sensors* **2017**, *17*, 420. [CrossRef] [PubMed]
8. Sani, M.F.; Abeywardena, S.; Psomopoulou, E.; Ascione, R.; Dogramadzi, S. Towards Finger Motion Tracking and Analyses for Cardiac Surgery. In Proceedings of the 15th Mediterranean Conference on Medical and Biological Engineering and Computing (MEDICON 2019), Coimbra, Porutugal, 26–28 September 2019; pp. 1515–1525.
9. Wang, L.; Meydan, T.; Williams, P.I. A two-axis goniometric sensor for tracking finger motion. *Sensors* **2017**, *17*, 770. [CrossRef]
10. Gumpp, T.; Azad, P.; Welke, K.; Oztop, E.; Dillmann, R.; Cheng, G. Unconstrained Real-time Markerless Hand Tracking for Humanoid Interaction. In Proceedings of the 6th IEEE-RAS International Conference on Humanoid Robots, Genova, Italy, 4–6 December 2006; pp. 88–93.
11. Ballan, L.; Cortelazzo, G.M. Marker-less motion capture of skinned models in a four camera set-up using optical flow and silhouettes. In Proceedings of the 3DPVT, Atlanta, GA, USA, 18–20 June 2008.
12. Shimada, N.; Shirai, Y.; Kuno, Y.; Miura, J. Hand gesture estimation and model refinement using monocular camera-ambiguity limitation by inequality constraints. In Proceedings of the IEEE International Conference on Automatic Face and Gesture Recognition, Nara, Japan, 14–16 April 1998; pp. 268–273.
13. Erol, A.; Bebis, G.; Nicolescu, M.; Boyle, R.D.; Twombly, X. Vision-based hand pose estimation: A review. *Comput. Vis. Image Underst.* **2007**, *108*, 52–73. [CrossRef]
14. Ahmad, N.; Ghazilla, R.A.R.; Khairi, N.M.; Kasi, V. Reviews on various inertial measurement unit (IMU) sensor applications. *Int. J. Signal Proc.* **2013**, *1*, 256–262. [CrossRef]
15. Esser, P.; Dawes, H.; Collett, J.; Howells, K. IMU: Inertial sensing of vertical CoM movement. *J. Biomech.* **2009**, *42*, 1578–1581. [CrossRef]
16. Bai, S.; Lai, J.; Lyu, P.; Xu, X.; Liu, M.; Huang, K. A System-Level Self-Calibration Method for Installation Errors in A Dual-Axis Rotational Inertial Navigation System. *Sensors* **2019**, *19*, 4005. [CrossRef]
17. Narasimhappa, M.; Mahindrakar, A.D.; Guizilini, V.C.; Terra, M.H.; Sabat, S.L. MEMS Based IMU Drift Minimization: Sage Husa Adaptive Robust Kalman Filtering. *IEEE Sens. J.* **2019**, *20*, 250–260. [CrossRef]
18. Wittmann, F.; Lambercy, O.; Gassert, R. Magnetometer-based drift correction during rest in IMU arm motion tracking. *Sensors* **2019**, *19*, 1312. [CrossRef] [PubMed]

19. Bravo-Illanes, G.P.; Halvorson, R.; Matthew, R.; Lansdown, D.; Ma, C.; Bajcsy, R. IMU Sensor Fusion Algorithm for Monitoring Knee Kinematics in ACL Reconstructed Patients. In Proceedings of the 41st Annual International Conference of the IEEE Engineering in Medicine and Biology Society (EMBC), Berlin, Germany, 23–27 July 2019; pp. 5877–5881.
20. Li, M.P.; Zhuo, Y.; He, B.; Liang, Z.; Xu, G.; Xie, J.; Zhang, S. A 3D-printed soft hand exoskeleton with finger abduction assistance. In Proceedings of the 16th International Conference on Ubiquitous Robots (UR), Jeju, Korea, 24–27 June 2019; pp. 319–322.
21. Lu, S.; Chen, D.; Liu, C.; Jiang, Y.; Wang, M. A 3-D finger motion measurement system via soft strain sensors for hand rehabilitation. *Sens. Actuators A* **2019**, *285*, 700–711. [CrossRef]
22. Li, X.; Wen, R.; Shen, Z.; Wang, Z.; Luk, K.D.K.; Hu, Y. A wearable detector for simultaneous finger joint motion measurement. *IEEE Trans. Biomed. Circuits Syst.* **2018**, *12*, 644–654. [CrossRef]
23. Sato, J.; Sekine, T.; Yi-Fei, W.; Takeda, Y.; Matsui, H.; Kumaki, D.; Dos Santos, F.D.; Miyabo, A.; Tokito, S. Ferroelectric polymer-based fully printed flexible strain rate sensors and their application for human motion capture. *Sens. Actuators A* **2019**, *295*, 93–98. [CrossRef]
24. Borghetti, M.; Sardini, E.; Serpelloni, M. Sensorized glove for measuring hand finger flexion for rehabilitation purposes. *IEEE Trans. Instrum. Meas.* **2013**, *62*, 3308–3314. [CrossRef]
25. Wise, S.; Gardner, W.; Sabelman, E.; Valainis, E.; Wong, Y.; Glass, K.; Drace, J.; Rosen, J.M. Evaluation of a fiber optic glove for se-automated goniometric measurements. *J. Rehabil. Res. Dev.* **1990**, *27*, 411–424. [CrossRef]
26. Najafi, B.; Aminian, K.; Paraschiv-Ionescu, A.; Loew, F.; Bula, C.J.; Robert, P. Ambulatory system for human motion analysis using a kinematic sensor: Monitoring of daily physical activity in the elderly. *IEEE Trans. Biomed. Eng.* **2003**, *50*, 711–723. [CrossRef]
27. Balogun, J.A.; Amusa, L.O.; Onyewadume, I.U. Factors affecting Caltrac®and Calcount®accelerometer output. *Phys. Ther.* **1988**, *68*, 1500–1504.
28. Forner-Cordero, A.; Mateu-Arce, M.; Forner-Cordero, I.; Alcántara, E.; Moreno, J.; Pons, J.L. Study of the motion artefacts of skin-mounted inertial sensors under different attachment conditions. *Physiol. Meas.* **2008**, *29*, N21. [CrossRef]
29. Gajdosik, R.L.; Bohannon, R.W. Clinical measurement of range of motion review of goniometry emphasizing reliability and validity. *Phys. Ther.* **1987**, *67*, 1867–1872. [CrossRef] [PubMed]
30. Hill, K.; Fujii, Y.; Johnson, D.C.; Kawasaki, B. Photosensitivity in optical fiber waveguides: Application to reflection filter fabrication. *Appl. Phys. Lett.* **1978**, *32*, 647–649. [CrossRef]
31. Erdogan, T. Fiber grating spectra. *J. Lightwave Technol.* **1997**, *15*, 1277–1294. [CrossRef]
32. Kuang, K.; Kenny, R.; Whelan, M.; Cantwell, W.; Chalker, P. Embedded fibre bragg grating sensors in advanced composite materials. *Compos. Sci. Technol.* **2001**, *61*, 1379–1387. [CrossRef]
33. Trutzel, M.N.P.; Wauer, K.; Betz, D.; Staudigel, L.; Krumpholz, O.; Muehlmann, H.-C.; Muellert, T.; Gleine, W. Smart sensing of aviation structures with fiber optic bragg grating sensors. In Proceedings of the Smart Structures and Materials 2000: Sensory Phenomena and Measurement Instrumentation for Smart Structures and Materials, Newport Beach, CA, USA, 6–9 March 2000; pp. 134–143.
34. Hopf, B.; Koch, A.W.; Roths, J. Temperature dependence of glue-induced birefringence in surface-attached FBG strain sensors. *J. Lightwave Technol.* **2015**, *34*, 1220–1227. [CrossRef]
35. Ramalingam, R.K.; Nast, R.; Neumann, H. Fiber bragg grating sensors for distributed torsional strain measurements in a (RE) BCO tape. *IEEE Sens. J.* **2014**, *15*, 2023–2030. [CrossRef]
36. Jin, L.; Zhang, W.; Zhang, H.; Liu, B.; Zhao, J.; Tu, Q.; Kai, G.; Dong, X. An embedded FBG sensor for simultaneous measurement of stress and temperature. *IEEE Photonics Technol. Lett.* **2005**, *18*, 154–156. [CrossRef]
37. Betz, D.C.; Thursby, G.; Culshaw, B.; Staszewski, W.J. Advanced layout of a fiber bragg grating strain gauge rosette. *J. Lightwave Technol.* **2006**, *24*, 1019–1026. [CrossRef]
38. Basumallick, N.; Chatterjee, I.; Biswas, P.; Dasgupta, K.; Bandyopadhyay, S. Fiber bragg grating accelerometer with enhanced sensitivity. *Sens. Actuators A* **2012**, *173*, 108–115. [CrossRef]
39. He, Y.; Zhang, X.; Zhu, L.; Sun, G.; Lou, X.; Dong, M. Optical Fiber Sensor Performance Evaluation in Soft Polyimide Film with Different Thickness Ratios. *Sensors* **2019**, *19*, 790. [CrossRef]

40. Ge, J.; James, A.E.; Xu, L.; Chen, Y.; Kwok, K.W.; Fok, M.P. Bidirectional Soft Silicone Curvature Sensor Based on Off-Centered Embedded Fiber bragg Grating. *IEEE Photonics Technol. Lett.* **2016**, *28*, 2237–2240. [CrossRef]
41. Park, Y.P.; Lee, J.; Bae, J. Development of a Finger Motion Measurement System using Linear Potentiometers. In Proceedings of the 2014 IEEE/ASME International Conference on Advanced Intelligent Mechatronics, Besacon, France, 8–11 July 2014; pp. 125–130.

sensors

MDPI

Article

Comparative Study of Damage Detection Methods Based on Long-Gauge FBG for Highway Bridges

Shi-Zhi Chen [1,*], De-Cheng Feng [2] and Wan-Shui Han [1]

1 Highway College, Chang'an University, Xi'an 710064, China; hws@gl.chd.edu.cn
2 Key Laboratory of Concrete and Prestressed Concrete Structures of the Ministry of Education, Southeast University, Nanjing 210096, China; dcfeng@seu.edu.cn
* Correspondence: szchen@chd.edu.cn; Tel.: +86-029-8233-4453

Received: 4 June 2020; Accepted: 26 June 2020; Published: 28 June 2020

check for updates

Abstract: Damage detection of highway bridges is a significant part of structural heath monitoring. Conventional accelerometers or strain gauges utilized for damage detection have many shortcomings, especially their monitoring gauge length being too short, which would result in poor damage detection results. Under this circumstance, long-gauge FBG sensors as a novel optical sensor were developed to measure the macro-strain response of the structure. Based on this sensor, many derived damage detection methods were proposed. These methods exhibit various characteristics and have not been systematically compared. As a result, it is difficult to evaluate the state of the art and also leads to confusion for users to select. Therefore, a strict comparative study on three representative methods using long-gauge FBG was carried out. First, these methods' theoretical backgrounds and formats were reformulated and unified for better comparison. Then, based on validated vehicle–bridge coupling simulation, these methods' performances were tested through a series of parametric studies including various damage scenarios, vehicle types, speeds, road roughness and noise levels. The precision and reliability of three methods have been thoroughly studied and compared.

Keywords: long-gauge FBG; damage detection; highway bridges; vehicle–bridge interaction; comparative study

1. Introduction

Nowadays, in-service highway bridges always suffer from damage caused by external effects like normal traffic and material degradation. Without appropriate maintenance actions, this damage would inevitably result in severe disaster, causing human and fortune losses. In order to detect potential damage and avoid these losses, the conception of structural health monitoring (SHM) was proposed. By installing sensors on a highway bridge, the SHM technique could identify potential damage and assess its condition through monitored structural response [1].

Damage detection methods, as a core part of SHM, have drawn a lot of attention. Currently, there have been plenty of methods proposed utilizing accelerometers or strain gauges [2]. For the methods based on accelerometers, the core part is extracting modal information of a highway bridge like intrinsic frequencies and modal shapes from acceleration response through modal identification algorithms, such as frequency domain decomposition (FDD), stochastic subspace identification (SSI) [3–5]. Owing to the intrinsic relationship between modal information and structural physical parameters, these methods could achieve the goal of damage detection. Salawu [6] and Doebling et al. [7] have systematically summarized this kind of method.

After being verified through numerical simulations and indoor experiments, these methods have also been tested in actual bridges. However, they were found to be insensitive to structural local damage. Huth et al. [8] conducted a modal test on a severely damaged prestressed concrete bridge.

Its modal shapes were found to be nearly unchanged compared with those when the bridge is intact. Through analyzing modal shape and other derivative parameters, such as modal curvature and modal strain energy, Alvandi and Cremona [9] found out that modal information would be sensitive to local damage only if there is no interference of noise. When introducing noise in monitored response, the modal information would no longer reflect the structural condition clearly. Chang et al. [10] carried out a destruction test on an actual bridge and discovered that in order to succeed in detecting damage, the order of modal information should be high enough. However, with current test technique, it is difficult to accurately obtain high order modal information. Due to these reasons, the performance of accelerometer based methods is not satisfying in practical application.

In contrast with modal information, strain response is directly related to structural local conditions. Hence, inherently, the methods based on strain gauge would be more sensitive to structural local damage [11]. Based on strain response, Cardini and Dewolf [12] deduced the neutral axis height of bridges in order to detect bridge damage. Catbas et al. [13] also proposed a damage detection method utilizing the correlations between strain time histories and verified this method in a bridge test. Li's [14] review has thoroughly collected this kind of method. Although strain gauge-based methods are more sensitive to local damage, in practice these methods have a common weakness that conventional strain gauge's monitoring range is too short. As a result, it is hard to cover all potential damaged area on a bridge with finite number of strain gauges. In the meantime, more sensors would be demanded to increase the possibility of damage identification, but corresponding problems of installation inconvenience would emerge [15]. Conventional strain gauge also has other shortcomings, like being fragile and sensitive to electromagnetic interference (EMI) [15,16].

To solve these problems, distributed optical fiber (DOF) sensors such as Brillouin optical time domain reflectometer (BOTDR) were used for bridge damage detection [17]. Normally they can achieve long-distance distributed sensing with a precision of 5 $\mu\varepsilon$, which is very suitable for large-scale structures like bridges. However, their sampling frequencies are rather low, less than 1 Hz, which is not capable of bridge dynamic monitoring. Moreover, the spatial resolution of these techniques is larger than 500 mm, making it quite discrepant to ideal distributed sensing [18]. Therefore, the advantage of these sensors would be diminished. In comparison with these distributed optical fiber sensors, the fiber Bragg grating (FBG) sensor as another kind of fiber optic sensor can realize high sampling frequency, fulfilling the demand of bridge SHM. Currently, the main drawback of FBG is its short monitoring gauge, which would lead to same problems met in the conventional strain gauge [19]. Moreover, the fiber-based sensors, especially FBG, are usually fragile, which is not durable under actual environment. These drawbacks restrained their application.

In order to expand monitoring gauge and increase durability of FBG, Li and Wu [20] designed a special packaging structure and invented the long-gauge FBG sensor. It is the only sensor which can be used to monitor the average strain response within its monitoring gauge, which is defined as macro-strain. Long-gauge FBG sensor's monitoring gauge length could vary from 100 mm to 1000 mm according to relevant monitoring demands. Meanwhile, due to the spatial resolution existing in DOF sensors, the performance of long-gauge FBG is quite similar with normal DOF sensors, so it could be treated as a quasi-DOF sensor. Owing to these advantages, long-gauge FBG has been gradually applied in bridge SHM [21–23]. Based on long-gauge FBG, some exclusive damage detection methods have also been proposed, which can be found in many studies [24,25]. These methods were developed separately and exhibited various characteristics. Although they were alleged to be feasible and reliable, there is not a comparative study among them, especially regarding their precision and reliability. Consequently, it is difficult to evaluate the state of the art, resulting in some confusion for users to select.

Therefore, this paper conducts a strict comparison on three representative damage detection methods based on long-gauge FBG. The methods employed in this study were proposed by Wu et al. [26], Hong et al. [27] and Chen et al. [28]. For better comparison, these methods' formats were firstly reformulated and unified. Then, a numerical simulation based on vehicle–bridge interaction was programmed and validated by an indoor experiment in order to carry out a series of parametric studies. The parameters considered include various damage scenarios, vehicle types, speeds, road surface roughness degrees and noise levels. The performances of different methods under various scenarios are tested and compared.

The paper is organized as follows: Section 2 introduces the unified theoretical background for each method. Section 3 presents the vehicle–bridge interaction theory used for simulation. To rectify the reliability of the simulation, an indoor experimental validation was also illustrated in Section 3. The numerical simulation and results discussion are given in Sections 4 and 5 draws the final conclusions.

2. Theoretical Background

2.1. Long-Gauge FBG

For better demonstration of each damage detection method based on long-gauge FBG, the basic structure and characteristics of long-gauge FBG are presented first.

The basic structure of long-gauge FBG is displayed in Figure 1a. It can be seen that the FBG as core sensing unit of long-gauge FBG is protected and isolated by a shield tube made by basalt fibers from the external environment. Hence, the durability of fragile FBG can be effectively enhanced. Meanwhile, the FBG is attached to the monitored structure only at two anchor points. At two anchor points, the inner fiber was solidified to the basalt fiber tube by epoxy resin. Owing to this special design, the long-gauge FBG can measure the average strain within the area between two anchor points. This area is defined as its monitoring gauge. In actual practice, in order to monitor large scale structures like bridges, long-gauge FBG sensors can also be multiplexed, forming a sensor sequence, as shown in Figure 1b. Compared with conventional strain gauge, one main merit of long-gauge FBG is that, by multiplexing, they can cover the whole span of a bridge while conventional strain gauge can only monitor some certain section of a bridge. As can be seen in Figure 1b, when the damage area was not covered by a strain gauge, the damage could not be detected. However, the long-gauge FBG can always successfully monitor the potential damage, even though damage would randomly occur.

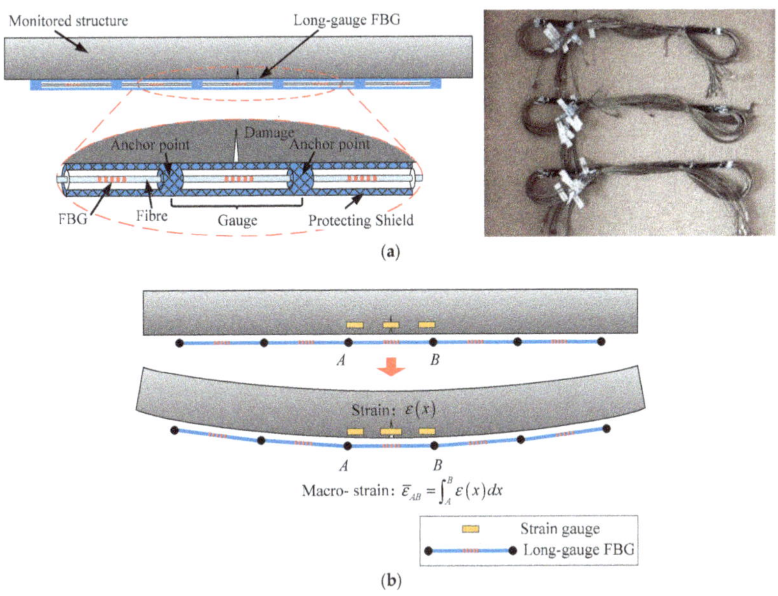

Figure 1. Schematic diagram of long-gauge fiber Bragg grating (FBG) sensor: (**a**) Basic structure and actual view of long-gauge FBG; (**b**) Comparison between long-gauge FBG and strain gauge.

As for its optical characteristics, they can be seen in Figure 2. The FBG within the gauge functions like a narrow-band reflective mirror. When a beam of light was inputted into long-gauge FBG, only the light with its wavelength equal to FBG's central wavelength λ_b would be reflected. The central wavelength λ_b is determined by fiber core's effective refractive index n_e and grating period Λ, as shown in Equation (1).

$$\lambda_b = 2n_e\Lambda \tag{1}$$

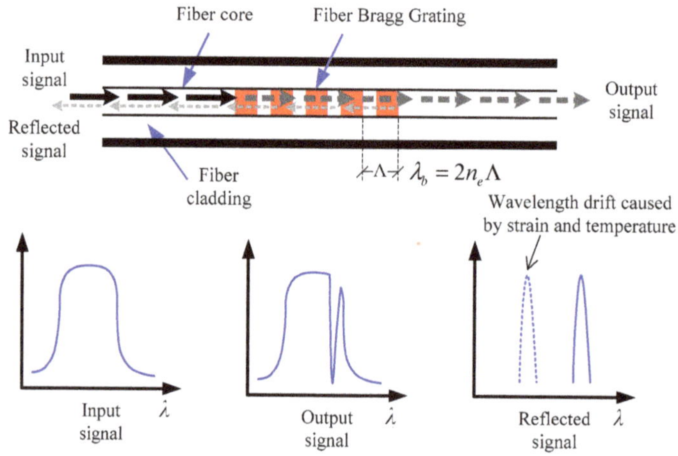

Figure 2. Optical characteristic of long-gauge FBG.

When the long-gauge FBG deforms with monitored structure, the grating period Λ would change accordingly, so as the central wavelength λ_b. The relationship between the change of central wavelength and deformation is:

$$\left[\frac{\Delta\lambda_b}{\lambda_b}\right] = C_S \cdot \Delta\varepsilon + C_T \cdot \Delta T \tag{2}$$

where $\Delta\lambda_b$ is the change of central wavelength, $\Delta\varepsilon$ is the strain monitored, ΔT is the change of temperature, C_S and C_T are strain factor and temperature factor, respectively, which can be calibrated by experiment.

According to the structure of long-gauge FBG, the relationship between measured structural average strain $\overline{\varepsilon}_{AB}$ and strain response at each point can be deduced, as shown in Figure 1b.

$$\overline{\varepsilon}_{AB} = \frac{1}{l_{AB}}\int_A^B \varepsilon(x)\,dx \tag{3}$$

in which $\varepsilon(x)$ represents the strain of point x, l_{AB} represents the distance between A and B. The average strain $\overline{\varepsilon}_{AB}$ is also defined as macro-strain. Currently, long-gauge FBG is the only way to measure the macro-strain response of the structure.

In Figure 1b, the main advantage of long-gauge FBG has also been explained. In order to capture the random damage feature of the structure, large amounts of conventional strain gauges are demanded, while long-gauge FBG can achieve covering large structural area with a finite number of sensors.

2.2. Macro-Strain Influence Line

After long-gauge FBG was developed, many derived damage detection methods have been carried out. Because moving vehicles are the main source of load on a bridge, all typical methods discussed in this study utilized the macro-strain influence line as their theoretical foundations. According to structural mechanics, the strain influence line response $\varepsilon_i(x)$ generated by a moving unit concentration force F at a certain section i of a simply supported beam (Figure 3) would be:

$$\varepsilon_i(x) = \begin{cases} \frac{(L-x_i)h}{(EI)_i L} \cdot x & 0 < x \le x_i \\ \frac{(L-x)h}{(EI)_i L} \cdot x_i & x_i < x \le L \end{cases} \tag{4}$$

in which L is the span length of a beam, x_i is the coordinate of section i, h is the height of neutral axis and $(EI)_i$ is the bending stiffness of section I.

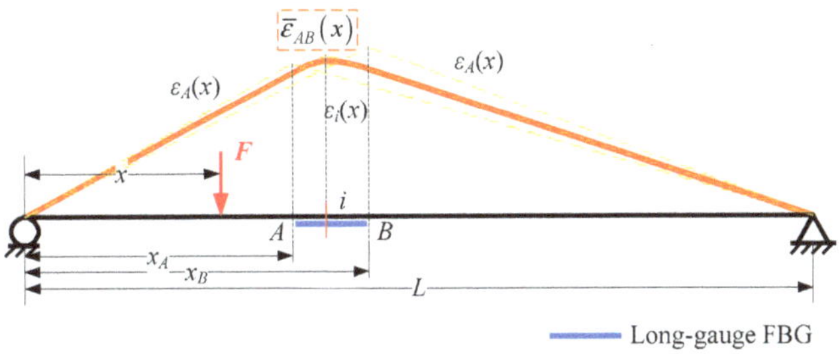

Figure 3. Strain and macro-strain influence line response.

Based on Equation (3), strain influence line response can be modified to macro strain influence line response $\overline{\varepsilon}_{AB}(x)$:

$$\overline{\varepsilon}_{AB}(x) = \begin{cases} \frac{(x_B-x_A)(2L-x_B-x_A)h}{2\left(\overline{EI}\right)_{AB} L l_g} \cdot x & 0 < x \le x_A \\ \frac{h}{2\left(\overline{EI}\right)_{AB} L l_g} \cdot \left[-Lx^2 + \left(2Lx_B + x_A^2 - x_B^2\right)x - Lx_A^2\right] & x_A < x \le x_B \\ \frac{\left(x_B^2-x_A^2\right)h}{2\left(\overline{EI}\right)_{AB} L l_g} \cdot (L-x) & x_B < x \le L \end{cases} \tag{5}$$

where x_A and x_B are two end coordinates of a long-gauge FBG, l_g is the gauge length of long-gauge FBG sensor and $\left(\overline{EI}\right)_{AB}$ represents the average bending stiffness within monitoring gauge AB.

For a vehicle passing through a bridge, its response would be the superposition of its axial weights multiplying macro-strain influence line response, as illustrated in Figure 4, because the bridge would be still in the elastic stage when a vehicle passes through. The macro-strain response caused by a vehicle could be expressed as:

$$\begin{aligned} \overline{\varepsilon}_{vehicle}(x) &= F_1 \cdot \overline{\varepsilon}_{AB}(x) + F_2 \cdot \overline{\varepsilon}_{AB}(x-d_1) + \cdots + F_n \cdot \overline{\varepsilon}_{AB}\left(x - \sum_{k=1}^{n-1} d_k\right) \\ &= \sum_{i=1}^{n} F_i \cdot \overline{\varepsilon}_{AB}\left(x - \sum_{k=1}^{i-1} d_k\right) \end{aligned} \tag{6}$$

where F_i is the ith axial weight, n is the total axle number of a vehicle, d_k is the kth wheelbase of a vehicle.

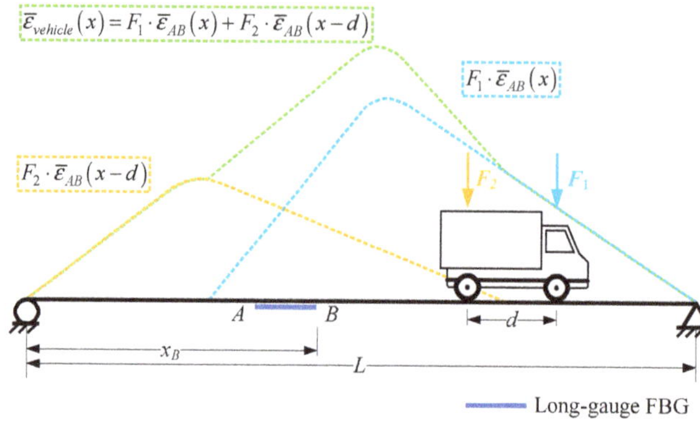

Figure 4. Macro-strain response caused by a vehicle.

Because the macro-strain response caused by a vehicle is just the superposition of a single influence line, they would exhibit the same characteristic. Hence, the damage detection method can be established by studying the basic features of single macro-strain influence line response. Then, with a long-gauge FBG sequence attached on a beam, a series of macro-strain responses would be obtained, which can be used for developing damage detection methods, as shown in Figure 5. Then, the damage detection methods discussed in this study were proposed.

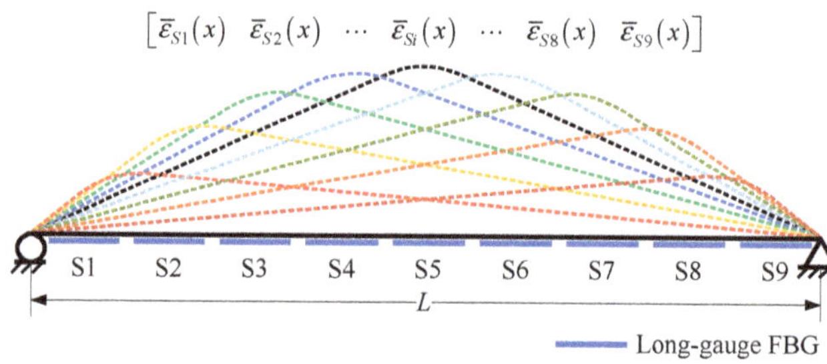

Figure 5. Macro-strain influence line responses obtained by long-gauge FBG sequence.

2.3. Damage Detection Method (M1)

The damage index proposed in Wu et al. [26] was called macro-strain influence line response envelope (MIE). It consists of the maximum of each macro-strain response obtained by a long-gauge FBG sensor: $\left[\ \max(\overline{\varepsilon}_{S1}) \ \cdots \ \max(\overline{\varepsilon}_{Si}) \ \cdots \ \max(\overline{\varepsilon}_{S9})\ \right]$. Based on Equation (5), as the geometric parameters such as L, l_g, x_A and x_B are fixed, the maximum of macro-strain response is only inversely proportional to the average bending stiffness $\left(\overline{EI}\right)_{AB}$ within each monitoring gauge, because when the sensor was installed these parameters would be invariant. Therefore, the maximum of macro-strain response can be expressed as:

$$\max(\overline{\varepsilon}_{Si}) = \frac{F\left(x_A, x_B, L, l_g, h\right)}{\left(\overline{EI}\right)_{Si}} \tag{7}$$

in which $F(x_A, x_B, L, l_g, h)$ is an implicit function of geometric parameters.

If there is damage causing β degree of stiffness degradation, the corresponding MIE would emerge as a peak at the relevant sensor, as displayed in Figure 6, because after damage the maximum of macro-strain response would be:

$$\max\left(\overline{\varepsilon}_{Si}^{D}\right) = \frac{F\left(x_A, x_B, L, l_g, h\right)}{\left(\overline{EI}\right)_{Si}^{D}} = \frac{F\left(x_A, x_B, L, l_g, h\right)}{(1-\beta)\left(\overline{EI}\right)_{Si}} = \frac{1}{(1-\beta)} \cdot \max(\overline{\varepsilon}_{Si}) \tag{8}$$

where superscript D represents the parameter under the damaged scenario.

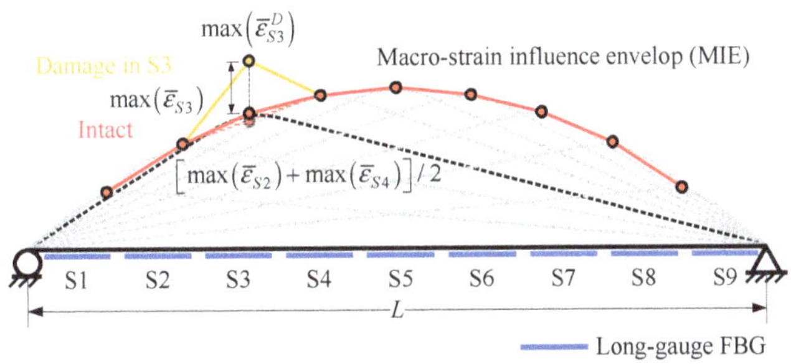

Figure 6. MIE obtained from long-gauge FBG sequence.

Due to this mechanism, the damage location can be detected, as shown in Figure 6. The damage degree can be calculated through:

$$\beta = 1 - \frac{\max(\overline{\varepsilon}_{Si})}{\max\left(\overline{\varepsilon}_{Si}^{D}\right)} \tag{9}$$

This is the core part of the damage detection method proposed by Wu et al. [26]. This method uses the MIE of the macro-strain response caused by a vehicle to locate and quantify the damage. However, in order to accurately quantify the damage extent, the MIE under intact conditions is needed for reference, which is often hard to get for an actual bridge. Therefore, Wu et al. [26] selected the average value of two nearby intact areas' MIE to approximate the intact MIE of the identified damage area:

$$\max\left(\overline{\varepsilon}_{Si}^{I}\right) \approx \left[\max(\overline{\varepsilon}_{Si-1}) + \max(\overline{\varepsilon}_{Si+1})\right]/2 \tag{10}$$

2.4. Damage Detection Method (M2)

Instead of MIE, the damage detection method proposed by Hong et al. [27] selected integrals of the macro-strain influence line response (IMIL) as the damage index. Through derivation based on Equation (5), IMIL would be:

$$\begin{aligned}
\int_0^L \overline{\varepsilon}_{AB}(x)\,dx &= \int_0^{x_A} \frac{(x_B - x_A)(2L - x_B - x_A)h}{2\left(\overline{EI}\right)_{AB} L l_g} \cdot x\,dx + \int_{x_A}^{x_B} \frac{h}{2\left(\overline{EI}\right)_{AB} L l_g} \cdot \left[-Lx^2 + \left(2Lx_B + x_A^2 - x_B^2\right)x - Lx_A^2\right] dx \\
&\quad + \int_{x_B}^{L} \frac{\left(x_B^2 - x_A^2\right)h}{2\left(\overline{EI}\right)_{AB} L l_g} \cdot (L - x)\,dx \\
&= -\frac{h}{12\left(\overline{EI}\right)_{AB} L l_g} \cdot (x_A - x_B) \cdot \left\{ \begin{array}{l} 3{x_a}^3 + (3x_b - 8L - 6) \cdot {x_a}^2 + \left[3L^2 + (12 - 2x_b) \cdot L - 6x_b\right] \cdot x_a \\ + L \cdot x_b \cdot (3L - 2x_b) \end{array} \right\} \\
&= \frac{G\left(x_A, x_B, L, l_g, h\right)}{\left(\overline{EI}\right)_{AB}}
\end{aligned} \tag{11}$$

Similarly, IMIL is also inversely proportional to the average bending stiffness $\left(\overline{EI}\right)_{AB}$. After the bending stiffness within ith sensor decreases, the corresponding IMIL would be:

$$IMIL_{Si}^{D} = \frac{G\left(x_A, x_B, L, l_g, h\right)}{\left(\overline{EI}\right)_{Si}^{D}} = \frac{G\left(x_A, x_B, L, l_g, h\right)}{(1-\beta)\left(\overline{EI}\right)_{Si}} = \frac{1}{(1-\beta)} \cdot IMIL_{Si} \tag{12}$$

Equation (12) means if there is a damage causing β degree of stiffness degradation, the corresponding IMIL would fluctuate, as shown in Figure 7. This is the basic theory of the damage detection methods proposed by Hong et al. [27]. The damage extent can be calculated through:

$$\beta = 1 - \frac{IMIL_{Si}}{IMIL_{Si}^{D}} \tag{13}$$

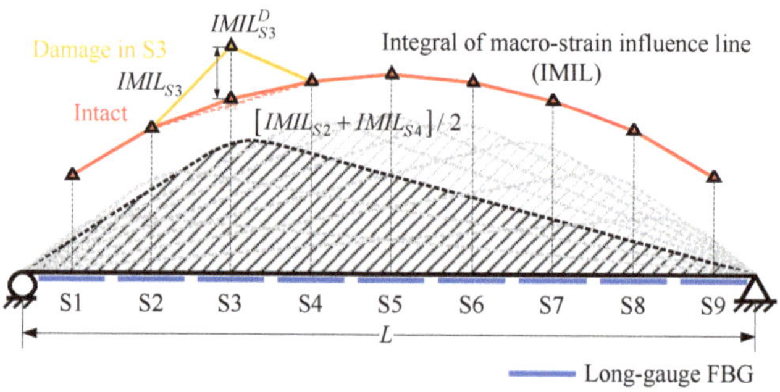

Figure 7. Influence line response (IMIL) obtained from long-gauge FBG sequence.

Likewise, Hong et al. [27] selected the average of two nearby intact areas' IMIL to substitute the intact IMIL reference needed for damage extent calculation.

$$\max\left(\overline{\varepsilon}_{Si}^{I}\right) \approx [\max(\overline{\varepsilon}_{Si-1}) + \max(\overline{\varepsilon}_{Si+1})]/2 \tag{14}$$

2.5. Damage Detection Method (M3)

Through analyzing the feature of macro-strain influence line response (Equation (5)), Chen et al. [28] realized that Equation (5) is a second order function of x only when x is between x_A and x_B. As a result, the second order difference of macro-strain influence line response (SODM) is derived as damage index:

$$\frac{d^2\overline{\varepsilon}_{AB}(x)}{dx^2} = \begin{cases} 0 & 0 < x \leq x_A \\ -\frac{h}{\left(\overline{EI}\right)_{AB} l_g} & x_A < x \leq x_B \\ 0 & x_B < x \leq L \end{cases} \tag{15}$$

Compared with the original macro-strain influence line response and two derivatives used in former methods, the expression of SODM is much more concise. The relationship between SODM and average bending stiffness is quite clear. The SODM obtained from long-gauge FBG sequence is illustrated in Figure 8. The peak of SODM would reflect the location of damage. The extent of damage would be:

$$\beta = 1 - \frac{SODM_{Si}}{SODM_{Si}^{D}} \tag{16}$$

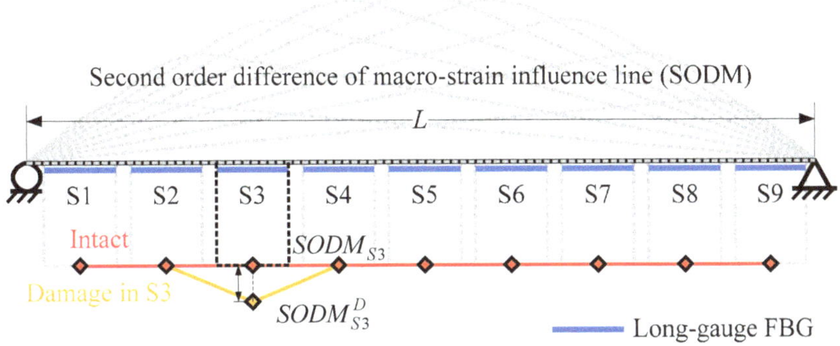

Figure 8. SODM obtained from long-gauge FBG sequence.

Because SODM forms a straight line, unlike a curve formed by MIE and IMIL, the method proposed by Chen et al. [28] can directly calculate the damage extent through Equation (16) without any approximate assumption.

According to the introductions mentioned above, it can be discovered that, theoretically, all these three methods would detect damage location and extent based on monitored macro-strain response through long-gauge FBG. All their theoretical foundations come from macro-strain influence line response. One evident difference among them is that methods proposed by Wu et al. [26] and Hong et al. [27] need to adopt an approximation for damage extent estimation because intact condition reference of the structure is hard to get, while the method proposed by Chen et al. [28] has no need. Except this, the difference among their characteristics cannot be revealed merely based on their theoretical backgrounds. In order to better compare these methods, a refined vehicle–bridge coupling simulation is needed to generate macro-strain response caused by moving vehicles under various parametric scenarios. The basic theory of vehicle–bridge coupling simulation would be introduced first in the next section.

3. Vehicle–Bridge Coupling Simulation

3.1. Vehicle–Bridge Coupling Simulation Theory

Compared with treating a vehicle as a group of moving loads, in vehicle–bridge coupling simulation, a vehicle is represented by a multi-degree of freedom model (Figure 9) containing the displacement of the vehicle body and suspension structures. The stiffness and damping of suspension structure and tire can also be effectively reflected. Besides these, the vehicle–bridge coupling simulation could reasonably consider the interaction effect between a vehicle and a bridge. Therefore, the bridge response simulated considering vehicle–bridge coupling would be more suitable for comparing the performance of these methods.

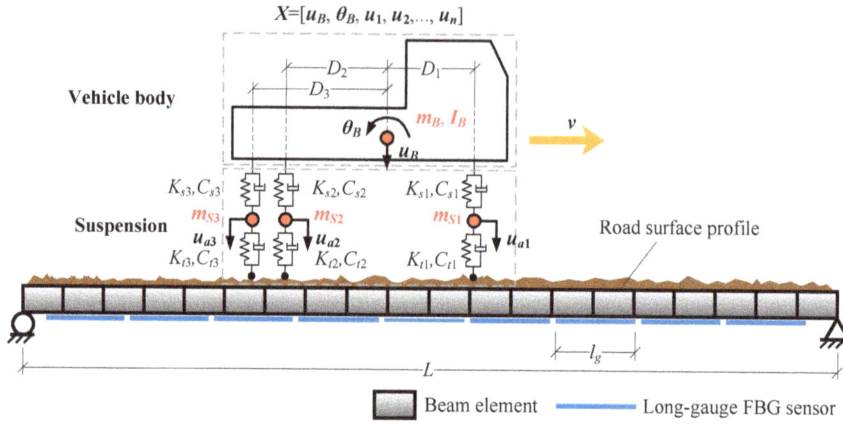

Figure 9. Vehicle–bridge coupling simulation.

Figure 9 demonstrates a schematic diagram of a vehicle–bridge coupling simulation. A two dimensional three-axle vehicle model moves at a velocity v over a bridge covered by long-gauge FBG sensors. According to D'Alembert's principle, the equation of motion of a vehicle would be:

$$\begin{bmatrix} \mathbf{M}_{V1} & 0 \\ 0 & \mathbf{M}_{V2} \end{bmatrix} \cdot \ddot{\mathbf{X}} + \begin{bmatrix} \mathbf{C}_{V11} & \mathbf{C}_{V12} \\ \mathbf{C}_{V21} & \mathbf{C}_{V22} \end{bmatrix} \cdot \dot{\mathbf{X}} + \begin{bmatrix} \mathbf{K}_{V11} & \mathbf{K}_{V12} \\ \mathbf{K}_{V21} & \mathbf{K}_{V22} \end{bmatrix} \cdot \mathbf{X} = -\begin{bmatrix} 0 \\ \mathbf{F}(t) \end{bmatrix} + \begin{bmatrix} 0 \\ \mathbf{F}_G \end{bmatrix} \quad (17)$$

in which $\mathbf{X} = [u_B, \theta_B, u_{a1}, u_{a2}, \cdots, u_{an}]^T$ is the response vector of a vehicle. $\mathbf{F}(t)$ is the interaction force between vehicle axle and bridge. $\mathbf{F}_G$ is vehicle's static axial weight. $\mathbf{M}_V$, $\mathbf{C}_V$, $\mathbf{K}_V$ are sub-matrices of vehicular mass, damping and stiffness matrices, respectively.

For a bridge, it can be simplified as an Euler-Bernoulli beam to simulate its mechanical behavior. A bridge's equation of motion would be:

$$\mathbf{M}_B \cdot \ddot{\mathbf{D}} + \mathbf{C}_B \cdot \dot{\mathbf{D}} + \mathbf{K}_B \cdot \mathbf{D} = \mathbf{I} \cdot \mathbf{F} \quad (18)$$

where $\mathbf{M}_B$, $\mathbf{C}_B$ and $\mathbf{K}_B$ are mass, damping and stiffness matrices of the bridge, respectively. $\mathbf{D}$ is the displacement vector, and $\mathbf{I}$ is an interpolating matrix, which would equivalently allocate the axial force of a vehicle to relevant nodes of a bridge.

According to the equilibrium of the interaction force between a bridge and a vehicle, a vehicle's equation of motion (Equation (17)) and a bridge's equation of motion (Equation (18)) can be coupled together forming a vehicle–bridge coupling system's equation of motion:

$$\mathbf{M}^C(t) \cdot \ddot{\mathbf{U}} + \mathbf{C}^C(t) \cdot \dot{\mathbf{U}} + \mathbf{K}^C(t) \cdot \mathbf{U} = \mathbf{F}(t) \quad (19)$$

where $\mathbf{U} = \{\mathbf{X}^T\ \mathbf{D}^T\}^T$ is the coupled displacement vector. $\mathbf{M}^C(t)$, $\mathbf{C}^C(t)$ and $\mathbf{K}^C(t)$ are mass, damping and stiffness matrices of vehicle–bridge coupling system.

Utilizing Newmark-β method, Equation (19) can be solved to obtain the dynamic response $D(t)$ of a bridge under a moving vehicle. Then, based on the plane section assumption, the dynamic strain response $\varepsilon(x,t)$ at point x under the bridge can be derived from $D(t)$:

$$\varepsilon(x,t) = h \cdot \frac{\partial^2 D(x,t)}{\partial x^2} \quad (20)$$

where $\varepsilon(x,t)$ and $D(x,t)$ are the strain and displacement response of a bridge at point x, respectively. h is the neutral axis height.

In order to simulate the macro-response caused by a long-gauge FBG sensor, simulated strain response $\varepsilon(x,t)$ needs to be expanded into macro-strain response $\overline{\varepsilon}_{AB}$. According to the definition of macro-strain in Equation (3), the macro-strain response caused by a vehicle can be deduced:

$$\begin{aligned} \overline{\varepsilon}_{AB} &= \frac{1}{l_{AB}}\int_A^B \varepsilon(x,t)\,dx \\ &= \frac{1}{l_{AB}}\int_A^B h\cdot\frac{\partial^2 D(x,t)}{\partial x^2}\,dx \\ &= \frac{\overline{h}}{l_{AB}}\mathbf{I}(x)\cdot\mathbf{D}(t) \end{aligned} \tag{21}$$

in which $\overline{h}$ is the average neutral axis height and $I(x) = \begin{bmatrix} 0 & \cdots & -1 & 0 & \cdots & 1 & 0 & \cdots & 0 \end{bmatrix}_{1\times n}$ is an interpolation vector.

Moreover, road surface roughness is an important factor, which would dramatically affect the bridge response caused by a moving vehicle (Figure 9). In order to consider this factor, a displacement power spectral density (PSD) defined by relevant specification is utilized to generate road surface roughness by inverse fast Fourier transformation [29]. By selecting different spectral roughness coefficient values, various classes of road surface roughness can be simulated. In this study, 5 classes A, B, C, D, E are selected representing very good, good, average, poor and very poor road surface roughness conditions, as shown in Table 1.

Table 1. Different classes of road surface roughness.

Class	Spectral Roughness Coefficient	Road Surface Roughness Sample
A	0 m^3/cycles	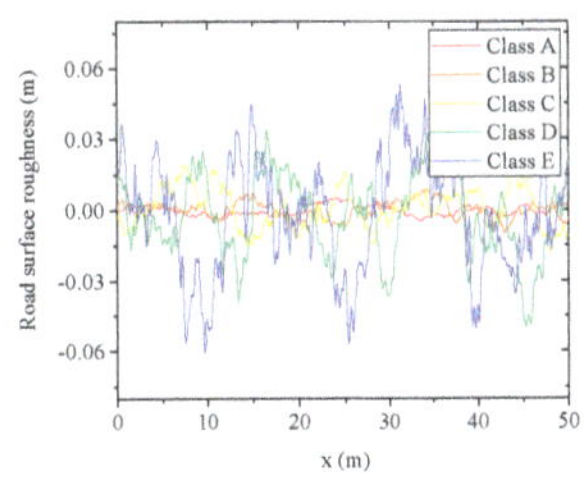
B	1×10^{-6} m^3/cycles	
C	6×10^{-6} m^3/cycles	
D	16×10^{-6} m^3/cycles	
E	64×10^{-6} m^3/cycles	

Then, according to basic theory of vehicle–bridge coupling simulation, the macro-strain response of a bridge measured by long-gauge FBG under the effect of a moving vehicle can be programmed and simulated in MATLAB. More specific explanations of the vectors and matrixes mentioned in this section can be found in Chen et al. [23]. In order to validate the correctness of the simulated macro-strain response, an indoor vehicle–bridge coupling experiment was carried out, which will be introduced in the next part.

3.2. Experimetental Validation

In order to validate the correctness of numerical results generated by vehicle–bridge coupling simulation, an indoor experiment was conducted. The whole experimental platform is displayed in Figure 10. The platform has three sections: the acceleration section, bridge model and braking section. The bridge model is fabricated by polymethyl-methacrylate and its span length is 3 m. It is simply supported by two supports. A total of 9 long-gauge FBG sensors with 300 mm gauge length were installed underneath the bridge model to capture the macro-strain response of the bridge model.

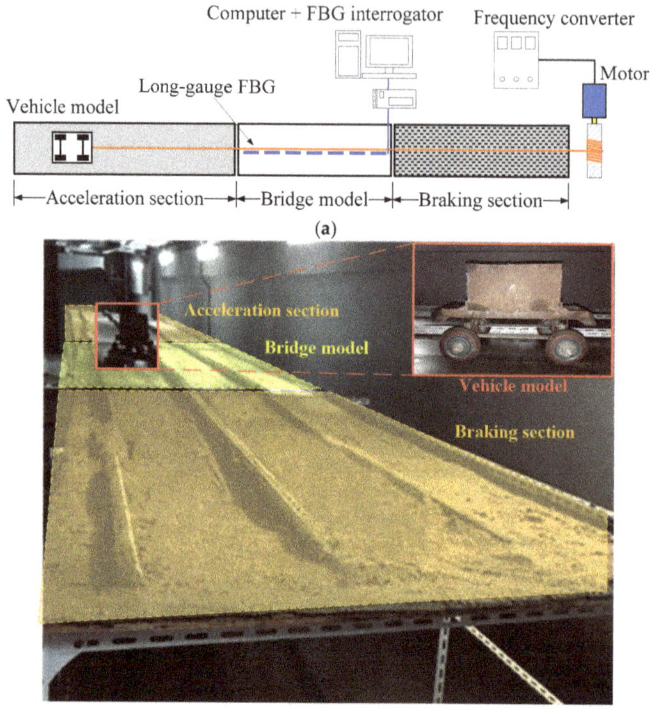

(a)

(b)

Figure 10. Vehicle–bridge coupling experimental platform: (**a**) Schematic diagram; (**b**) Actual view.

A vehicle model is dragged by a motor and moves through a bridge model. The macro-strain response generated by the moving vehicle model is collected through a FBG interrogator fabricated by Micro Optics Inc., whose type is SM530 (Micro-optics, Inc., Hackettstown NJ, USA). In experiments, the velocity of the vehicle model can be precisely controlled by changing the rotation frequency of the motor using a frequency converter.

In the experiment, macro-strain responses under different vehicle moving scenarios were collected by long-gauge FBG. They were utilized to validate the vehicle–bridge coupling simulation programmed. The comparison between typical experimental and simulation results is presented in Figure 11. In the legend of Figure 11, "E-S*i*" and "S-S*i*" represent the corresponding experimental and simulated macro-strain responses of sensor S*i*, respectively. It can be found that the macro-strain response simulated agrees well with the response captured in the experiment. Under various vehicle velocities, the difference between experimental and simulated results varied little, which implies that the vehicle–bridge coupling simulation is suitable for conducting comparative studies of damage detection methods. Therefore, in next section, a series of numerical simulations were introduced to compare the performance of three damage detection methods introduced in Section 2 based on the model parameters validated through this indoor experiment.

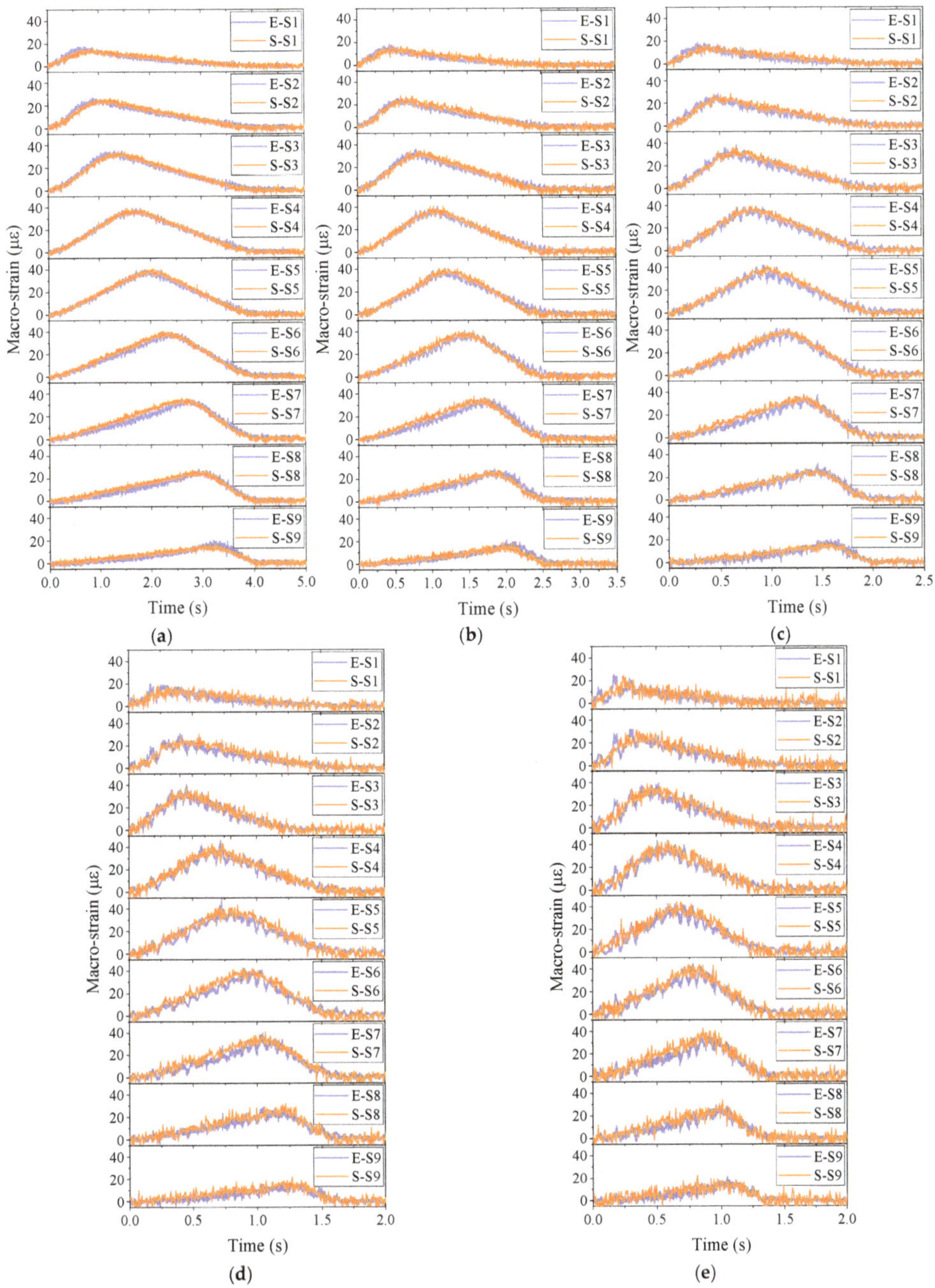

Figure 11. Comparison between typical experimental and simulated macro-strain results: vehicle velocity: (**a**) 10 km/h; (**b**) 15 km/h; (**c**) 20 km/h; (**d**) 25 km/h; (**e**) 30 km/h.

4. Numerical Simulation

4.1. Simulation Scenario

For better comparing the performances of three damage detection methods mentioned in Section 2, a series numerical simulations were carried out with the model parameters validated by an indoor experiment. The corresponding model parameters validated from experiment are listed in Table 2. These parameters also referred to relevant studies [30–32]. Three damage scenarios with various impact factors are designed in this simulation, as shown in Figure 12, containing single damage and multiple damage conditions. In addition, three typical damage extents: 5%, 10% and 15%, are considered. Moreover, the potential impact factors for damage detection are taken into account, such as vehicle type, speed, road surface roughness and signal-to-noise ratio.

Table 2. Parameters of vehicle–bridge coupling simulation.

Bridge Model		Vehicle Model			
L	30 m	m_B	260 kg	I_B	500 kg·m^2
E	3.023×10^4 MPa	D_1	1.8 m	D_2	1.8 m
I	5.697×10^{11} mm^4	m_{S1}	10 kg	m_{S2}	10 kg
A	1.138×10^6 mm^2	k_{S1}	4.15×10^3 N/m	k_{S2}	4.15×10^3 N/m
h	2000 mm	k_{T1}	3.89×10^3 N/m	K_{T2}	3.89×10^3 N/m
ρ	1.171×10^4 g/mm^3	C_{S1}	39.5 N/m/s	C_{S2}	39.5 N/m/s
ξ	0.02	C_{D1}	42.0 N/m/s	C_{D2}	42.0 N/m/s

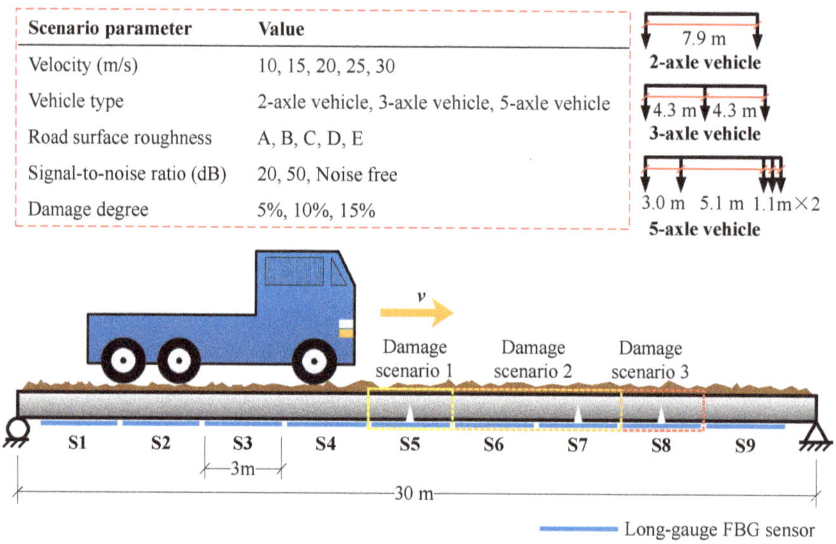

Figure 12. Damage scenarios designed in numerical simulation.

The typical macro-strain responses obtained are shown in Figure 13. It can be seen that macro-strain response varies dramatically under various parametric scenarios, which would be very suitable for testing the method's performance. Their performance would be impacted more or less. Then, inputting these data into three damage detection methods, corresponding results would be obtained for comparison. The discussion of results is given in the next section.

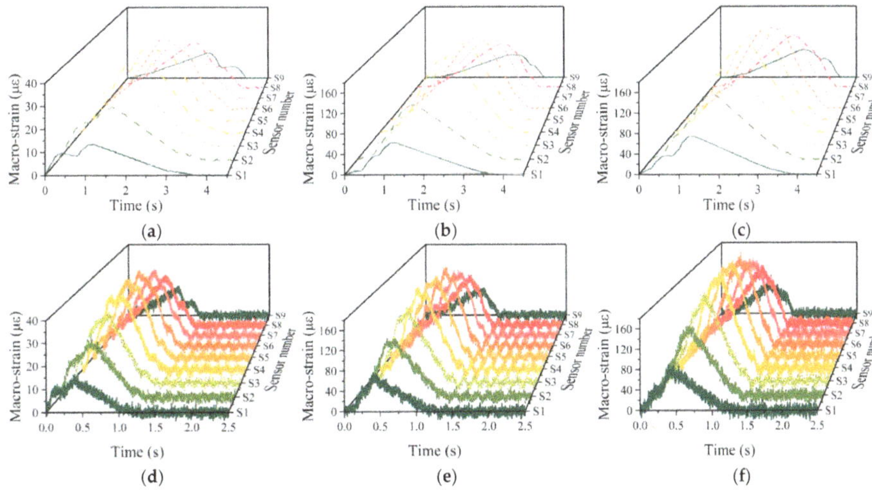

Figure 13. Typical macro-strain response obtained through vehicle-coupling simulation: Noise-free signal of (**a**) 2-axle vehicle; (**b**) 3-axle vehicle; (**c**) 5-axle vehicle with 10 m/s on A road surface; 20 signal-to-noise ratio signal of (**d**) 2-axle vehicle; (**e**) 3-axle vehicle; (**f**) 5-axle vehicle with 30 m/s on E road surface.

4.2. Results Discussion

4.2.1. Comparison under Different Vehicle Types

Firstly, in the numerical simulation, three different typical vehicle types are considered to investigate its influence. For brevity, the methods proposed by Wu et al. [26], Hong et al. [27] and Chen et al. [28] are represented by M1, M2 and M3. The corresponding damage indexes of three methods MIE, IMIL and SODM under three vehicle types are displayed in Figure 14. They were calculated based on Equations (8), and (12) and (15). The letters 'D' and 'V', with numbers behind in legends, represent the damage scenario and vehicle type. The other parameters like speed, road surface roughness, noise level in Figure 14 are set as 10 m/s and A class without noise.

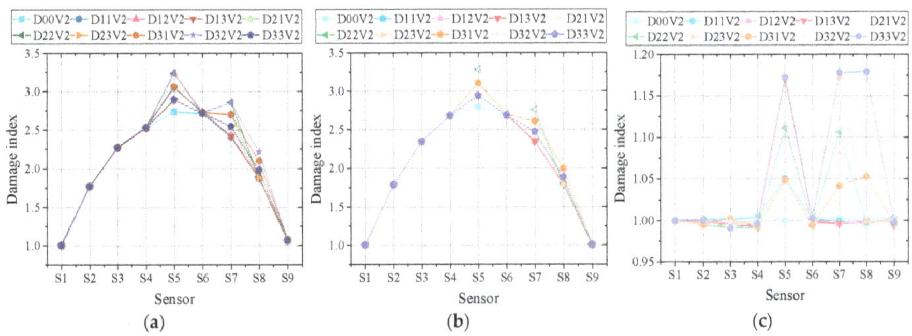

Figure 14. *Cont.*

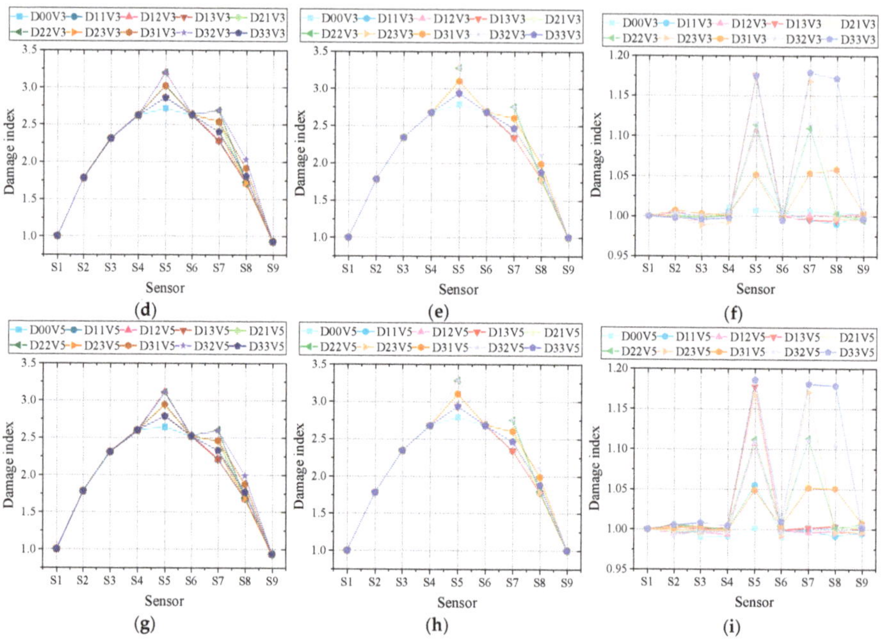

Figure 14. Damage detection results: (**a**) M1 with 2-axle vehicle; (**b**) M2 with 2-axle vehicle; (**c**) M3 with 2-axle vehicle; (**d**) M1 with 3-axle vehicle; (**e**) M2 with 3-axle vehicle; (**f**) M3 with 3-axle vehicle; (**g**) M1 with 5-axle vehicle; (**h**) M2 with 5-axle vehicle; (**i**) M3 with 5-axle vehicle.

From Figure 14, it can be seen that the overall performance of three methods under different vehicle types is rather good. The damage location in each damage scenario can be reflected by the corresponding damage index. However, it can be discovered that the damage index of M1 showed some fluctuations under three vehicle types. Moreover, the shape of M1's damage index is not ideally symmetric as introduced in its basic theory (Figure 6), while M2's damage index remains more stable and symmetric. In addition, compared with two curved lines of damage indexes obtained in M1 and M2, M3's damage index is more convenient for calculating damage extent. The damage index of M3 still has some tiny fluctuations, which would influence the accuracy of damage extent calculation. Then, using Equations (9), (13) and (16), the damage extent measured through the three methods can be calculated, which are listed in Table 3. The value within brackets is the relative error.

Table 3. Damage extent detection results of three methods with different vehicles.

Designed Damage Extent (%)		Detected Damage Extent (%)								
		M1			M2			M3		
		V2	V3	V5	V2	V3	V5	V2	V3	V5
5	S5	9.11 (82.20%)	8.16 (63.20%)	8.10 (62.00%)	8.73 (74.60%)	8.73 (74.60%)	8.73 (74.60%)	4.65 (7.00%)	5.20 (4.00%)	4.76 (4.80%)
	S7	9.58 (91.60%)	9.45 (89.00%)	9.68 (93.60%)	9.37 (87.40%)	9.37 (87.40%)	9.37 (87.40%)	4.39 (12.20%)	5.28 (5.60%)	5.27 (5.40%)
	S8	36.23 (624.60%)	34.56 (591.20%)	34.99 (599.80%)	34.69 (593.80%)	34.69 (593.80%)	34.69 (593.80%)	5.02 (0.40%)	5.49 (9.80%)	4.82 (3.60%)
10	S5	13.91 (39.10%)	12.93 (29.30%)	12.89 (28.90%)	13.45 (34.50%)	13.45 (34.50%)	13.45 (34.50%)	10.24 (2.40%)	9.66 (3.40%)	9.68 (3.20%)
	S7	14.44 (44.40%)	14.17 (41.70%)	14.30 (43.00%)	13.96 (39.60%)	13.96 (39.60%)	13.96 (39.60%)	9.58 (4.20%)	9.84 (1.60%)	10.15 (1.50%)
	S8	39.56 (295.60%)	37.99 (279.90%)	38.50 (285.00%)	38.15 (281.50%)	38.15 (281.50%)	38.15 (281.50%)	9.96 (0.40%)	10.12 (1.20%)	9.52 (4.80%)
15	S5	18.68 (24.53%)	17.77 (18.47%)	17.68 (17.87%)	18.16 (21.07%)	18.16 (21.07%)	18.16 (21.07%)	14.56 (2.93%)	14.96 (0.27%)	15.10 (0.67%)
	S7	19.27 (28.47%)	18.86 (25.73%)	18.88 (25.87%)	18.54 (23.60%)	18.54 (23.60%)	18.54 (23.60%)	14.65 (2.33%)	14.37 (4.20%)	14.60 (2.67%)
	S8	42.90 (186.00%)	41.49 (176.60%)	42.07 (180.47%)	41.65 (177.67%)	41.65 (177.67%)	41.65 (177.67%)	15.20 (1.33%)	14.65 (2.33%)	15.19 (1.27%)

According to Table 3, it can be found that the performance of M1 and M2 is very poor with the lowest relative error larger than 15%. The main reason is when the intact condition of the bridge is unknown, an approximation is adopted that uses the average of nearby intact gauges' indexes approximate intact condition, as given in Equations (10) and (14). This assumption itself would introduce huge error, because the damage index curve of M1 and M2 is a quadratic curve rather than a straight line. In contrast, M3's performance is rather good owing to its not needing an intact condition reference. As for their performance under different vehicle types, M2's results are the most stable, which seems to be immune to the change of vehicle type. Meanwhile, M1 and M3's results change under different vehicle types. Between these two methods, M1's performance is steadier, though its accuracy is much poorer than that of M3. In general, the performance with 2-axle vehicles is a little bit weaker than those with the other two vehicle types. This might be the result of the response signal being more easily polluted if its amplitude is smaller. In this study, the weight of two-axle vehicle is the lightest, leading to the smallest response amplitude.

In addition, their performance under different damage scenarios is also given in Table 3. Under nine designed damage scenarios, all three methods exhibit a similar trend in that, with the damage extent rising, the detection results become more accurate. Meanwhile, the relative error in sensor S5 is the lowest, exceeding 100% under different damage scenarios. This might be the result of the response amplitude in S5 being higher than those in S7 and S8. Besides this trend, it can be seen that M1 and M2's relative error in S8 is far beyond those in S5 and S7. The reason is that in damage scenarios concerned with sensor S8, the nearby sensor S7 also suffered from damage. Under this situation, when calculating the damage extent of S8, the damage index of S6 is selected as S6 is the nearest intact gauge. Consequently, more approximation error is introduced into the results of S8.

Moreover, once the intact condition is known for reference, the damage extent of M1 and M2 would become much more ideal as listed in Table 4. Under this circumstance, the accuracy of M1 and M2 is better than that of M3. M2 exhibits the most precise and stable performance. Nevertheless, it still needs to be clarified that the excellent performance of these two methods is based on the precondition that bridge's intact condition is known beforehand.

Table 4. Damage extent detection results of M1 and M2 with intact reference.

Designed Damage Extent (%)		Detected Damage Extent (%)					
		M1			M2		
		V2	V3	V5	V2	V3	V5
5	S5	5.18 (3.60%)	4.97 (0.60%)	5.12 (2.40%)	5.00 (0.00%)	5.00 (0.00%)	5.00 (0.00%)
	S7	5.42 (8.40%)	5.17 (3.40%)	5.17 (3.40%)	5.00 (0.00%)	5.00 (0.00%)	5.00 (0.00%)
	S8	5.42 (8.40%)	5.22 (4.40%)	5.20 (4.00%)	5.21 (4.20%)	5.21 (4.20%)	5.20 (4.20%)
10	S5	10.38 (3.80%)	10.00 (0.00%)	10.13 (1.30%)	10.00 (0.00%)	10.00 (0.00%)	10.00 (0.00%)
	S7	10.79 (7.90%)	10.44 (4.40%)	10.13 (1.30%)	10.00 (0.00%)	10.00 (0.00%)	10.00 (0.00%)
	S8	10.68 (6.80%)	10.37 (3.70%)	10.40 (4.00%)	10.44 (4.40%)	10.44 (4.40%)	10.44 (4.40%)
15	S5	15.54 (3.60%)	15.07 (0.47%)	15.08 (0.53%)	15.00 (0.00%)	15.00 (0.00%)	15.00 (0.00%)
	S7	15.78 (5.20%)	15.29 (1.93%)	15.02 (0.13%)	15.00 (0.00%)	15.00 (0.00%)	15.00 (0.00%)
	S8	15.52 (3.47%)	15.53 (3.53%)	15.62 (4.13%)	15.71 (4.73%)	15.71 (4.73%)	15.70 (4.67%)

Overall, under different damage scenarios and vehicle types, all three methods can successfully detect the damage location. However, for damage extent calculation, unless intact condition reference is available, the performance of M1 and M2 is unacceptable. As for the impact of vehicle type, M3's performance is a little bit weak, while M2's is the most stable. In addition, the impact of damage location on three damage detection methods is more obvious than that of vehicle type. The damage detection results in S7 and S8 are worse than that in S5 because the corresponding response amplitude is smaller. Due to the same reason, the relative errors in 15% damage scenarios are the lowest.

4.2.2. Comparison under Different Speeds and Road Surface Roughness

Based on the findings in last section, the parameter vehicle type and damage scenario is fixed as two-axle vehicle and scenario 3 to test the performance of three methods under various speeds and road surface roughness. The reason why these two factors are considered together is that there might be some couple effect within these two factors impacting the performance of damage detection methods. Corresponding damage indexes of three methods under various speeds and road surface roughness are displayed in Figure 15. The letters 'S' and 'R', with numbers behind in legends, represent speed and roughness degree. The noise level is set as noise-free.

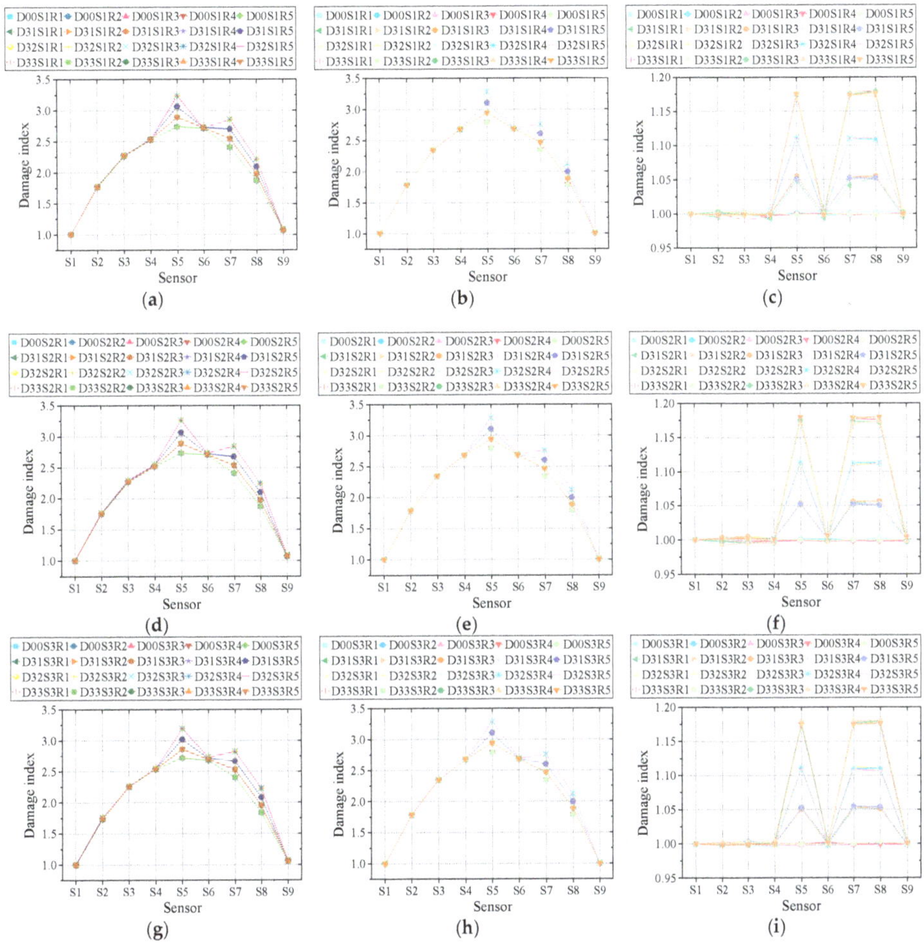

Figure 15. Damage detection results: (**a**) M1 with 2-axle vehicle; (**b**) M2 with 2-axle vehicle; (**c**) M3 with 2-axle vehicle; (**d**) M1 with 3-axle vehicle; (**e**) M2 with 3-axle vehicle; (**f**) M3 with 3-axle vehicle; (**g**) M1 with 5-axle vehicle; (**h**) M2 with 5-axle vehicle; (**i**) M3 with 5-axle vehicle.

According to the results presented in Figure 15, it can be revealed that vehicle speed and road surface roughness' influence on the performance of three methods is limited. Moreover, the coupling effect of these two factors was not discovered. These findings proved these methods' reliability in actual stochastic traffic flow to some degree, which accords with the findings in previous works [24–28]. Among three methods, M2 behaves the best, whose results remained stable under various speeds and road surface roughness. Similar to the findings in the last section, M3's behavior is less ideal with some fluctuations in results. Similarly, through Equations (7), (11) and (14), the damage extent can be calculated, as listed in Table 5. It needs to be clarified that in Table 5 the results of M1 and M2 are based on intact condition reference, because the results without intact condition proved to be incomparable in the last section.

Table 5. Damage extent detection results of three methods under various speed and roughness.

Designed Damage Extent (%)		Detected Damage Extent (%)								
		M1			M2			M3		
		S1	S2	S3	S1	S2	S3	S1	S2	S3
5	R1	5.18 (3.60%)	5.32 (6.40%)	4.37 (12.60%)	5.00 (0.00%)	5.01 (0.20%)	5.00 (0.00%)	5.02 (0.40%)	4.76 (4.80%)	4.62 (7.60%)
	R2	5.18 (3.60%)	5.32 (6.40%)	4.37 (12.60%)	5.00 (0.00%)	5.01 (0.20%)	5.00 (0.00%)	5.06 (1.20%)	4.81 (3.80%)	5.10 (2.00%)
	R3	5.18 (3.60%)	5.32 (6.40%)	4.37 (12.60%)	5.00 (0.00%)	5.01 (0.20%)	5.00 (0.00%)	5.11 (2.20%)	4.98 (0.40%)	5.01 (0.20%)
	R4	5.18 (3.60%)	5.32 (6.40%)	4.37 (12.60%)	5.00 (0.00%)	5.01 (0.20%)	5.00 (0.00%)	4.91 (1.80%)	4.79 (4.20%)	4.86 (2.80%)
	R5	5.18 (3.60%)	5.32 (6.40%)	4.37 (12.60%)	5.00 (0.00%)	5.01 (0.20%)	5.00 (0.00%)	5.01 (0.20%)	5.13 (2.60%)	4.62 (7.60%)
10	R1	10.38 (3.80%)	10.73 (7.30%)	9.35 (6.50%)	10.00 (0.00%)	10.01 (0.10%)	10.00 (0.00%)	10.24 (2.40%)	10.44 (4.40%)	9.90 (1.00%)
	R2	10.38 (3.80%)	10.73 (7.30%)	9.35 (6.50%)	10.00 (0.00%)	10.01 (0.10%)	10.00 (0.00%)	9.98 (0.20%)	10.02 (0.20%)	9.90 (1.00%)
	R3	10.38 (3.80%)	10.73 (7.30%)	9.35 (6.50%)	10.00 (0.00%)	10.01 (0.10%)	10.00 (0.00%)	9.94 (0.60%)	10.00 (0.00%)	10.34 (3.40%)
	R4	10.38 (3.80%)	10.73 (7.30%)	9.35 (6.50%)	10.00 (0.00%)	10.01 (0.10%)	10.00 (0.00%)	10.18 (1.80%)	10.08 (0.80%)	10.00 (0.00%)
	R5	10.38 (3.80%)	10.73 (7.30%)	9.35 (6.50%)	10.00 (0.00%)	10.01 (0.10%)	10.00 (0.00%)	10.08 (0.80%)	9.78 (2.20%)	10.06 (0.60%)
15	R1	15.54 (3.60%)	16.13 (7.53%)	14.33 (4.47%)	15.00 (0.00%)	15.01 (0.07%)	15.00 (0.00%)	14.56 (2.93%)	14.72 (1.87%)	14.81 (1.27%)
	R2	15.54 (3.60%)	16.13 (7.53%)	14.33 (4.47%)	15.00 (0.00%)	15.01 (0.07%)	15.00 (0.00%)	14.98 (0.13%)	14.94 (0.40%)	14.93 (0.47%)
	R3	15.54 (3.60%)	16.13 (7.53%)	14.33 (4.47%)	15.00 (0.00%)	15.01 (0.07%)	15.00 (0.00%)	15.16 (1.07%)	14.91 (0.60%)	15.16 (1.07%)
	R4	15.54 (3.60%)	16.13 (7.53%)	14.33 (4.47%)	15.00 (0.00%)	15.01 (0.07%)	15.00 (0.00%)	14.91 (0.60%)	14.48 (3.47%)	15.24 (1.60%)
	R5	15.54 (3.60%)	16.13 (7.53%)	14.33 (4.47%)	15.00 (0.00%)	15.01 (0.07%)	15.00 (0.00%)	15.10 (0.67%)	15.13 (0.87%)	14.89 (0.73%)

According to Table 5, all three methods' performance under various speeds and road surface roughness conditions are good. The accuracy of M2 is the best. Similar to the findings in the last section, the performance of M3 is relatively unstable. The detection error varies under different scenarios. This demonstrates that, as for stability, M3's performance is not ideal, while M1 and M2's excellent performance is based on the precondition that intact condition was known in advance. Therefore, although M3's performance is not very stable, its not needing an intact condition reference still made it an alternative method.

In summary, all three methods showed good performance under different vehicle speeds and road surface roughness, except that M3's behavior exhibited little fluctuations.

4.2.3. Comparison under Different Signal-to-Noise Ratios

At last, the influence of noise level on the three methods was studied in order to illustrate the three methods' performance under adverse environments. The noise level was controlled by signal-to-noise level, which was set as 20, 50 and noise-free to mimic various noise conditions, as designed in previous studies. According to the definition of signal-to-noise level, with the value decreasing, the noise level would rise accordingly. For better comparison, at this part the other factors such as vehicle type, speed and road surface roughness are set as two-axle vehicle with 10 m/s under A class road surface roughness. Corresponding damage indexes of three methods under various noise levels are displayed in Figure 16.

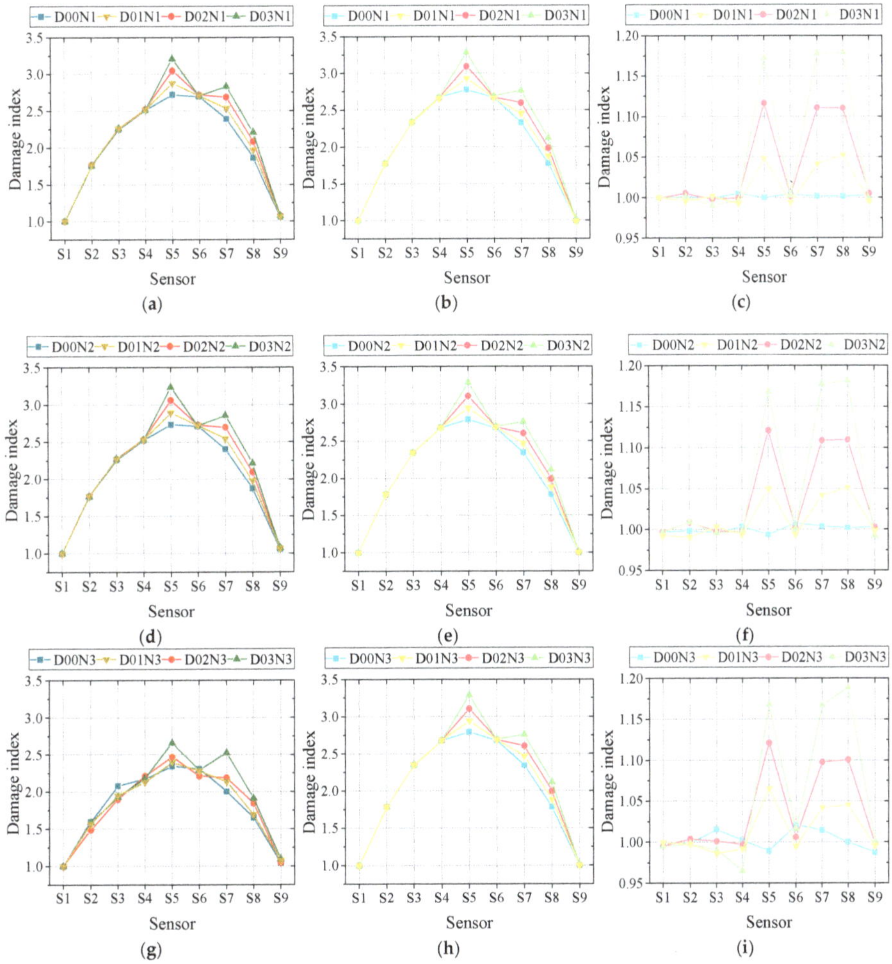

Figure 16. Damage detection results: (**a**) M1 noise-free; (**b**) M2 noise-free; (**c**) M3 noise-free; (**d**) M1 with 50 signal-to-noise ratio; (**e**) M2 with 50 signal-to-noise ratio; (**f**) M3 with 50 signal-to-noise ratio; (**g**) M1 with 20 signal-to-noise ratio; (**h**) M2 with 20 signal-to-noise ratio; (**i**) M3 with 20 signal-to-noise ratio.

As can be seen in Figure 16, the influence of noise level on the three damage detection methods varied dramatically. M2 is the most robust one facing different noise levels, while M1 and M3's performance is less ideal. Especially in noise level 3, with signal-to-noise level reaching 20, evident abnormal fluctuations were observed in M1 and M3's damage detection results. Then, utilizing Equations (7), (11) and (14), the damage extent can be obtained as given in Table 6. It also needs to be elucidated that the results of M1 and M2 in Table 6 are on the basis of intact condition reference due to the same reason explained in Section 4.2.2.

Table 6. Damage extent detection results of three methods under various speeds and roughness.

Designed Damage Extent (%)	Detected Damage Extent (%)								
	M1			M2			M3		
	Noise-Free	50	20	Noise-Free	50	20	Noise-Free	50	20
5	5.18 (3.60%)	5.38 (7.60%)	2.18 (56.40%)	5.00 (0.00%)	5.01 (0.20%)	5.11 (2.20%)	5.02 (0.40%)	5.31 (6.20%)	7.07 (41.40%)
10	10.38 (3.80%)	10.73 (7.30%)	5.21 (47.90%)	10.00 (0.00%)	10.02 (0.20%)	10.27 (2.70%)	10.24 (2.40%)	11.32 (2.91%)	11.73 (6.64%)
15	15.54 (3.60%)	15.98 (6.53%)	11.86 (20.93%)	15.00 (0.00%)	15.01 (0.07%)	15.38 (2.53%)	14.56 (2.93%)	14.92 (0.53%)	15.25 (1.67%)

According to Table 6, the findings in Figure 16 are certified. M2's are the most stable and accurate under three noise levels with the largest error less than 3%. Meanwhile, the performance of M1 and M3 degrades intensely as noise level rises. From noise-free to 20 signal-to-noise ratio, the relative error of M1 is raised from 3.6% to 56.40% for the 5% damage extent scenario. For the more severe 15% damage extent scenario, the relative error dropped to around 20%, which is a little bit lower but still unacceptable. For M3, the overall performance is better than M1, while its relative error still reached 40% for the 5% damage extent scenario with a 20 signal-to-noise ratio. Similarly, the accuracy under more severe damage scenarios is much better for M3. In summary, M2's robustness under different noise levels is the best among the three methods. M3 and M1's performance under different noise levels is very unstable, while M1 is the worst one.

5. Conclusions

Long-gauge FBG, as a novel optical sensor with many advantages, has been applied in SHM. Many exclusive damage detection methods have been proposed utilizing this sensor and exhibited various characteristics. Although they were alleged to be feasible and reliable, there is not a comparative study among them, especially regarding their precision and reliability. As a result, it is difficult to evaluate the state of the art, which also results in some confusion for users to select. Therefore, a strict comparison of three typical damage detection methods based on long-gauge FBG was conducted. A numerical simulation based on vehicle–bridge interaction was programmed and validated by an indoor experiment. The performances of different methods under various scenarios were tested and compared through a series of parametric numerical studies. The following conclusions can be drawn:

(1) The damage detection methods proposed by Wu et al. [26] (M1), Hong et al. [27] (M2) and Chen et al. [28] (M3) as three representative methods were selected to conduct this comparative study. Firstly, these three methods' theoretical backgrounds were reformulated and unified for comparison. It can be discovered that, theoretically, all these three methods would detect damage location and extent based on monitored macro-strain response. All their theoretical foundations come from macro-strain influence line, while the damage indexes they selected are derivatives of the macro-strain influence line. One evident difference among them is that M1 and M2 need to adopt a approximation for damage extent estimation because the intact condition of the structure is hard to get, while M3 has no need.

(2) In order to compare the performances of the three methods under various critical parameters, a parametric study based on vehicle–bridge interaction was programmed and validated. The parameters considered in this study include damage scenario, vehicle type, speed, road surface roughness and noise level. Under different damage scenarios and vehicle types, all of them can successfully detect the damage location. However, for damage extent calculation, unless intact condition reference is available, the accuracy of M1 and M2 is very poor. As for the impact of vehicle type, although M3's precision is a little bit weak, it is still the most stable one. In addition, the impact of damage location on three damage detection methods is more obvious than that of vehicle type.

(3) As for the influence of vehicle speed and road surface roughness, all three methods showed good performance under different vehicle speeds and road surface roughness, except that M3's behavior exhibited little fluctuations. At last, as for the impact of noise level, M2's robustness under different

noise levels is the best among the three methods. M3 and M1's performance under different noise levels is very unstable, while M1 is the worst one. All in all, the M2's performance is the most eminent only if the intact condition reference is available. Meanwhile M3 is a little bit unstable but it has a merit that the intact condition reference is not needed.

Author Contributions: Conceptualization, S.-Z.C. and D.-C.F.; methodology, S.-Z.C.; project administration, W.-S.H.; resources, W.-S.H.; software, S.-Z.C.; validation, S.-Z.C.; formal analysis, S.-Z.C. and D.-C.F.; investigation, S.-Z.C.; writing—original draft preparation, S.-Z.C.; writing—review and editing, D.-C.F. and W.-S.H.; visualization, S.-Z.C.; funding acquisition, D.-C.F. and W.-S.H. All authors have read and agreed to the published version of the manuscript.

Funding: This research was funded by National Key Research and Development Project of China, grant number 2019YFB1600700, National Natural Science Foundation of China, grant number 51878058, Natural Science Foundation of Shaanxi Province, 2019JZ-02.

Acknowledgments: The present work benefits from the financial support of National Key Research and Development Project of China, National Natural Science Foundation of China and Natural Science Foundation of Shaanxi Province.

Conflicts of Interest: The authors declare no conflict of interest.

References

1. Brownjohn, J.M.W. Structural health monitoring of civil infrastructure. *Philos. Trans. R. Soc. A* **2007**, *365*, 589–622. [CrossRef] [PubMed]
2. Das, S.; Saha, P.; Patro, S.K. Vibration-based damage detection techniques used for health monitoring of structures: A review. *J. Civil. Struct. Health Monit.* **2016**, *6*, 477–507. [CrossRef]
3. Peeters, B.; de Roeck, G. Reference-based stochastic subspace identification for output-only modal analysis. *Mech. Syst. Signal Process.* **1999**, *13*, 855–878. [CrossRef]
4. Malekjafarian, A.; Brien, E.J. Identification of bridge mode shapes using Short Time Frequency Domain Decomposition of the responses measured in a passing vehicle. *Eng. Struct.* **2014**, *81*, 386–397. [CrossRef]
5. He, W.; He, J.; Ren, W. Damage localization of beam structures using mode shape extracted from moving vehicle response. *Measurement* **2018**, *121*, 276–285. [CrossRef]
6. Salawu, O.S. Detection of structural damage through changes in frequency: A review. *Eng. Struct.* **1997**, *19*, 718–723. [CrossRef]
7. Doebling, S.W.; Farrar, C.R.; Prime, M.B. A summary review of vibration-based damage identification methods. *Shock Vib. Dig.* **1998**, *30*, 91–105. [CrossRef]
8. Huth, O.; Feltrin, G.; Maeck, J.; Kilic, N.; Motavalli, M. Damage identification using modal data: Experiences on a prestressed concrete bridge. *J. Struct. Eng.* **2005**, *131*, 1898–1910. [CrossRef]
9. Alvandi, A.; Cremona, C. Assessment of vibration-based damage identification techniques. *J. Sound Vib.* **2006**, *292*, 179–202. [CrossRef]
10. Chang, K.; Kim, C. Modal-parameter identification and vibration-based damage detection of a damaged steel truss bridge. *Eng. Struct.* **2016**, *122*, 156–173. [CrossRef]
11. Glisic, B.; Inaudi, D. Development of method for in-service crack detection based on distributed fiber optic sensors. *Struct. Health Monit.* **2012**, *11*, 161–171. [CrossRef]
12. Cardini, A.J.; DeWolf, J.T. Long-term Structural Health Monitoring of a Multi-girder Steel Composite Bridge Using Strain Data. *Struct. Health Monit.* **2009**, *8*, 47–58. [CrossRef]
13. Catbas, F.N.; Gokce, H.B.; Gul, M. Nonparametric analysis of structural health monitoring data for identification and localization of changes: Concept, lab, and real-life studies. *Struct. Health Monit.* **2012**, *11*, 613–626. [CrossRef]
14. Li, Y.Y. Hypersensitivity of strain-based indicators for structural damage identification: A review. *Mech. Syst. Signal Process.* **2010**, *24*, 653–664. [CrossRef]
15. Torres Gorriz, B.; Rinaudo, P.; Calderon Garcia, P.A. Comparison between point and long-gage FBG-based strain sensors during a railway bridge load test. *Strain* **2017**, *53*, e122304. [CrossRef]
16. Chen, S.; Wu, G.; Xing, T.; Feng, D. Prestressing force monitoring method for a box girder through distributed long-gauge FBG sensors. *Smart Mater. Struct.* **2018**, *27*, 0150151. [CrossRef]

17. Barrias, A.; Casas, J.R.; Villalba, S. A Review of Distributed Optical Fiber Sensors for Civil Engineering Applications. *Sensors* **2016**, *16*, 748. [CrossRef]
18. Bao, X.; Chen, L. Recent Progress in Brillouin Scattering Based Fiber Sensors. *Sensors* **2011**, *11*, 4152–4187. [CrossRef]
19. Chen, S.; Wu, G.; Xing, T. Deflection monitoring for a box girder based on a modified conjugate beam method. *Smart Mater. Struct.* **2017**, *26*, 0850348. [CrossRef]
20. Li, S.; Wu, Z. Development of distributed long-gage fiber optic sensing system for structural health monitoring. *Struct. Health Monit.* **2007**, *6*, 133–143. [CrossRef]
21. Wu, B.; Lu, H.; Chen, B.; Gao, Z. Study on Finite Element Model Updating in Highway Bridge Static Loading Test Using Spatially-Distributed Optical Fiber Sensors. *Sensors* **2017**, *17*, 1657. [CrossRef]
22. Chen, S.; Wu, G.; Feng, D. Development of a bridge weigh-in-motion method considering the presence of multiple vehicles. *Eng. Struct.* **2019**, *191*, 724–739. [CrossRef]
23. Chen, S.; Wu, G.; Feng, D.; Zhang, L. Development of a Bridge Weigh-in-Motion System Based on Long-Gauge Fiber Bragg Grating Sensors. *J. Bridge Eng.* **2018**, *23*, 040180639. [CrossRef]
24. Chen, S.; Wu, G.; Feng, D.; Wang, Z.; Cao, X. Multi-cross reference method for highway bridge damage identification based on long-gauge fiber Bragg grating sensors. *J. Bridge Eng.* **2020**, *25*, 04020023. [CrossRef]
25. Chen, S.; Wu, G.; Feng, D. Damage detection of highway bridges based on long-gauge strain response under stochastic traffic flow. *Mech. Syst. Signal Process.* **2019**, *127*, 551–572. [CrossRef]
26. Wu, B.; Wu, G.; Yang, C.; He, Y. Damage identification and bearing capacity evaluation of bridges based on distributed long-gauge strain envelope line under moving vehicle loads. *J. Intel. Mat. Syst. Struct.* **2016**, *27*, 2344–2358. [CrossRef]
27. Hong, W.; Cao, Y.; Wu, Z. Strain-Based Damage-Assessment Method for Bridges under Moving Vehicular Loads Using Long-Gauge Strain Sensing. *J. Bridge Eng.* **2016**, *21*, 04016059. [CrossRef]
28. Chen, S.; Wu, G.; Wu, B. *Simultaneous Identification Method of Damage and Vehicle Parameters on Bridges Utilizing Long-Gauge Strain Influence Line under Moving Vehicle Loads*; The Society of Photo-Optical Instrumentation Engineers (SPIE): Portland, OR, USA, 2017.
29. Wu, S.Q.; Law, S.S. Vehicle axle load identification on bridge deck with irregular road surface profile. *Eng. Struct.* **2011**, *33*, 591–601. [CrossRef]
30. Zhang, Y.; Cai, C.; Shi, X.; Wang, C. Vehicle-induced dynamic performance of FRP versus concrete slab bridge. *J. Bridge Eng.* **2006**, *4*, 410–419. [CrossRef]
31. Harris, N.; Obrien, E.; González, A. Reduction of bridge dynamic amplification through adjustment of vehicle suspension damping. *J. Sound Vib.* **2007**, *302*, 471–485. [CrossRef]
32. He, W.; Deng, L.; Shi, H.; Cai, C.; Yu, Y. Novel virtual simply supported beam method for detecting the speed and axles of moving vehicles on bridges. *J. Bridge Eng.* **2017**, *22*, 04016141. [CrossRef]

sensors

MDPI

Article

Multi-Addressed Fiber Bragg Structures for Microwave-Photonic Sensor Systems

Oleg Morozov [1], Airat Sakhabutdinov [1,*], Vladimir Anfinogentov [2], Rinat Misbakhov [3], Artem Kuznetsov [1] and Timur Agliullin [1]

1 Department of Radiophotonics and Microwave Technologies, Kazan National Research Technical University named after A.N. Tupolev-KAI, 10, Karl Marx st., 420111 Kazan, Tatarstan, Russia; microoil@mail.ru (O.M.); aakuznetsov@kai.ru (A.K.); taagliullin@mail.ru (T.A.)

2 Department of Special Mathematics, Kazan National Research Technical University named after A.N. Tupolev-KAI, 10, Karl Marx st., 420111 Kazan, Tatarstan, Russia; v.anfinogentov@yandex.ru

3 Engineering Center "Computer Modeling and Engineering in the Field of Energy and Power Engineering", Kazan State Power Engineering University, 51, Krasnoselskaya st., 420066 Kazan, Tatarstan, Russia; energy@zerdex.pro

* Correspondence: azhsakhabutdinov@kai.ru

Received: 21 April 2020; Accepted: 7 May 2020; Published: 9 May 2020

check for updates

Abstract: The new theory and technique of Multi-Addressed Fiber Bragg Structure (MAFBS) usage in Microwave Photonics Sensor Systems (MPSS) is presented. This theory is the logical evolution of the theory of Addressed Fiber Bragg Structure (AFBS) usage as sensors in MPSS. The mathematical model of additive response from a single MAFBS is presented. The MAFBS is a special type of Fiber Bragg Gratings (FBG), the reflection spectrum of which has three (or more) narrow notches. The frequencies of narrow notches are located in the infrared range of electromagnetic spectrum, while differences between them are located in the microwave frequency range. All cross-differences between optical frequencies of single MAFBS are called the address frequencies set. When the additive optical response from a single MAFBS, passed through an optic filter with an oblique amplitude–frequency characteristic, is received on a photodetector, the complex electrical signal, which consists of all cross-frequency beatings of all optical frequencies, which are included in this optical signal, is taken at its output. This complex electrical signal at the photodetector's output contains enough information to determine the central frequency shift of the MAFBS. The method of address frequencies analysis with the microwave-photonic measuring conversion method, which allows us to define the central frequency shift of a single MAFBS, is discussed in the work.

Keywords: microwave-photonic sensor systems; Fiber Bragg Gratings; Addressed Fiber Bragg Structures; Multi-Addressed Fiber Bragg Structures

1. Introduction

Common problems of Fiber Bragg Gratings (FBG) array interrogation in optical sensor systems are their complexity and the high cost of interrogators due to the technique of interrogation and FBG multiplexing [1–5]. Wavelength [1], time [2], frequency [3], polarizing [4], and spatial [5] division multiplexing requires using complex devices, such as spectrum analyzers, spectrometers with tunable Fabry–Perot interferometers, diffraction gratings, etc. All of them use the technique of optic signal receiving on charge-coupled devices with its further complex analysis. The complexity is also described by the fact that these sensors are not addressable per se, and therefore, any spectrum overlapping leads to interrogation errors [6–8].

In parallel with multiplexing methods and microwave photonics methods, the optical pulse-coding, phase-coding, low coherent interferometer with the cascaded FBGs methods is developed. The coding methods allow us to recognize two or more FBGs with the same spectral range [9–12]. Spectral-coding sensors are based on code-multiplexing technology [11,12], where the interrogation is produced in real time according to autocorrelation between sensor spectra and its code signature. In the series of works, it was demonstrated that the spectral-amplitude-coding method with super-structured FBG sensors, based on discrete prolate spheroidal sequences, can be useful even in the case of overlap of their optical ranges [13–16].

An easier solution was found in the Addressed Fiber Bragg Structure (AFBS) usage with the microwave photonics interrogation method [17]. The AFBS is a special type of FBG, the reflection spectrum of which has two narrow notches. The light, passed through AFBS, has two narrow optical frequencies, the difference between which is much less than an optical frequency (THz) and located in microwave range (GHz). The differential frequency is called the address frequency of an AFBS. The address frequency is invariant to stress or temperature fields; moreover, it is invariant to AFBS central frequency shifting. The AFBS in sensor systems is used both as a two-frequency source (due to the fact that it has two narrow optical frequencies with the difference between them being in the microwave range) and as a sensor of measurement system (due to the fact that its address frequency is invariant to measurable fields) simultaneously. It allows us to design a microwave-photonic sensor system based on arrays of AFBSs, on condition that the set of address frequencies in the array is orthogonal [17,18].

The evolution from AFBS to Multi-Addressed Fiber Bragg Structures (MAFBS) is logical and self-consistent. In spectral response of MAFBS, three (or more) frequency carriers are configured, and their beatings on a photodetector form three (or more) address frequencies. The combining of address frequencies allows us to expand the sensor capacity of the measurement system; moreover, it allows us to increase the accuracy of central wavelength determination.

2. Multi-Addressed Fiber Bragg Structure

The MAFBS (as well as AFBS) is a special type of FBG, the reflection spectrum of which has three (or more) narrow notches. The light, passed through MAFBS, has three (or more) narrow optical frequencies, the difference between which is much less than an optical frequency (THz) and is located in the microwave range (GHz). The set of all differential frequencies is named the address frequencies set of a MAFBS. The address frequencies set is invariant to strain or temperature fields, and it is also invariant to central frequency shifting [17–20]. The MAFBS is both a multi-frequency source and a sensor of the measurement system at the same time. It is necessary to require the additional conditions to use a MAFBS as a sensor in the microwave sensor system, namely: The light only from MAFBS's narrow-band frequencies must trap into the light analysis area for the whole MAFBS central frequency shift range, which corresponds to the measurement range [17–20].

There are at least two approaches to MAFBS (as well as AFBS) forming: The first of them is the introduction of phase π-shifts into the classic FBG periodic structure [21] (Figure 1a), and the second one is the MAFBS forming as a set of ultra-narrowband FBGs [17], (Figure 1b).

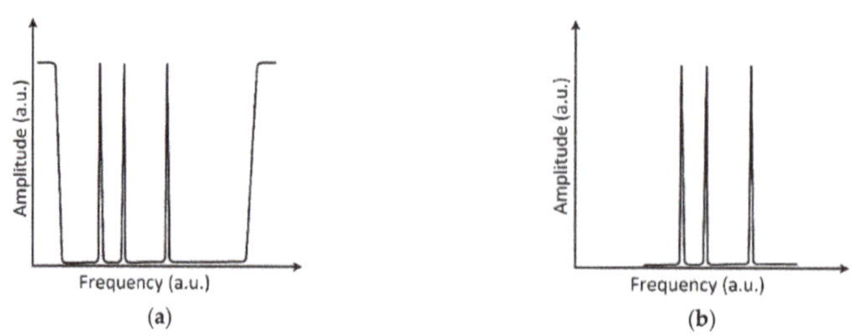

Figure 1. Amplitude–frequency diagrams: (**a**) transmitted through Multi-Addressed Fiber Bragg Structure (MAFBS), formed using Fiber Bragg Gratings (FBG) with π-shifts; (**b**) reflected from MAFBS, formed as a set of ultra-narrowband FBGs.

3. Mathematical Model

The optoelectronic schemata, designed for light reflection and propagation, differ from each other a little, but both of them are based on the common idea. The complex light response from MAFBS, passed through optic filter with an oblique amplitude–frequency characteristic, is received on a photodetector with the subsequent analysis of output electric signal on the address frequencies set. An example of the optoelectronic scheme for light propagation is presented in Figure 2a. The optic source (1) forms a finite band light (a), which passes through MAFBS (2) and forms a multi-frequency (in this case three-frequency) signal (c); the three-frequency optic signal, passing through the optic filter with an oblique amplitude–frequency characteristic, forms an asymmetric three-frequency optic signal (d), which is received on a photodetector (4); after the photodetector, signal is received on an analog-to-digital converter (5), its subsequent analysis is performed. The system also includes the reference channel, in which the optic signal is received on a photodetector (7) directly after the MAFBS without its asymmetrical deformation in the filter (3), and then the signal is processed using the analog-to-digital converter (8). All subsequent calculations are produced with the ratios of values in the measuring and reference channels. This allows us to avoid influence of light power fluctuation, which is not connected with the MAFBS central frequency shifting.

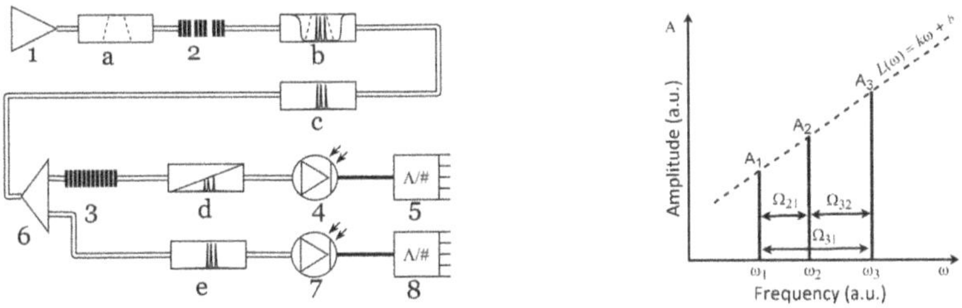

(a) (b)

Figure 2. Multi-Addressed Fiber Bragg Structure: (**a**) a microwave-photonic interrogation scheme; (**b**) a spectral response of Multi-Addressed Fiber Bragg Structure.

The shape of a MAFBS spectral response at the output end of the optic filter with an oblique amplitude–frequency characteristic is shown in Figure 2b, where the following notation is used: ω_1, ω_2, and ω_3 are the frequencies of optical carriers; Ω_{21} and Ω_{32} are the address frequencies; k and b are the predefined parameters of the optic filter with an oblique amplitude–frequency characteristic.

The central frequency shift of the MAFBS leads to a change of the mutual relation of the optical carriers' amplitudes, which causes a change of the beating parameters at the address frequencies Ω_{21}, Ω_{32}, and their sum $\Omega_{31} = \Omega_{21} + \Omega_{32}$. The task is to determine the MAFBS central frequency (any of the frequencies ω_1, ω_2, or ω_3), using the known parameters of the beating at the address frequencies set.

The light response from the MAFBS, which passes through the filter with an oblique amplitude–frequency response, can be written as:

$$S(t) = A_1 \sin(\omega_1 t + \varphi_1) + A_2 \sin(\omega_2 t + \varphi_2) + A_3 \sin(\omega_3 t + \varphi_3), \tag{1}$$

where A_1, A_2, and A_3 are the amplitudes, φ_1, φ_2, and φ_3 are the initial phases of the signal at the optical frequencies ω_1, ω_2, and ω_3. The output current of the photodetector $F(t)$ is proportional to the square of the optic response:

$$\begin{aligned} F(t) \sim\ & A_1^2 + A_2^2 + A_3^2 + 2A_1A_2\cos(\Omega_{21}t + \varphi_2 - \varphi_1) + \\ & + 2A_2A_3\cos(\Omega_{32}t + \varphi_3 - \varphi_2) + 2A_1A_3\cos(\Omega_{31}t + \varphi_3 - \varphi_1), \end{aligned} \tag{2}$$

in which the oscillations at optical (terahertz) frequencies are excluded [17,18]. The known values in (2) are the address frequencies Ω_{21} and Ω_{32}. The constant signal level in (2), the amplitudes at the address frequencies Ω_{21}, Ω_{32}, and their sum Ω_{31} give four independent equations for the determination of three unknown amplitudes, A_1, A_2, and A_3:

$$\begin{cases} D_0 = A_1^2 + A_2^2 + A_3^2 \\ D_{21} = 2A_1A_2 \\ D_{32} = 2A_2A_3 \\ D_{31} = 2A_1A_3 \end{cases}, \quad (3)$$

where D_0, D_{21}, D_{32}, and D_{31} are the measured values of constant signal level, amplitudes of the address frequencies Ω_{21}, Ω_{32}, and their sum Ω_{31}, respectively.

The resulting system of four equations is overdetermined, since the number of equations exceeds the number of unknowns. Moreover, Equation (3) must be supplemented by the requirements that the points (ω_1, A_1), (ω_2, A_2), and (ω_3, A_3) belong to the same line:

$$L(\omega) = k{\cdot}\omega + b, \quad (4)$$

which describes the filter with an oblique amplitude–frequency characteristic. Moreover, it is necessary to require that the differences $\omega_2 - \omega_1$ and $\omega_3 - \omega_2$ are equal to the address frequencies Ω_{21} and Ω_{32}, respectively, and the condition $\omega_3 - \omega_1 = \Omega_{32} + \Omega_{21}$ would also be automatically satisfied. Thus, it is necessary to add the additional relation to Equation (3):

$$\frac{A_2 - A_1}{\Omega_{21}} = \frac{A_3 - A_2}{\Omega_{32}}, \quad (5)$$

which binds the task parameters, imposing restrictions on finding a solution, while simultaneously describing the mutual relations between the frequencies. Having the amplitudes A_1, A_2, and A_3 from Equation (3), supplemented by Equation (5), and using the known values of the parameters k and b of the filter with an oblique linear amplitude–frequency characteristic, one can calculate the MAFBS frequencies ω_1, ω_2, and ω_3.

There are different ways to define the MAFBS central frequency, since MAFBS has three narrow resonances in addition to its main Bragg resonance. We used the definition of the central frequency of MAFBS as an average, according to the formula:

$$\omega_{Br} = \frac{1}{2}\left(\frac{\omega_1 + \omega_2}{2} + \frac{\omega_2 + \omega_3}{2}\right) = \frac{1}{4}(\omega_1 + 2\omega_2 + \omega_3), \quad (6)$$

which uniquely determines the position of the MAFBS.

The overdetermined equations system can be solved by searching the conditional extremum of the function:

$$\Phi(A_1, A_2, A_3, \lambda) = \left(A_1^2 + A_2^2 + A_3^2 - D_0\right)^2 + (D_{12} - 2A_1A_2)^2 + (D_{23} - 2A_2A_3)^2 + (D_{13} + 2A_1A_3)^2, \quad (7)$$

relatively to the restriction:

$$f(A_1, A_2, A_3) = (A_2 - A_1)\Omega_{32} - (A_3 - A_2)\Omega_{21} = 0, \quad (8)$$

requiring a minimum of the Lagrange function, expressed in the form:

$$\Psi(A_1, A_2, A_3, \lambda) = \Phi(A_1, A_2, A_3) + \lambda{\cdot}f(A_1, A_2, A_3) \rightarrow \min, \quad (9)$$

where λ is the Lagrange multiplier.

Equation (9) is equivalent to the requirement that all partial derivatives with respect to the variables A_1, A_2, A_3, and λ are equal to zero, which leads to a set of four non-linear equations:

$$\begin{cases} \partial\Psi/\partial A_1 = \partial\Phi/\partial A_1 - \lambda\Omega_{32} = 0 \\ \partial\Psi/\partial A_2 = \partial\Phi/\partial A_2 + \lambda(\Omega_{21} + \Omega_{32}) = 0 \\ \partial\Psi/\partial A_3 = \partial\Phi/\partial A_3 - \lambda\Omega_{21} = 0 \\ \partial\Psi/\partial\lambda = (A_2 - A_1)\Omega_{32} - (A_3 - A_2)\Omega_{21} \end{cases}, \tag{10}$$

where the partial derivatives $\partial\Phi/\partial A_i$ are not expressed due to their obvious simplicity.

Non-linear Equation (10) (due to the nonlinearity of the partial derivatives $\partial\Phi/\partial A_i$, i = 1, 2, 3) can be solved only numerically. The values A_1, A_2, and A_3, which are the solution of Equation (3) with the exception of the first equation, can be taken as the initial conditions, and λ can be taken as the initial value equal to zero:

$$A_{01} = \sqrt{\frac{D_{31}D_{21}}{2D_{32}}},\ A_{02} = \sqrt{\frac{D_{21}D_{32}}{2D_{31}}},\ A_{03} = \sqrt{\frac{D_{31}D_{32}}{2D_{21}}},\ \lambda_0 = 0. \tag{11}$$

After that, Equation (10), supplemented by the initial values in Equation (11), is solved by any well-converging iterative method, for example, the Levenberg–Marquardt [22,23] or Newton–Raphson [24] methods.

Equation (10) solution gives the values of the amplitudes A_1, A_2, and A_3, each of which can be used to determine the MAFBS central frequency position relative to the filter with an oblique amplitude–frequency characteristic. Substituting the found values of the amplitudes A_1, A_2, and A_3 in Equation (4), and combining them in Equation (6), we obtain the expression for the central frequency of the MAFBS:

$$\omega_{Br}(D_0, D_{12}, D_{13}, D_{32}) = \frac{1}{4k}(A_1 + A_2 + A_3 - 3b), \tag{12}$$

as a function of the measured values of D_0, D_{21}, D_{32}, and D_{31}—a constant signal level, and amplitudes at address frequencies Ω_{21}, Ω_{32}, and their sum Ω_{31}, respectively.

4. Generalized Modulation Factor

An alternative solution for MAFBS central frequency determination can be obtained through a generalization of the modulation factor of the output current after the photodetector (Equation (2)), which for a two-address MAFBS can be written as follows:

$$M(\omega) = \frac{A_1(\omega)A_2(\omega) + A_1(\omega)A_3(\omega) + A_2(\omega)A_3(\omega)}{A_1^2(\omega) + A_2^2(\omega) + A_3^2(\omega)}. \tag{13}$$

Note that in the particular case, when one of the amplitudes is equal to zero, the modulation factor coincides with the modulation factor of two-frequency beatings within a constant factor [18–20]. Equation (13) is a monotonic function of the MAFBS central frequency shifting relative to a filter with an oblique amplitude–frequency characteristic. The monotonicity of the generalized modulation factor in Equation (13) allows us to use it as single measured parameter for central frequency determination.

5. Results of Numerical Modeling

There is no detection limit of the central wavelength shift of the MAFBS in the numerical model, but it depends on an analog-to-digital converter accuracy. We use the classical approach to transform equations system into dimensionless quantities, and the characteristic quantities are defined as dimensional frequency Ω^0 and dimensional amplitude A^0. As the characteristic dimensional frequency of task Ω^0, we set the frequency corresponding to 125 GHz (in wavelength terms, it is 1 nm). The characteristic dimensional amplitude A^0 depends on the maximum output current of the photodetector, which can be independently amplified or attenuated to any value. We normalize all the task variables, so that the maximum central frequency shifting of the MAFBS does not exceed 125 GHz or Ω^0, which is equal to 1 conventional unit. We normalize the maximum signal amplitude, so that in dimensionless quantities the maximum signal level does not exceed 1000 conventional units. Based on this, we define the parameters of optic filter with an oblique linear amplitude–frequency characteristic as $k = 1000$ conventional units and $b = 100$ conventional units. As the MAFBS model, we choose a structure with address frequencies $\Omega_{21} = 0.01$ conventional units (1.25 GHz) and $\Omega_{32} = 0.02$ conventional units (2.50 GHz), with a range of changes in the MAFBS central frequency to 1 conventional unit.

The relative error in determining the MAFBS central frequency is determined by the formula:

$$\varepsilon(D_0, D_{12}, D_{13}, D_{23}, E_F) = \frac{|\hat{\omega}_{Br} - \omega_{Br}|}{\omega_{Br}}. \tag{14}$$

where $\hat{\omega}_{Br} = \hat{\omega}_{Br}(D_0, D_{12}, D_{13}, D_{32})$ is the central frequency of the MAFBS, calculated without error, and $\omega_{Br} = \omega_{Br}(D_0, D_{12}, D_{13}, D_{32}, E_F)$ is the central frequency, calculated with the error E_F of determining the amplitudes D_0, D_{21}, D_{32} and D_{31}.

All simulations were made for two values of an amplitude determination error—with the error E_F of 0.01% and 0.001% of the full scale. The dependence of the modulation factor on the central frequency shift of the MAFBS at amplitude determination error equal to 0.01% is shown in Figure 3a, and at 0.001% is shown in Figure 3b. The blue line indicates the modulation factor, and the red line is the spectral response of the filter with an oblique linear amplitude–frequency characteristic. As can be seen from the Figure 3, only high-precision amplitude measurement of the photodetector output current leads to acceptable accuracy of the MAFBS central frequency shift determination. Due to this fact, the subsequent development of microwave-photonic measuring systems based on the generalized modulation factor is a very difficult task, since it requires high accuracy in the amplitude determination of the output current of the photodetector. The relative error of MAFBS central frequency shifting determination does not exceed 10^{-1} and 10^{-2} for $E_F = 0.01\%$ and 0.01%, respectively. These values cannot be considered acceptable for high-precision measurements.

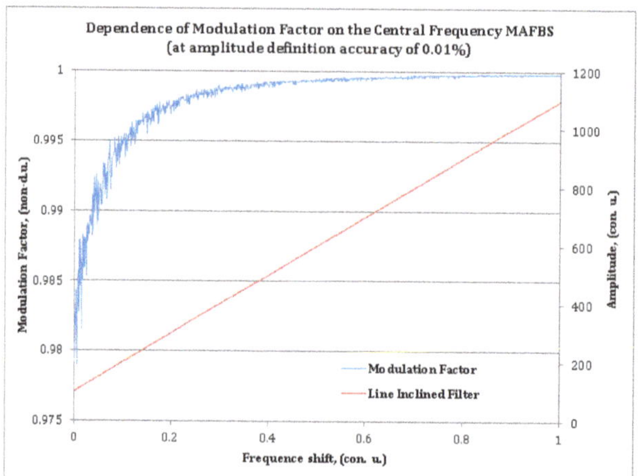

(a)

Figure 3. *Cont.*

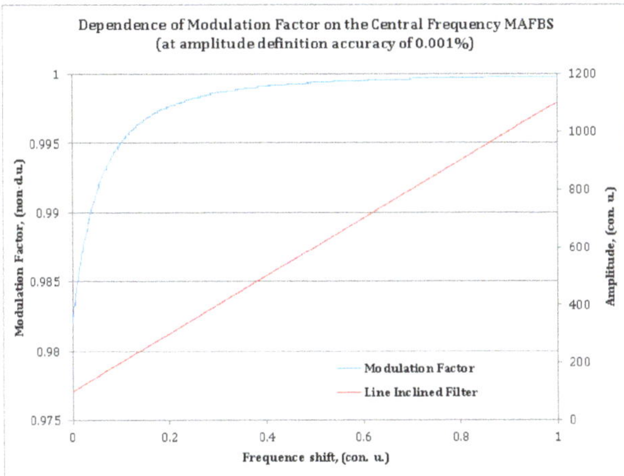

(b)

Figure 3. Modulation factor as the function of MAFBS central frequency shifting: The blue line is the dependence of the modulation factor; the red line is the spectral characteristic of the optic linear oblique filter, for an amplitude definition accuracy of (**a**) 0.01% and (**b**) 0.001%.

The second conclusion that follows from the modulation factor dependence is that this dependence does not provide uniformity of the measurement scale in the whole measurement range. The task of regularizing the measurement scale can be solved [17]; however, it requires additional complication of the optoelectronic scheme.

In Figure 4, violet, red, and brown lines show the calculated amplitudes of the carrier frequencies of the MAFBS obtained by numerically solving Equation (10) with the initial values in Equation (11). The blue line shows the relative error of central frequency shift determination of the MAFBS. Two calculation sets for E_F = 0.01% (Figure 4a) and E_F = 0.001% (Figure 4b) were made. The relative error of the central frequency shift determination of MAFBS, calculated via Equation (13), does not exceed 10^{-3} (for E_F = 0.01%) and 10^{-4} (for E_F = 0.001%), almost in the whole measurement range. The only exception is a small sector, where the amplitudes (A_1, A_2, and A_3) are close to zero.

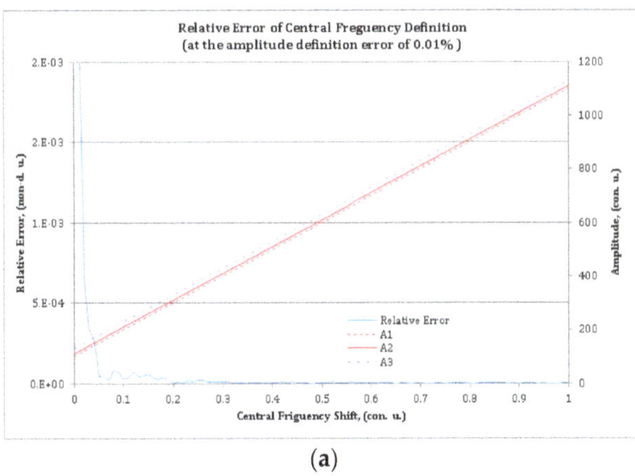

(a)

Figure 4. *Cont.*

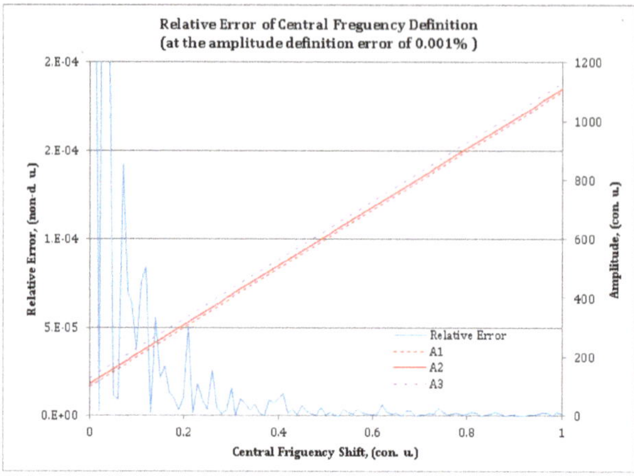

(**b**)

Figure 4. Relative error of the MAFBS central frequency definition for the amplitude definition error of (**a**) 0.01% and (**b**) 0.001%: The thick line is the relative error; the thin lines are the calculated amplitudes of the frequency components.

The simulation results clearly demonstrate that the proposed method of the MAFBS central frequency determination is two orders of magnitude more precise than the method based on the generalized modulation factor.

6. Conclusions

Studies have confirmed the perspectivity of MAFBS usage instead of AFBS for Microwave Photonics Sensor Systems (MPSS). The MAFBS multi-addressing idea is that in FBG structure, three (or more) narrow optical frequencies are formed, the difference between which is much less than an optical frequency (THz) and is located in the microwave range (GHz). The task of strict definition of central frequency shifting of MAFBS with the given address frequencies set is specified and solved. The equations system, which describes photodetector output current dependence from MAFBS central frequency shifting relative to the optic filter with oblique characteristics parameters, is written. The additional restrictions, connected with a microwave-photonic interrogation method and an optoelectronic scheme, were added to this equations system. We suggest the mathematical model based on this equations system, and the method of its solution, which allows us to define the central frequency shift unequivocally, according to the output current of a photodetector in MPSS, and as a result to define the values of applied physical fields. We made an attempt to use the generalization of modulation factor as a single measuring parameter to correlate it with the MAFBS central frequency shifting. It was found that generalization of modulation factor usage has a lot of disadvantages, since it requires high measurement accuracy of photodetector output current parameters, thus making this method unattractive.

Based on the mathematical model and its simulation, the task of strict definition of central frequency shifting of MAFBS, when the parameters of output photodetector current are measured with errors, was solved. It was shown that the suggested method of MAFBS usage in MPSS satisfies accuracy requirements, has potential for development, and allows us to move to an experimental research step.

Author Contributions: Conceptualization, O.M.; Formal analysis, A.K.; Funding acquisition, O.M. and A.K.; Investigation, A.S., V.A., R.M. and T.A.; Methodology, V.A.; Software, A.S.; Validation, A.K.; Writing—original draft, T.A.; Writing—review & editing, A.S. All authors have read and agreed to the published version of the manuscript.

Funding: This work was supported by Grant from the President of the Russian Federation for state support of young Russian scientists—candidates of sciences MK-3421.2019.8 (agreement No. 075-15-2019-309) and by RFBR, DST, NSFC, and NRF according to the research project 19-57-80006 BRICS_t.

Conflicts of Interest: The authors declare no conflict of interest.

References

1. Melle, S.; Liu, K. Wavelength demodulated Bragg grating fiber optic sensing systems for addressing smart structure critical issues. *Smart Mater. Struct.* **1992**, *1*, 36.
2. Davis, M.A.; Bellemore, D.G.; Kersey, A.D. Structural strain mapping using a wavelength/time division addressed fiber Bragg grating array. *Proc. SPIE* **1994**, *2361*, 342–345.
3. Matveenko, V.P.; Shardakov, I.N.; Voronkov, A.A.; Kosheleva, N.A.; Lobanov, D.S.; Serovaev, G.S.; Spaskova, E.M.; Shipunov, G.S. Measurement of strains by optical fiber Bragg grating sensors embedded into polymer composite material. *Struct. Control Health Monit.* **2017**, *25*, 1–11. [CrossRef]
4. Qiao, X.; Shao, Z.; Bao, W.; Rong, Q. Fiber Bragg grating sensors for the oil industry. *Sensors* **2017**, *17*, 429. [CrossRef]
5. Ma, Z.; Chen, X. Fiber Bragg gratings sensors for aircraft wing shape measurement: Recent applications and technical analysis. *Sensors* **2019**, *19*, 55. [CrossRef] [PubMed]
6. Karim, F. *Full Matlab Code for Synthesis and Optimization of Bragg Gratings*; Cambridge Scholars Publishing: Newcastle upon Tyne, UK, 2019; p. 24.
7. Cormier, G.; Boudreau, R.; Thériault, S. Real-coded genetic algorithm for Bragg grating parameter synthesis. *J. Opt. Soc. Am. B* **2001**, *18*, 1771–1776. [CrossRef]
8. Li, K. Review of the strain modulation methods used in fiber Bragg grating sensors. *J. Sens.* **2016**, *9*, 1284520. [CrossRef]
9. Koo, K.P.; LeBlanc, M.; Tsai, T.E.; Vohra, S.T. Fiber-chirped grating Fabry–Perot sensor with multiple-wavelength-addressable free-spectral ranges. *IEEE Photonics Technol. Lett.* **1998**, *10*, 1006–1008. [CrossRef]
10. Wei, Z.; Ghafouri-Shiraz, H.; Shalaby, H.M.H. New code families for fiber-Bragg-grating-based spectral-amplitude-coding optical CDMA systems. *IEEE Photonics Technol. Lett.* **2001**, *13*, 890–892.
11. Kataoka, N.; Wada, N.; Terada, Y.; Sakamoto, A.; Nishiwaki, K.; Nishide, K. Phase-shifted superstructured fiber Bragg grating. *Fujikura Tech. Rev.* **2011**, *40*, 6–11.
12. Triana, C.A.; Pastor, D.; Varon, M. Optical code division multiplexing in the design of encoded fiber Bragg grating sensors. *Pura Y Apl.* **2016**, *49*, 17–28. [CrossRef]
13. Triana, A.; Pastor, D. Interrogation of super-structured FBG sensors based on discrete prolate spheroidal sequences. *Proc. SPIE* **2017**, *10231*, 102310H.
14. Djordjevic, I.B.; Saleh, A.H.; Küppers, F. Design of DPSS based fiber bragg gratings and their application in all-optical encryption, OCDMA, optical steganography, and orthogonal-division multiplexing. *Opt. Express* **2014**, *22*, 10882–10897. [CrossRef]
15. Kim, Y.; Jeon, S.W.; Kwon, W.B.; Park, C.S. A wide dynamics and fast scan interrogating method for a fiber Bragg grating sensor network implemented using code division multiple access. *Sensors* **2012**, *12*, 5888–5895. [CrossRef]
16. Triana, A. A code division design strategy for multiplexing fiber Bragg grating sensing networks. *Sensors* **2017**, *17*, 2508. [CrossRef] [PubMed]
17. Sakhabutdinov, A.J. Radio-Photon Sensor Systems Based on Address Fiber Bragg Structures and their Application for Solving Practical Problems. Ph.D. Thesis, Kazan National Research Technical University, Kazan, Russia, 2018. (In Russian).
18. Morozov, O.G.; Sakhabutdinov, A.J. Addressed fiber Bragg structures in quasidistributed microwave-photonic sensor systems. *Comput. Opt.* **2019**, *43*, 535–543. [CrossRef]
19. Morozov, O.G.; Sakhabutdinov, A.J.; Nureev, I.I.; Misbakhov, R.S. Modelling and record technologies of address fibre Bragg structures based on two identical ultra-narrow gratings with different central wavelengths. *J. Phys.* **2019**, *1368*, 022049. [CrossRef]

20. Morozov, O.G.; Sakhabutdinov, A.J.; Nureev, I.I.; Misbakhov, R.S. Modelling and record technologies of address fiber Bragg structures based on gratings with two symmetrical pi-phase shifts. *J. Phys.* **2019**, *1368*, 022048. [CrossRef]
21. Novikova, V.A.; Konnov, K.A.; Gribaev, A.I.; Vargel, S.V. The Method of Forming a Fiber Bragg Grating with a Phase Shift. Patent RF No. 2676191, 24 January 2018. (In Russian).
22. Levenberg, K. A Method for the Solution of Certain Non-Linear Problems in Least Squares. *Q. Appl. Math.* **1944**, *2*, 164–168. [CrossRef]
23. Marquardt, D. An Algorithm for Least-Squares Estimation of Nonlinear Parameters. *Siam J. Appl. Math.* **1963**, *11*, 431–441. [CrossRef]
24. Ypma, T.J. Historical development of the Newton–Raphson method. *Siam Rev.* **1995**, *37*, 531–551. [CrossRef]

sensors

MDPI

Article

Addressed Fiber Bragg Structures in Load-Sensing Wheel Hub Bearings

Timur Agliullin [1,*], Robert Gubaidullin [1], Airat Sakhabutdinov [1], Oleg Morozov [1], Artem Kuznetsov [1] and Valentin Ivanov [2]

[1] Department of Radiophotonics and Microwave Technologies, Kazan National Research Technical University named after A.N. Tupolev-KAI, 420111 Kazan, Russia; diablogrr@gmail.com (R.G.); kazanboy@yandex.ru (A.S.); microoil@mail.ru (O.M.); aakuznetsov@kai.ru (A.K.)

[2] Automotive Engineering Group, Technische Universität Ilmenau, 98693 Ilmenau, Germany; valentin.ivanov@tu-ilmenau.de

* Correspondence: taagliullin@mail.ru

Received: 7 September 2020; Accepted: 28 October 2020; Published: 30 October 2020

Abstract: The work presents an approach to instrument the load-sensing bearings for automotive applications for estimation of the loads acting on the wheels. The system comprises fiber-optic sensors based on addressed fiber Bragg structures (AFBS) with two symmetrical phase shifts. A mathematical model for load–deformation relation is presented, and the AFBS interrogation principle is described. The simulation includes (i) modeling of vehicle dynamics in a split-mu braking test, during which the longitudinal wheel loads are obtained, (ii) the subsequent estimation of bearing outer ring deformation using a beam model with simply supported boundary conditions, (iii) the conversion of strain into central frequency shift of AFBS, and (iv) modeling of the beating signal at the photodetector. The simulation results show that the estimation error of the longitudinal wheel force from the strain data acquired from a single measurement point was 5.44% with a root-mean-square error of 113.64 N. A prototype load-sensing bearing was instrumented with a single AFBS sensor and mounted in a front right wheel hub of an experimental vehicle. The experimental setup demonstrated comparable results with the simulation during the braking test. The proposed system with load-sensing bearings is aimed at estimation of the loads acting on the wheels, which serve as input parameters for active safety systems, such as automatic braking, adaptive cruise control, or fully automated driving, in order to enhance their effectiveness and the safety of the vehicle.

Keywords: microwave photonic sensor system; numerical simulation; addressed fiber Bragg structures; load-sensing bearings; vehicle dynamics control

1. Introduction

Various active safety systems such as the antilock braking system (ABS) and electronic stability control (ESC) have been used to increase driving safety for several decades [1,2]. However, their efficiency depends on the correctness of tire–road traction parameters that are currently estimated by means of indirect methods [3,4] which do not always provide sufficient accuracy, therefore limiting the capabilities of such systems for vehicle dynamics control.

In order to mitigate this issue, a significant research effort is aimed at the development of the corresponding direct methods, in particular, based on the real-time measurement of loads acting on the wheels, using a variety of sensor types. Theoretically, every component of the vehicle transferring the loads from the tire contact patch to the vehicle body can be used to estimate the wheel loads. Thus, four main categories of wheel force determination methods can be distinguished in accordance with the vehicle components they are based on: tire-based methods (including the ones that use piezoelectric [5–7], fiber-optic [8–10], or optical [11–13] sensors; electrical resistance strain gauges [14,15]; or accelerometers [16]); wheel disc-based methods (primarily used in testing equipment [7]); suspension-based methods (including strain measurement of suspension bushings [17], displacement measurement of knuckle [18]), and wheel hub bearing based methods (load-sensing bearings) [19–24]. Wheel force measurements based on load-sensing hub bearings have a number of advantages in comparison with the other approaches. Firstly, the sensors are located on a nonrotating ring of the bearing, therefore significantly simplifying data acquisition. Secondly, the hub bearings generally have a much longer lifespan than tires. Thirdly, the methods based on load-sensing bearings are robust in case of tire or wheel change. Finally, load-sensing hub bearings are located closer to the contact patch than other suspension components, therefore providing a more precise load estimation.

The current work aims to present a novel approach to instrument the load-sensing hub bearings with fiber-optic sensors suitable for automotive applications. The paper introduces a comprehensive model of the load measurement process and an experimental demonstration of the system operation.

2. Load-Sensing Bearings in Automotive Applications

Figure 1 illustrates the dependence of wheel loads on road conditions by presenting examples of the longitudinal wheel force as the function of the wheel slip ratio k for various conditions of the road surface [25]. As can be seen from the figure, the magnitude of the wheel force varies for different road conditions in the wide range.

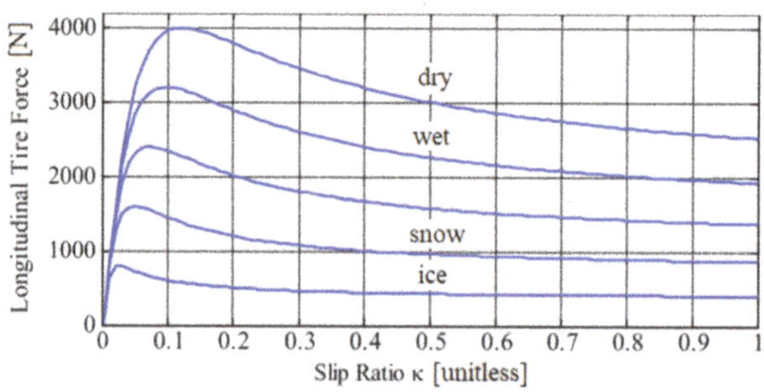

Figure 1. Dependence of longitudinal wheel force on the wheel slip ratio for different conditions of the road surface [25].

The wheel slip ratio k is determined using the Expressions (1) and (2) in case of braking and acceleration, respectively:

$$k = \frac{v_x - \omega_w r_w}{v_x}, \tag{1}$$

$$k = \frac{\omega_w r_w - v_x}{\omega_w r_w}, \tag{2}$$

where v_x is the longitudinal velocity of the vehicle, ω_w is the angular velocity of the wheel, and r_w is the wheel dynamic radius.

As it is demonstrated in Figure 1, the peak value of the longitudinal wheel force is reached at a certain wheel slip ratio, which corresponds to the most efficient braking or acceleration [26]. It must be noted that the wheel slip coinciding with the maximum wheel force is different for various road conditions, which further complicates its indirect estimation. When the vehicle brakes or accelerates, the load-sensing wheel hub bearings identify the maximum wheel force, and the slip ratio is corrected in such a way as to maintain the peak value of the wheel force, therefore enhancing the effectiveness of the wheel–road interaction [24]. It should be noted that some low-friction roads as well as deformable surfaces have no clearly pronounced extremum of the longitudinal force–slip-curve. In this case, the analysis of the curve slope should be considered by seeking the peak value. However, the presented paper does not include case studies with such surfaces.

In this work, we consider wheel hub bearings with nonrotating outer rings. The load caused by the wheel–road interaction is applied to the inner ring of the bearing, and then it is translated to the outer ring by means of the rolling elements (balls) and induces strain of the outer ring. This tangential deformation detected by a single sensor is periodic and sinusoidal even at constant load due to the passage of the rolling elements during bearing rotation [27]. Therefore, the sensor detects the maximum strain when its location coincides with the ball position, and the minimum strain is detected when the two adjacent rolling elements are equidistant from the sensor [22] (see Figure 2, where P_r is the radial load applied to the inner ring of the bearing).

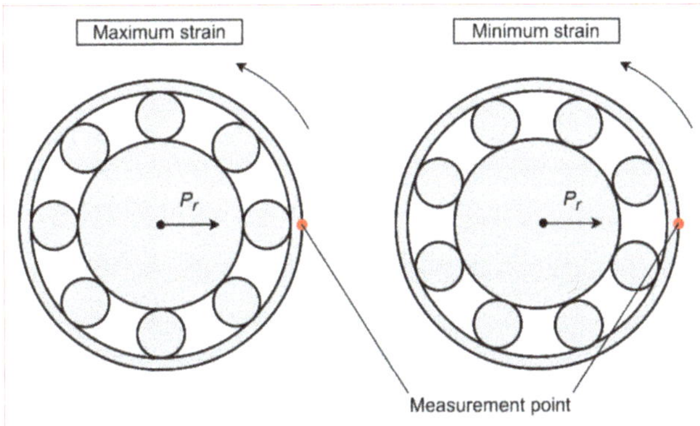

Figure 2. Sensor position relative to the rolling elements at maximum and minimum of measured strain.

The current work proposes the usage of fiber-optic sensors for strain measurement of the bearing outer ring. With all the intrinsic advantages of fiber-optic sensors based on fiber Bragg gratings (FBGs), such as low weight, small footprint, immunity to electromagnetic interference, and absence of electric power supply, a significant disadvantage still persists that is related to the usage of electro-optic interrogators. The complexity and high cost of such devices, which are designed to define the FBG central frequency by scanning its spectral response, constrain the areas of application of FBG-based sensor systems [28,29].

In order to solve the abovementioned issue, the addressed fiber Bragg structures (AFBSs) have been developed [24,27,30–35] and further expanded into multi-addressed fiber Bragg structures [29]. An AFBS is a fiber Bragg grating that has the optical frequency response with two narrowband components. The frequency spacing between the components is called the address frequency and lies in the radio frequency range. The system can include several AFBSs with the same Bragg frequency, and the sensors are distinguished using their address frequencies, which are unique for each sensor in the system and do not change when the sensors are subjected to strain or temperature fields. Two approaches for AFBS design have been proposed to date: 2λ-FBG and 2π-FBG. The 2λ-FBG is based on two sequentially formed ultra-narrowband FBGs with different Bragg frequencies [30,31]. An addressed structure with two symmetrical π-phase shifts (2π-FBG) consists of three FBGs divided by the discrete symmetrical phase shifts [24,27,32–34]. The 2π-FBGs are generally shorter in comparison with 2λ-FBGs, which makes them more suitable for applications where minimal sensor length is required, for instance, in load-sensing bearings, due to their relatively small diameter [30].

Thus, the application of 2π-FBGs as sensing elements of load-sensing hub bearings allows mitigating the disadvantages of the traditional FBGs, such as the high cost of interrogation devices, while maintaining the benefits of fiber-optic sensor technology and making the system suitable for automotive applications.

3. AFBS Interrogation Principle

Figure 3 presents the optic-electrical scheme of the interrogation system for two AFBSs of 2π-FBG type. A wideband light source (1) generates light radiation (diagram a) corresponding to the range of the frequency shift of the sensors. The light passes through a fiber-optic splitter (9) and enters the 2π-FBG-sensors (2.1 and 2.2), both of which form two-frequency radiations that are merged into a four-frequency optical response (diagrams b, c) by a fiber-optic coupler (10). After that, the optical signal is divided into two channels (the measuring one and the reference one) via a fiber-optic splitter (6). The measuring channel is equipped with an optical filter (3) with a predefined linear inclined frequency response, which converts the two-frequency radiation into the asymmetrical one (diagram d). After that, the signal is guided to the photodetector (4) and is processed by the measuring analog-to-digital converter (ADC) (5). The output signal of the ADC (5) is used to determine the central frequencies of the AFBS sensors. In the reference channel, the unmodified signal (diagram e) is transmitted directly to the reference photodetector (7) and then undergoes processing by the reference ADC (8). All the subsequent calculations for AFBS central frequency determination are carried out using the relations of the signal amplitudes from the outputs of the measuring and reference ADCs [27]. Therefore, the intensity of the output signal is normalized, eliminating the influence of the light source power fluctuations on the amplitudes of the AFBSs' optical spectrum response.

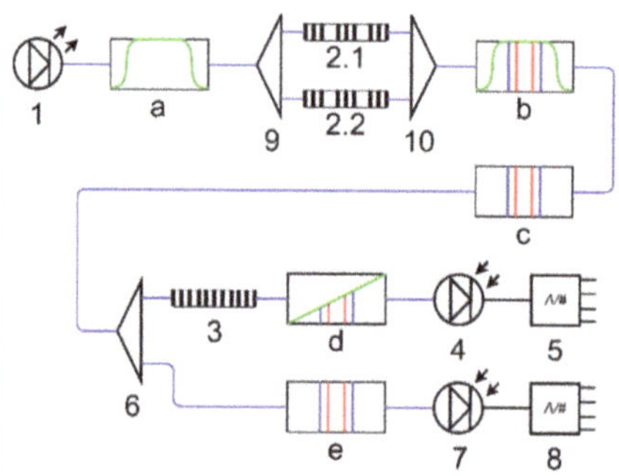

Figure 3. Structure of the interrogation system for two addressed fiber Bragg structure (AFBS)-based sensors: (1) wideband light source, (2.1, 2.2) AFBS sensors, (3) optical filter with a linear inclined frequency response, (4) photodetector of the measuring channel, (5) analog-to-digital converter (ADC) of measuring channel, (6,9) fiber-optic splitters, (7) photodetector of the reference channel, (8) ADC of the reference channel, (10) fiber-optic coupler, (a) spectrum of the wideband light source, (b,c) are the spectra of light propagated through the AFBS sensors, (d) spectra of AFBS sensors at the output of the optical filter, (e) spectra of AFBSs in the reference channel; blue connection lines denote optical fibers, black connection lines represent electrical wires.

The optical response from the i-th AFBS is represented as follows:

$$E_i(t) = A_i e^{j\omega_i t + \varphi_{Ai}} + B_i e^{j(\omega_i + \Omega_i)t + \varphi_{Bi}}, \tag{3}$$

where A_i and B_i are the amplitudes of the AFBS spectral components, ω_i is the frequency of the left spectral component of the i-th AFBS, Ω_i is the address frequency, and φ_{Ai} and φ_{Bi} are the phases. It must be noted that the proposed mathematical representation of the AFBS spectrum does not take into account the spectral shape of the transparency windows, which can be described using the Gaussian (in case of 2λ-FBG [31]) or Lorentz (in case of 2π-FBG [27]) functions.

The luminous power received by the photodetector from N addressed structures can be described using Expression (4) by multiplying Expression (3) with its complex conjugate:

$$\begin{aligned} P(t) = & \left(\sum_{i=1}^{N} E_i(t)\right)\left(\overline{\sum_{i=1}^{N} E_i(t)}\right) = \left(\sum_{i=1}^{N}\left(A_i e^{j\omega_i t+\varphi_{Ai}} + B_i e^{j(\omega_i+\Omega_i)t+\varphi_{Bi}}\right)\right)\left(\sum_{k=1}^{N}\left(A_k e^{-(j\omega_k t+\varphi_{Ak})} + B_k e^{-(j(\omega_k+\Omega_k)t+\varphi_{Bk})}\right)\right) = \\ & = \sum_{i=1}^{N}\left(A_i^2 + B_i^2\right) + 2\sum_{i=1}^{N} A_i B_i \cos(\Omega_i t + \varphi_{Ai} - \varphi_{Bi}) + \\ & + 2\sum_{i=1}^{N}\sum_{k=i+1}^{N}\left(\begin{array}{c} A_i A_k \cos((\omega_i - \omega_k)t + \varphi_{Ai} - \varphi_{Ak}) + \\ A_i B_k \cos((\omega_i - \omega_k - \Omega_k)t + \varphi_{Ai} - \varphi_{Bk}) + \\ B_i A_k \cos((\omega_i - \omega_k + \Omega_i)t + \varphi_{Bi} - \varphi_{Ak}) + \\ B_i B_k \cos((\omega_i - \omega_k + \Omega_i - \Omega_k)t + \varphi_{Bi} - \varphi_{Bk}) \end{array}\right). \end{aligned} \quad (4)$$

As can be seen, the oscillation of the output electrical signal of the photodetector at the address frequency Ω_i is proportional to the amplitudes of the AFBS optical spectral components A_i and B_i. The amplitudes A_i and B_i are defined by the parameters u and v of the linear function describing the frequency response of the optical filter with inclined frequency response ((3) in Figure 3):

$$A_i = L_0(u\omega_i + v), B_i = L_0(u(\omega_i + \Omega_i) + v), \quad (5)$$

where u is the slope ratio, v is the free term of the equation describing the frequency response of the optical filter, and L_0 is the amplitude of the optical spectral component of the AFBS prior to entering the filter with inclined linear frequency response. By measuring the amplitude of the photodetector output signal at the address frequency Ω_i, it is possible to define the central frequency shift (or the frequency of the left spectral component ω_i) of the AFBS. However, due to the appearance of the additional frequency components in the last sum of Expression (4), the filtering of the signal at the address frequencies is required.

4. Modeling of Vehicle Dynamics

As mentioned before, load-sensing bearings are used to assess the tire–road friction characteristics based on the wheel forces. In order to define the forces acting on the wheels in various conditions, consider a split-mu braking test, which was simulated using the CarSim software. A split-mu braking test is a common vehicle testing procedure that presupposes straight braking in a lane with significantly different frictional coefficients for the left and the right wheel paths. The visualization of the maneuver is shown in Figure 4a.

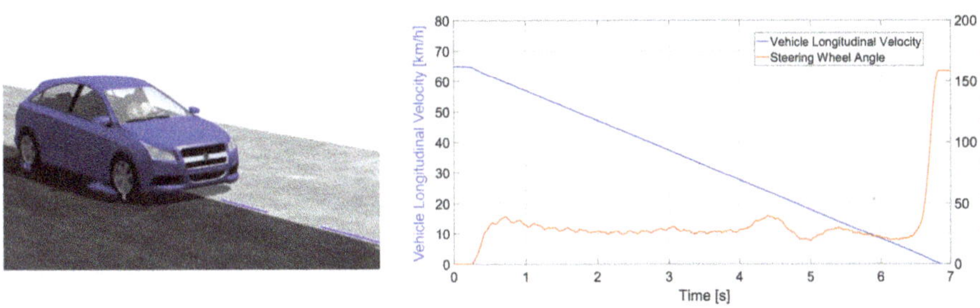

(a) (b)

Figure 4. (**a**) Visualization of the split-mu braking test. (**b**) Vehicle longitudinal velocity (blue line) and steering wheel angle (orange line) during the split-mu braking test.

The vehicle under test was a generic C-class hatchback with the following main parameters: a sprung mass of 1270 kg, a tire size of 215/55 R17, and a generic braking system with ABS. The friction of the left side of the road was $\mu_L = 0.2$, and the friction of the right side was $\mu_R = 0.5$, which imitates the possible road conditions at low temperatures. The maneuver included braking from the initial velocity of 65 km/h to a standstill with the constant pressure of 15 MPa applied to the brake master cylinder. The ABS was activated during the test in order to eliminate skidding. The profiles of the vehicle longitudinal velocity and steering wheel angle are presented in Figure 4b. As can be seen, the virtual driver applied a certain steering input in order to compensate for the yaw moment generated due to the inequality of left and right wheel forces.

The longitudinal wheel forces obtained from the simulation during the split-mu braking test are presented in Figure 5. Using these values, the corresponding deformation of the hub bearing outer ring can be calculated, which is described in the next section.

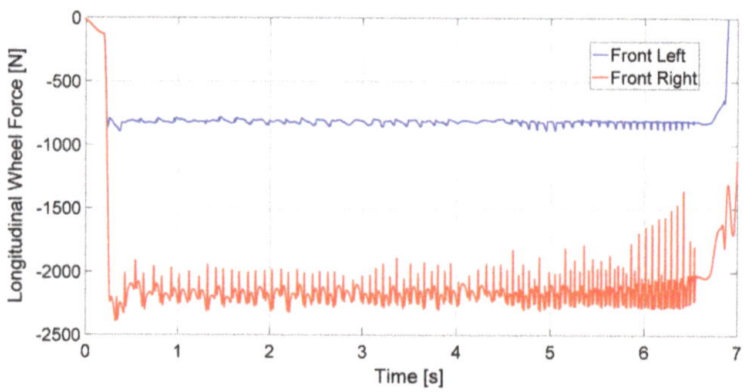

Figure 5. Longitudinal force of the front left wheel (blue line) and front right wheel (red line) during the split-mu braking test.

5. Modeling of Bearing Outer Ring Deformation

In order to estimate the tangential deformation of the outer ring of the load-sensing bearing caused by the loads acting on the wheels during the maneuver, consider a beam model with simply supported boundary conditions [23]. The model provides sufficient accuracy for preliminary estimations while maintaining simplicity of calculations [23]. A schematic representation of the beam model is shown in Figure 6, where the following denotations are used: P is the load transmitted to the outer ring by a ball, P_r is the radial load applied to the inner ring of the bearing, x is the position of the sensor (i.e., the point at which the strain is estimated), and a and b are the load positions relative to the left end and the right end of the beam, respectively.

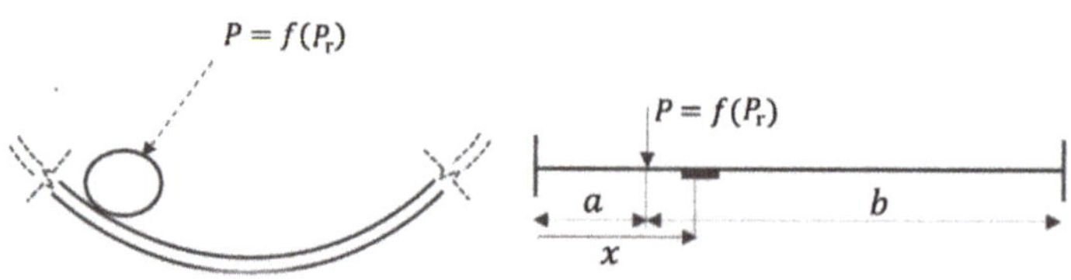

Figure 6. Beam model for a single-load case for estimation of the bearing outer ring deformation [23]: P is the load transmitted by a ball, P_r is the load applied to the inner ring of the bearing, x is the sensor position, and a and b are the load positions relative to the left end and the right end of the beam, respectively.

As mentioned in Section 2, the strain of the outer ring measured by a single sensor is periodic due to the passage of rolling elements. The maximum values of strain are achieved when the position of the ball coincides with the position of the sensor, and the sensor is located in the middle of a beam, i.e., $a = b = x$, while the minimum strain is detected when the sensor is positioned in the middle between two adjacent balls, i.e., $x - a = \beta/2$, where β is the arc length between the positions of two subsequent balls. In order to define the load applied by the rolling element to the beam, we use a generally accepted relation found by Stribeck in 1900, according to which the load on the most-loaded element is 4.37 times higher than the average load distributed on the bearing balls [36]:

$$P = 4.37\frac{P_r}{n}, \tag{6}$$

where n is the number of balls in a single row of a bearing.

A single load acting on the bearing ball in a beam model with simply supported boundary conditions is expressed as follows [23]:

$$P = -\frac{L}{bx}\frac{S_x EI}{y}, \tag{7}$$

where S_x is the tangential strain at 'x' distance from the left end of a beam; E is Young's modulus; I is the area moment of inertia of a cross-section of the beam; y is distance from the neutral axis, where the strain is calculated (i.e., half of the outer ring thickness); and $L = a + b$ is the beam length (i.e., half of the perimeter of the outer ring circumference).

Considering $a = b = x = L/2$ for the maximum case and $x = L/2$, $b = L/2 - L/n$ for the minimum case, the corresponding values of strain can be estimated based on Expression (7):

$$S_{x,\max} = -\frac{(L/2)^2}{L}\frac{yP}{EI}, \tag{8}$$

$$S_{x,\min} = -\frac{(n-2)L}{4n}\frac{yP}{EI}. \tag{9}$$

Using Expressions (8) and (9), the tangential deformation of the bearing outer ring was calculated for the whole duration of the maneuver. The results are presented in Figure 7a for the whole duration of the simulated testing procedure and in Figure 7b for a shorter time interval. As can be seen from the figure, the sensor on the front right hub bearing is subjected to significantly higher strain in comparison with the left one due to higher loads acting on the right wheel caused by higher road traction.

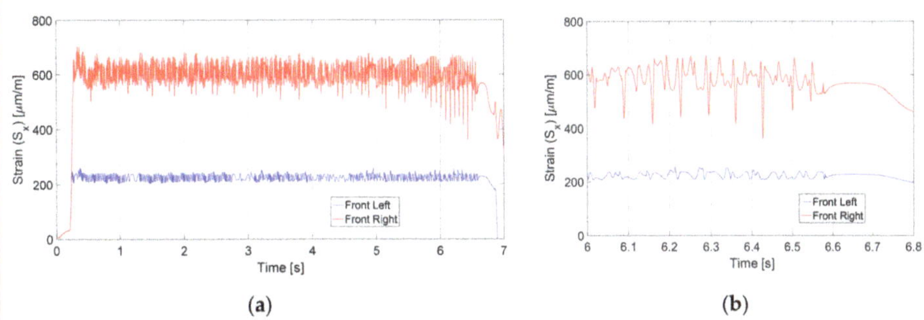

Figure 7. Strain simulated for the front left wheel hub bearing (blue line) and front right wheel hub bearing (red line): (**a**) for the whole duration of the split-mu braking test; (**b**) for t = [6, 6.8] s.

A system based on load-sensing bearings is aimed at solving the inverse problem, i.e., the calculation of force using the measured values of strain. In order to assess the estimation accuracy for a system with a single strain-sensor, an upper envelope of the simulated strain was obtained for the front right wheel hub of the simulated vehicle, which is shown in Figure 8a. After that, the longitudinal wheel force was estimated using Expressions (6) and (7) considering only the maximum strain case. The results are illustrated in Figure 8b, and the estimation error found as the difference between simulated and estimated forces is presented in Figure 8c. The high overshoot of error at the beginning of simulation results from linear interpolation of strain due to the relatively long time interval as well as rapid strain raise between the subsequent strain peaks. The root-mean-square error for the whole duration of the simulation was 113.64 N, which indicates a deviation of 5.44% from the average force of 2087.15 N.

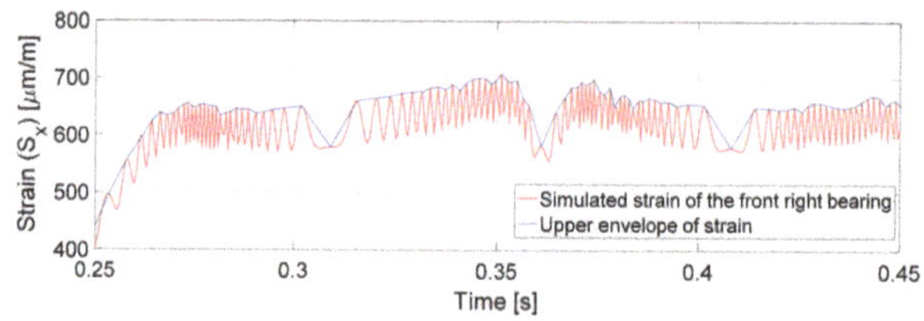

(a)

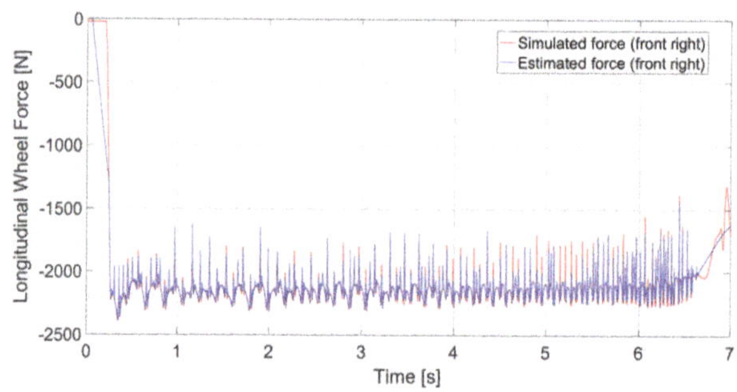

(b)

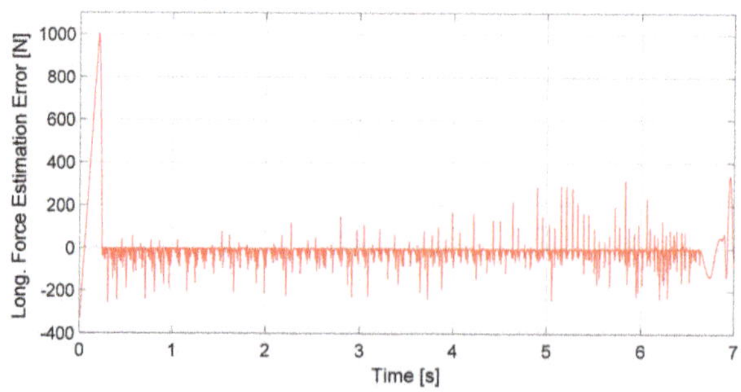

(c)

Figure 8. Assessment of the force estimation accuracy by solving the inverse problem: (**a**) strain simulated for the front right hub bearing (red line) and its upper envelope (blue line) shown for a short time interval; (**b**) longitudinal wheel force obtained from the simulation (red line) and estimated force (blue line) for the whole duration of the maneuver; (**c**) force estimation error.

6. Modeling of AFBS Interrogation

Consider a 2π-FBG structure with the address frequency Ω = 3.75 GHz as the sensing element for strain measurement of the bearing outer ring. The simulations presented in this section were performed using the OptiSystem 7.0 software. The optoelectronic interrogation scheme implemented for the simulation is shown in Figure 9. The model does not include the reference channel shown in Figure 3, since the optical source is ideal and does not induce fluctuations of signal intensity. The AFBS spectrum was obtained using the OptiGrating software, and the left slope of the frequency response of the trapezoidal optical filter with the central frequency of 193.215 THz and bandwidth of 136.979 GHz was utilized as a filter with inclined frequency response. The abovementioned filter parameters provide a linear frequency response of the filter in the whole range of AFBS sensor frequencies.

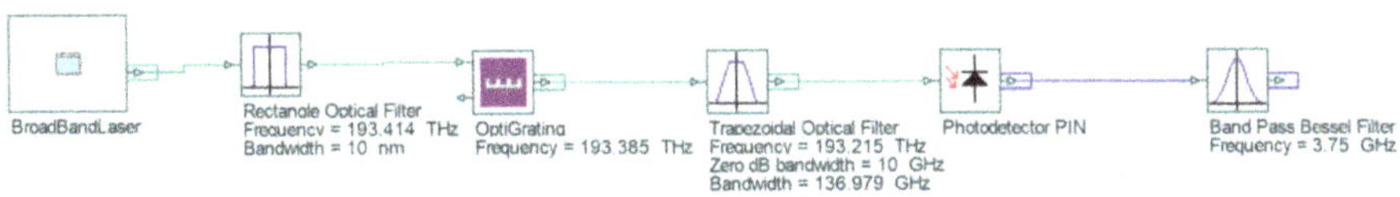

Figure 9. Simulation of the optoelectronic interrogation scheme in the OptiSystem software.

It is known that the central frequency shift of AFBS, similarly to conventional FBG, depends both on the applied strain and temperature. According to [37], the strain measurement can be represented as a function of the central frequency shift for the strain sensor and the temperature sensor:

$$\begin{gathered} g(\Delta f_{\mathrm{T}}, \Delta f_{\mathrm{P}}) = c_{2,3} \cdot \Delta f_{\mathrm{T}}^2 \cdot \Delta f_{\mathrm{P}}^3 + c_{2,2} \cdot \Delta f_{\mathrm{T}}^2 \cdot \Delta f_{\mathrm{P}}^2 + c_{2,1} \cdot \Delta f_{\mathrm{T}}^2 \cdot \Delta f_{\mathrm{P}} + \\ + c_{2,0} \cdot \Delta f_{\mathrm{T}} + c_{1,2} \cdot \Delta f_{\mathrm{T}} \cdot \Delta f_{\mathrm{P}}^2 + c_{1,1} \cdot \Delta f_{\mathrm{T}} \cdot \Delta f_{\mathrm{P}}^2 + c_{1,1} \cdot \Delta f_{\mathrm{T}} \cdot \Delta f_{\mathrm{P}} + \\ + c_{1,0} \cdot \Delta f_{\mathrm{T}} + c_{3,0} \cdot \Delta f_{\mathrm{P}}^3 + c_{2,0} \cdot \Delta f^2 + c_{0,1} \cdot \Delta f_{\mathrm{P}} + c_{0,0}, \end{gathered} \quad (10)$$

where Δf_{P} is the shift of the central frequency due to the deformation, Δf_{T} is the shift of the central frequency due to temperature exposure, and $c_{i,j}$ are calibration coefficients.

In the current simulation, the temperature of the sensor is assumed to be constant and its influence on the central frequency shift is neglected. The software used for modeling the vehicle dynamics is unable to estimate the temperature of bearings; therefore, further studies are required to assess their temperature. Several papers have reported thermal modeling of bearings [38,39]. Nevertheless, due to the relatively short duration of the testing procedure, the temperature variation of the bearing is not expected to exceed several degrees Celsius, which results in the central frequency shift that is at least by one order of magnitude smaller than the one caused by the bearing strain (considering typical FBG temperature sensitivity of approximately 10 pm/°C in terms of wavelength [40]). If the temperature influence is excluded from Equation (10), then the function relating the shift of the central frequency and deformation can be represented as follows:

$$g(\Delta f_{\mathrm{P}}) = c_3 \cdot \Delta f_{\mathrm{P}}^3 + c_2 \cdot \Delta f_{\mathrm{P}}^2 + c_1 \cdot \Delta f_{\mathrm{P}} + c_0, \quad (11)$$

where c_i are calibration coefficients. Applying Equation (11) to the simulation and taking into account a typical gauge factor of FBGs equal to 1.2 pm of wavelength shift per microstrain applied to the fiber [40], the central frequency of the AFBS is calculated for five cases: without strain, at t = 6.35 s from the beginning of the maneuver for both front left and front right bearings, and at t = 6.8 s for both front left and front right bearings. The diagram showing the relative positions of the AFBS spectra and the filter with an inclined frequency response for the abovementioned cases is presented in Figure 10a.

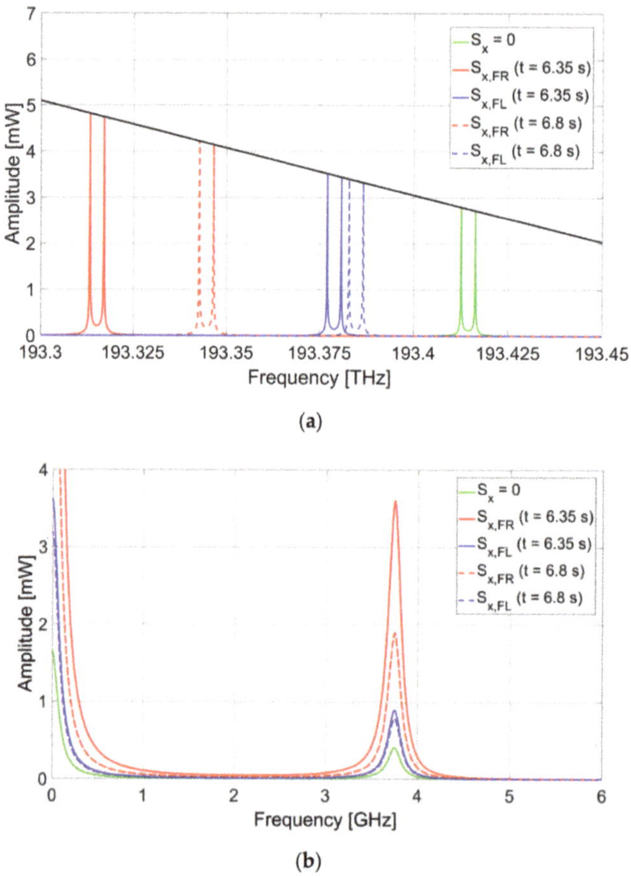

Figure 10. (**a**) AFBS spectra and the optical filter frequency response for five cases of applied strain: AFBS without strain (green line), front right bearing at $t = 6.35$ s (red solid line), front right bearing at $t = 6.8$ s (red dashed line), front left bearing at $t = 6.35$ s (blue solid line), and front left bearing at $t = 6.8$ s (blue dashed line); (**b**) Corresponding spectra of electrical signal at the photodetector.

For each of the five cases of the AFBS central frequency, the spectra of the electrical signal at the photodetector output were simulated. Figure 10b presents the spectra for each corresponding case. As demonstrated in Figure 10b, the amplitude of the electrical signal of the photodetector at the address frequency monotonically increases with the increase of the AFBS central frequency shift.

7. Experimental Results

The prototype bearing depicted in Figure 11a was instrumented with a single AFBS-sensor with the address frequency of 6.05 GHz. In order to ensure the uniformity of sensor strain and to eliminate lateral deformation of the sensor (since large lateral deformations applied to FBG impair the accuracy of the standard frequency-shift–strain relation of sensors [41]), a notch corresponding to the sensor length was made on the outer ring of the bearing. For preliminary testing, a static load was applied to the bearing by means of a mechanical press shown in Figure 11b, and the inner ring of the bearing was rotated using an electric screwdriver with the rotational speed of 360 rpm. The resulting amplitude of the measuring channel relative to the reference channel is presented in Figure 11c.

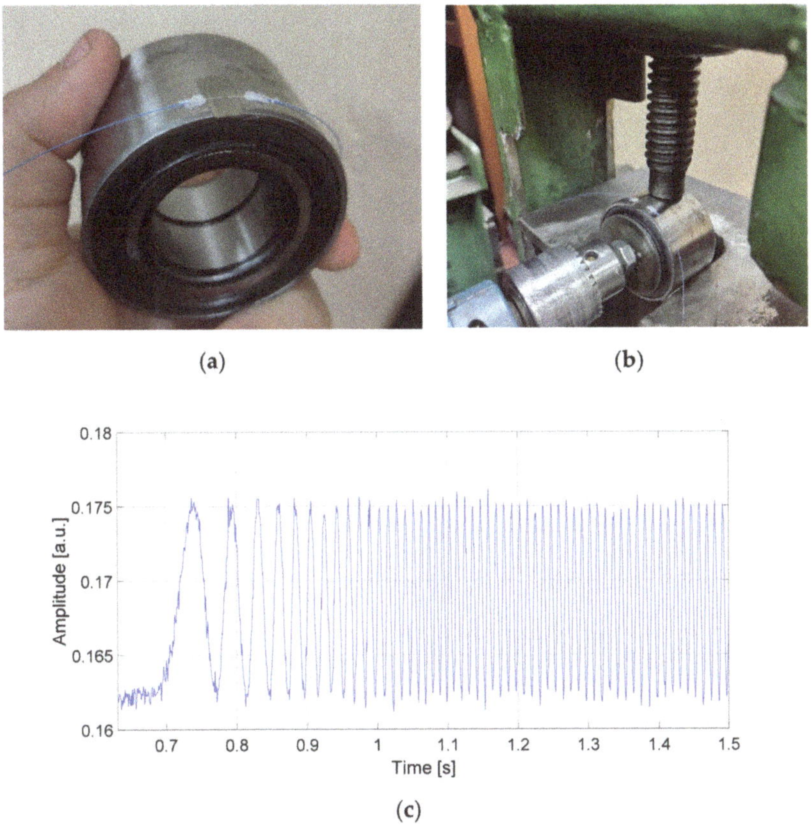

Figure 11. Static load test: (**a**) prototype load-sensing bearing with AFBS sensor; (**b**) experimental setup with a mechanical press for static load test; (**c**) relative amplitude of the resulting beating signal at the photodetector during the static load test.

For the dynamic testing, the experimental setup included a B-class passenger vehicle with a prototype load-sensing bearing installed in the modified front right wheel hub with a notch to accommodate the sensor. The sprung mass of the testing vehicle was approximately 1200 kg, which is close to the one of the simulated vehicle.

The conditions of the split-mu braking test were unavailable at the time when experimental studies were carried out; therefore, straight-line braking to a standstill from the initial velocity of 30 km/h was chosen as a testing procedure. The testing was performed on a private driveway with dry asphalt pavement and air temperature of 21 °C.

Figure 12 illustrates the results of the tangential strain estimation for the whole duration of the testing maneuver. As can be seen, the obtained experimental values are comparable with the simulation data. Higher experimental values of strain are mainly caused by higher road traction during the testing in comparison with the simulated conditions.

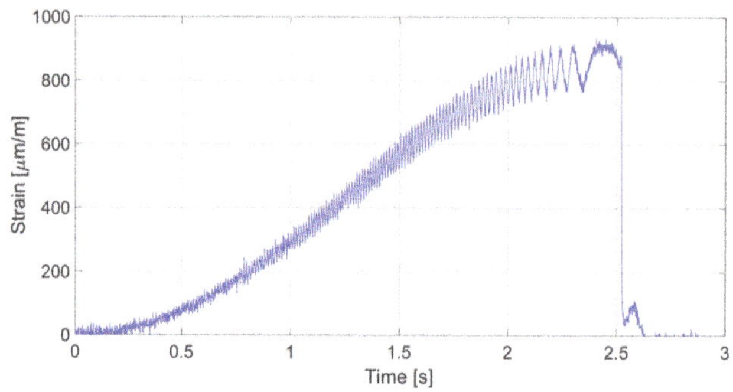

Figure 12. Tangential strain of the bearing measured by AFBS sensor for the whole duration of the dynamic testing procedure.

The experimental results confirm the applicability of the proposed measurement approach for automotive load-sensing bearings. Further studies are planned to increase the number of sensors used in the system for more comprehensive wheel force measurements. Moreover, due to the temperature sensitivity of the AFBS-based strain sensors, the temperature compensation is required for prolonged testing procedures and real-world operation of the system. This can be realized by means of at least one additional AFBS sensor located in proximity to the other sensors and isolated from strain, which will be considered in further studies.

8. Conclusions

The work presents a theoretical and experimental investigation of a novel usage of AFBS in load-sensing bearings for automotive applications. The two-frequency optical spectral response of the AFBS results in a beating signal at the output of the photodetector, the amplitude of which can be used to explicitly define the central frequency shift of the AFBS sensor. According to the simulation results, the estimation of the longitudinal wheel force had an error of 5.44% when the strain data were acquired by a single sensor, with the root-mean-square error of 113.64 N. The experimental results of the strain measurement are fully comparable with the simulation data. Further research will be conducted with an increased number of sensors providing measurements of wheel forces in various directions as well as ensuing temperature compensation for prolonged testing procedures.

Author Contributions: Conceptualization, T.A., O.M. and V.I.; methodology, A.S. and A.K.; software, R.G.; validation, T.A., R.G. and A.K.; formal analysis, A.S.; investigation, T.A. and R.G.; resources, R.G.; data curation, A.S.; writing—original draft preparation, T.A.; writing—review and editing, A.S., O.M. and V.I.; visualization, R.G.; supervision, O.M.; project administration, A.S.; funding acquisition, O.M. and A.K. All authors have read and agreed to the published version of the manuscript.

Funding: T.A., R.G., A.S. and O.M. were funded by the Ministry of Science and Higher Education of the Russian Federation (Agreement No. 075-03-2020-051, Topic No. fzsu-2020-0020), A.K. was funded by a grant from the President of the Russian Federation for state support of young Russian scientists—candidates of sciences MK-3421.2019.8 (Agreement No. 075-15-2019-309).

Conflicts of Interest: The authors declare no conflict of interest.

References

1. Pretagostini, F.; Ferranti, L.; Berardo, G.; Ivanov, V.; Shyrokau, B. Survey on Wheel Slip Control Design Strategies, Evaluation and Application to Antilock Braking Systems. *IEEE Access* **2020**, *8*, 10951–10970. [CrossRef]
2. Aksjonov, A.; Augsburg, K.; Vodovozov, V. Design and Simulation of the Robust ABS and ESP Fuzzy Logic Controller on the Complex Braking Maneuvers. *Appl. Sci.* **2016**, *6*, 382. [CrossRef]

3. Viehweger, M.; Vaseur, C.; Van Aalst, S.; Acosta, M.; Regolin, E.; Alatorre, A.; Desmet, W.; Naets, F.; Ivanov, V.; Ferrara, A.; et al. Vehicle State and Tyre Force Estimation: Demonstrations and Gudelines. *Veh. Syst. Dyn.* **2020**, *232*, 1883–1930. [CrossRef]
4. Canudas-de-Wit, C.; Tsiotras, P.; Efstathios, V.; Michel, B.; Gissinger, G.L. Dynamic Friction Models for Road/Tire Longitudinal Interaction. *Veh. Syst. Dyn.* **2003**, *39*, 189–226. [CrossRef]
5. Jousimaa, O.J.; Xiong, Y.; Niskanen, A.J.; Tuononen, A.J. Energy harvesting system for intelligent tyre sensors. In Proceedings of the 2016 IEEE Intelligent Vehicles Symposium (IV), Gothenburg, Sweden, 19–22 June 2016. [CrossRef]
6. Hopping, K.; Augsburg, K.; Buchner, F. Extending the HSRI tyre model for large inflation pressure changes. In Proceedings of the Engineering for a Changing World: 59th IWK, Technische Universität Ilmenau, Ilmenau, Germany, 11–15 September 2017.
7. Vehicle Dynamics, Durability and Tire Testing. Kistler Group (2020). Available online: https://www.kistler.com/en/applications/automotive-research-test/vehicle-dynamics-durability/tire-testing/ (accessed on 24 August 2020).
8. Coppo, F.; Pepe, G.; Roveri, N.; Carcaterra, A. A Multisensing Setup for the Intelligent Tire Monitoring. *Sensors* **2017**, *17*, 00576. [CrossRef] [PubMed]
9. Carcaterra, A.; Roveri, N.; Gianluca, P. OPTYRE—A new technology for tire monitoring: Evidence of contact patch phenomena. *Mech. Syst. Signal Process* **2015**, *66–67*, 793–810. [CrossRef]
10. Roveri, N.; Pepe, G.; Mezzani, F.; Carcaterra, A.; Culla, A.; Milana, S. OPTYRE—Real Time Estimation of Rolling Resistance for Intelligent Tyres. *Sensors* **2019**, *19*, 5119. [CrossRef]
11. Xiong, Y.; Juhani, A. A multi-laser sensor system to measure rolling deformation for truck tyres. *Veh. Perform.* **2017**, *3*, 115–126. [CrossRef]
12. Xiong, Y.; Juhani, A. Rolling deformation of truck tires: Measurement and analysis using a tire sensing approach. *J. Terramechanics* **2015**, *61*, 33–42. [CrossRef]
13. Matsuzaki, R.; Hiraoka, N.; Todoroki, A.; Mizutani, Y. Strain Monitoring and Applied Load Estimation for the Development of Intelligent Tires Using a Single Wireless CCD Camera. *J. Solid Mech. Mater. Eng.* **2012**, *6*, 935–949. [CrossRef]
14. Mendoza-Petit, M.F.; Garcia-Pozuelo, D.; Olatunbosun, O.A. A Strain-Based Method to Estimate Tire Parameters for Intelligent Tires under Complex Maneuvering Operations. *Sensors* **2019**, *19*, 2973. [CrossRef] [PubMed]
15. Mendoza-Petit, M.F.; Garcia-Pozuelo, D.; Díaz, V.; Olatunbosun, O.A. A Strain-Based Intelligent Tire to Detect Contact Patch Features for Complex Maneuvers. *Sensors* **2020**, *20*, 1750. [CrossRef] [PubMed]
16. Singh, K.B.; Taheri, S. Accelerometer Based Method for Tire Load and Slip Angle Estimation. *Vibration* **2019**, *2*, 11. [CrossRef]
17. Suzuki, M.; Nakano, K.; Miyoshi, A.; Katagiri, A.; Kunii, M. Method for Sensing Tire Force in Three Directional Components and Vehicle Control Using This Method. *SAE Tech. Paper 2007-01-0830 2007* **2007**. [CrossRef]
18. Ohkubo, N.; Horiuchi, T.; Yamamoto, O.; Inagaki, H. Brake Torque Sensing for Enhancement of Vehicle Dynamics Control Systems. *SAE Tech. Paper 2007-01-0867* **2007**. [CrossRef]
19. Den Engelse, J. Estimation of the Lateral Force, Acting at the Tire Contact Patch of a Vehicle Wheel, Using a Hub Bearing Unit Instrumented with Strain Gauges and Eddy-current Sensors. Master's Thesis, Delft University of Technology, Delft, The Netherlands, 2013.
20. Kerst, S.; Shyrokau, B.; Holweg, E. Reconstruction of Wheel Forces Using an Intelligent Bearing. *SAE Int. J. Passeng. Cars–Mech. Syst.* **2016**, *9*, 196–203. [CrossRef]
21. Nishikawa, K. Hub Bearing with Integrated Multi-axis Load Sensor. *Tech. Rev.* **2011**, *79*, 58–63.
22. Kerst, S.; Shyrokau, B.; Holweg, E. A model-based approach for the estimation of bearing forces and moments using outer-ring deformation. *IEEE Trans. Ind. Electron.* **2019**, *67*, 461–470. [CrossRef]
23. Gandhi, N. Load Estimation and Uncertainty Analysis Based on Strain Measurement: With Application to Load Sensing Bearing. Master's Thesis, Delft University of Technology, Delft, The Netherlands, 2013.
24. Sakhabutdinov., A.Z.; Agliullin, T.A.; Gubaidullin, R.R.; Morozov, O.G.; Ivanov, V.; Sakhabutdinov, A.Z.; Agliullin, T.A.; Gubaidullin, R.R.; Morozov, O.G.; Ivanov, V. Numerical Modeling of Microwave-Photonic Sensor System for Load Sensing Bearings. In Proceedings of the 2020 Wave Electronics and its Application in Information and Telecommunication Systems (WECONF), Saint-Petersburg, Russia, 1–5 June 2020. [CrossRef]

25. Dincmen, E.; Güvenç, B.A.; Acarman, T. Extremum-Seeking Control of ABS Braking in Road Vehicles with Lateral Force Improvement. *IEEE Trans. Control Syst. Technol.* **2014**, *22*, 230–237. [CrossRef]
26. Pacejka, H. *Tire and Vehicle Dynamics*, 3rd ed.; Butterworth-Heinemann: Oxford, UK, 2012; pp. 87–147.
27. Sahabutdinov, A.Z.; Morozov, O.G.; Agliullin, T.A.; Gubaidullin, R.R.; Ivanov, V. Modeling of Spectrum Response of Addressed FBG-Structures in Load Sensing Bearings. In Proceedings of the 2020 Systems of Signals Generating and Processing in the Field of on Board Communications, Moscow, Russia, 19–20 March 2020. [CrossRef]
28. Sakhabutdinov, A.Z.; Morozov, O.G.; Morozov, G.A. Universal Microwave Photonics Approach to Frequency-Coded Quantum Key Distribution. In *Advanced Technologies of Quantum Key Distribution*; Gnatyuk, S., Ed.; IntechOpen: London, UK, 2018. [CrossRef]
29. Morozov, O.; Sakhabutdinov, A.; Anfinogentov, V.; Misbakhov, R.; Kuznetsov, A.; Agliullin, T. Multi-Addressed Fiber Bragg Structures for Microwave-Photonic Sensor Systems. *Sensors* **2020**, *20*, 2693. [CrossRef]
30. Gubaidullin, R.R.; Sahabutdinov, A.Z.; Agliullin, T.A.; Morozov, O.G.; Ivanov, V. Application of Addressed Fiber Bragg Structures for Measuring Tire Deformation. In Proceedings of the 2019 Systems of Signal Synchronization, Generating and Processing in Telecommunications (SYNCHROINFO), Yaroslavl, Russia, 1–3 July 2019. [CrossRef]
31. Gubaidullin, R.R.; Agliullin, T.A.; Nureev, I.I.; Sahabutdinov, A.Z.; Ivanov, V. Application of Gaussian Function for Modeling Two-Frequency Radiation from Addressed FBG. In Proceedings of the 2020 Systems of Signals Generating and Processing in the Field of on Board Communications, Moscow, Russia, 19–20 March 2020. [CrossRef]
32. Agliullin, T.A.; Gubaidullin, R.R.; Ivanov, V.; Morozov, O.G.; Sakhabutdinov, A.Z. Addressed FBG-structures for tire strain measurement. In Proceedings of the SPIE 11146, Optical Technologies for Telecommunications 2018, Ufa, Russian Federation, 20–22 November 2018; p. 111461. [CrossRef]
33. Morozov, O.G.; Sakhabutdinov, A.J. Addressed fiber Bragg structures in quasidistributed microwave-photonic sensor systems. *Comput. Opt.* **2019**, *43*, 535–543. [CrossRef]
34. Morozov, O.G.; Sakhabutdinov, A.Z.; Nureev, I.I.; Misbakhov, R.S. Modelling and record technologies of address fibre Bragg structures based on gratings with two symmetrical pi-phase shifts. *J. Phys. Conf. Ser.* **2019**, *1368*, 022048. [CrossRef]
35. Gubaidullin, R.R.; Agliullin, T.A.; Morozov, O.G.; Sahabutdinov, A.Z.; Ivanov, V. Microwave-Photonic Sensory Tire Control System Based on FBG. In Proceedings of the 2019 Systems of Signals Generating and Processing in the Field of on Board Communications, Moscow, Russia, 20–21 March 2019. [CrossRef]
36. Oswald, F.B.; Zaretsky, E.V.; Poplawski, J.V. Effect of Internal Clearance on Load Distribution and Life of Radially Loaded Ball and Roller Bearings. *Tribol. Trans.* **2012**, *55*, 245–265. [CrossRef]
37. Sakhabutdinov, A.Z.; Nureev, I.I.; Morozov, O.G.; Kuznetsov, A.A.; Faskhutdinov, L.M.; Petrov, A.V.; Kuchev, S.M. Calibration of combined pressure and temperature sensors. *Int. J. Appl. Eng. Res.* **2015**, *10*, 44948–44957.
38. Gupta, P.K.; Taketa, J.I.; Price, C.M. Thermal interactions in rolling bearings. *Proc. Inst. Mech. Eng. Part J J. Eng. Tribol.* **2019**, *234*, 1233–1253. [CrossRef]
39. Kumaran, S.S.; Velmurugan, P.; Tilahun, S. Effect on stress and thermal analysis of tapered roller bearings. *J. Crit. Rev.* **2020**, *7*, 492–501. [CrossRef]
40. Fajkus, M.; Nedoma, J.; Martinek, R.; Vasinek, V.; Nazeran, H.; Siska, P. A Non-Invasive Multichannel Hybrid Fiber-Optic Sensor System for Vital Sign Monitoring. *Sensors* **2017**, *17*, 111. [CrossRef]
41. Lai, M.; Karalekas, D.; Botsis, J. On the Effects of the Lateral Strains on the Fiber Bragg Grating Response. *Sensors* **2013**, *13*, 2631–2644. [CrossRef]

Publisher's Note: MDPI stays neutral with regard to jurisdictional claims in published maps and institutional affiliations.

sensors

MDPI

Article

Generation of Vortex Optical Beams Based on Chiral Fiber-Optic Periodic Structures

Azat Gizatulin *, Ivan Meshkov, Irina Vinogradova, Valery Bagmanov, Elizaveta Grakhova and Albert Sultanov

Department of Telecommunications, Ufa State Aviation Technical University, 450008 Ufa, Russia; mik.ivan@bk.ru (I.M.); vil-4@mail.ru (I.V.); bagmanov.valeriy@yandex.ru (V.B.); eorlingsbest@mail.ru (E.G.); tks@ugatu.ac.ru (A.S.)

* Correspondence: azat_poincare@mail.ru

Received: 22 August 2020; Accepted: 17 September 2020; Published: 18 September 2020

check for updates

Abstract: In this paper, we consider the process of fiber vortex modes generation using chiral periodic structures that include both chiral optical fibers and chiral (vortex) fiber Bragg gratings (ChFBGs). A generalized theoretical model of the ChFBG is developed including an arbitrary function of apodization and chirping, which provides a way to calculate gratings that generate vortex modes with a given state for the required frequency band and reflection coefficient. In addition, a matrix method for describing the ChFBG is proposed, based on the mathematical apparatus of the coupled modes theory and scattering matrices. Simulation modeling of the fiber structures considered is carried out. Chiral optical fibers maintaining optical vortex propagation are also described. It is also proposed to use chiral fiber-optic periodic structures as sensors of physical fields (temperature, strain, etc.), which can be applied to address multi-sensor monitoring systems due to a unique address parameter—the orbital angular momentum of optical radiation.

Keywords: fiber Bragg gratings; chiral structures; orbital angular momentum; apodization; chirp; coupled modes theory

---

1. Introduction

Nowadays, the demand for broadband multimedia services is still growing, which leads to an increase of transmitted data volume as part of the development of the digital economy and the expansion of the range of services (video conferencing, telemedicine, online broadcasting, streaming, etc.). This circumstance requires the development of broadband access technologies in both wired (fiber optic, copper) and wireless networks, however, the existing ways to increase the throughput of communication systems (based on the use of time, frequency, polarization, space as a multiplexing domains) face the theoretical Shannon's throughput limit, which forces researchers to look for alternative physical parameters of electromagnetic (EM) waves that can be used to transmit information. For example, within the framework of 5G technology, the possibility of using non-orthogonal multiple access (NOMA) is being explored [1,2], which implies ranging and differentiating subscriber signals by power. However, from the point of view of the physical properties of electromagnetic waves, in recent years, the orbital angular momentum (OAM) of EM waves, which defines the vortex dislocation of the signal wavefront, has been of great interest to researchers. This physical parameter was determined 30 years ago and has since found many applications, including also telecommunication applications in both optical and radio communication systems.

As it is known from classical and quantum physics, EM waves (and photons) carry both energy and momentum. The total momentum consists of momentum P and angular momentum L. In particular, angular momentum has an additional component related to polarization, spin angular momentum

(SAM), and the other component related to a spatial field distribution is called orbital angular momentum (OAM). In an optical vortex, the planes of the constant phase of electric and magnetic vector fields form twists or spirals that move in the direction of propagation. That is, the wavefront of the signal obtains a continuous helical structure rather than a set of parallel planes. Wave vorticity is not related to polarization and is another physical property of EM waves. Vortex is characterized by a number called a topological charge (or order) that represents the amount of turns per wavelength of the signal. The application fields of OAM in communication are vast, although there are some issues that need to be solved before full deployment of OAM-based systems. There are some unresolved problems in OAM generation and reception, as well as in practical implementation of related technologies, e.g., mode-hopping spread spectrum (MHSS), which is now under consideration in several research projects [3]. Taking full advantage of the OAM photon dimension to modulate or multiplex data can significantly improve the information capacity of a single photon, which can increase the bandwidth of single wavelength and single mode fiber.

One of the major challenges in the development of OAM-based communication systems is the generation of waves with helical wavefronts. The two main approaches for OAM optical signal generation are free-space optics and fiber-optic technology. Regarding free-space solutions, we can distinguish the following approaches: cylindrical lenses, spiral phase plates (SPP), holographic gratings, spatial light modulators (SLM), metamaterials, liquid crystal q-plates, computer-generated holograms etc. If you want to use these methods to generate OAM beams, you need to use supplementary tools; this is related to the spatial conversion of light and the coupling of optical beams from free space to optical fibers, which leads to problems arising from the complicated procedure of stabilization and alignment of the applied optomechanical devices, as well as their sensitivity to vibration, high cost, and high precision equipment. It is also worth noting that these tools can only be used in laboratory conditions. These methods are described in detail for example in [4,5]. Thus, the task of generating OAM beams in free space optics is generally solved, but it is not clear how to apply free space solutions to already deployed fiber links or future fiber systems.

In addition to free-space optics, several methods have been developed to generate OAM beams directly in the optical fiber. Four main approaches of the research in this area include (1) chiral fibers (including chiral fibers with Bragg gratings), (2) microstructured fibers, (3) photonic lanterns and (4) long-period fiber gratings (LPFGs). We will discuss these methods in more detail.

Helical Bragg grating is one of the simplest and obvious technical ways to create an optical fiber OAM beam. In [6] an optical vortex (OV) generation method is proposed using a helical fiber Bragg grating (H-FBG), which is written directly in a few-mode fiber (FMF). Spiral modulation of the refractive index is achieved by rotating the fiber under one-sided ultra violet radiation using a single-phase mask (in theory). The reflective properties of H-FBG are analyzed analytically. It has been shown that stable OVs can be achieved with precise fiber construction and that the OV order can be tuned by adjusting the resonant wavelength and controllability of H-FBG.

Bragg gratings consisting of a multimode fiber can excite a higher-order linear polarized (LP) mode ($l > 1$). Due to the helical structure of these grating arrays, higher-order modes can be generated with a simple laser model described in [6]. It is also shown in [7] that when using multimode fiber, the conversion of optical fibers with different topological charges: $0 \rightarrow \pm1$, $\pm1 \rightarrow 0$, $0 \rightarrow \pm2$, $0 \rightarrow \pm3$ with efficiency up to 97% can be achieved. According to [7], one can simply increase the number of fiber modes by changing the fiber parameters in order to excite a higher-order topological charge. For example, an increase in the diameter shown in [7] can enlarge the fiber core size providing the support of the higher-order modes. However, it is not theoretically clear from [6] and [7] how this grating provides proper mode coupling between fundamental and high-order modes.

OAM modes generation methods using conventional quartz fibers typically use additional devices such as polarization controllers (PCs) and mode splitters [8]. However, achieving a specific OAM mode by configuring PCs in all-fiber optic systems is often too difficult. It is even more convenient if you can use a dedicated fiber-optic device to couple the main (fundamental) mode to the desired

OAM mode directly. Several studies have shown that the possibility of exciting OAM modes directly with the help of a special microstructured optical fiber construction that is under investigation.

The fiber design utilizes two basic operating principles to convert the fundamental mode to higher-order OAM modes. One of them is developed on the basis of the coupled modes theory [9]. The authors proposed and investigated a tunable fiber optic microstructure to create different OAM modes through modeling based on the coupled modes theory. Microstructured optical fiber consists of a high refractive index ring and a hollow core surrounded by four small air holes as shown in [9]. The hollow core and surrounding four air holes are infiltrated by optical functional material whose refractive index can be physically changed (e.g., by voltage), causing conversion between the polarized fundamental mode and different OAM modes in different cycles of a high refractive index ring with normal operating wavelength.

Another way to generate the OAM mode in-fiber is to mimic the element of the repetitive spiral phase [10,11]. This method typically uses multicore fiber (MCF) [10] or photonic crystal fiber (PCF) [11]. For MCF, phase transformation is a function of the core refractive index. The phase difference between the nearest cores is exactly $2\pi l/N$, where l is the order of the required OAM order and N is the number of MCF cores. With such a phase differentiation, when spatial phase modulated beams are repeatedly connected to a ring core fiber (RCF), the OAM with the order of l is effectively generated.

The third approach is based on the use of so-called photonic lanterns (PLs)—spatial mode converters that connect unidirectional single-mode signals from multiple individual waveguide cores into a single multimode waveguide, fabricated either by using optical fibers or planar waveguides [12]. PL can be classified as a spatial multiplexer for spatial multiplexing systems because PL can be considered as an N-mode multiplexer, where the N is the SMF number used to manufacture the PL, which can actually provide the communication system capacity multiplication by factor N. For example, in [13] the authors show a method of creating OAM modes, based on a PL with a center of all threaded rings. The device consists of a 5-mode selective photonic lantern (MSPL), with an effective refractive index profile that is arranged to a ring shape. It has been shown that high-quality OAM beams are generated up to the second-order by simultaneous excitation of the degenerate linear polarized mode pairs of the MSPL.

Regarding fibers maintaining vortex propagation, several approaches have been developed. The earliest was based on a fiber proposed by Ramachandran [14] having a refractive index resembling a coaxial structure. The next approach of the OAM fiber-generation methods was based on the inverse parabolic profile of refractive index [15]. The reverse parabolic reverse grading index (IPGIF) fiber has been proposed for robust transmission of cylindrical vector modes as well as integer OAM carrying modes. Large isolation of an efficient index between vector modes in the mode group LP_{11} $\{TE_{01}, HE_{21}, TM_{01}\}$ ($>2.1 \times 10^{-4}$) is numerically and experimentally confirmed in this fiber in the whole C-band as well as the possibility of OAM transmission of +/−1 orders for distances up to 1.1 km. The authors also provide simple design optimization rules for redefining fiber parameters.

Finally, the last method considered was based on the LPFGs [16]. It has been shown that the OAM signal can be generated via twisting a strong modulated LPFG written in a four-mode fiber (4MF). With a special design and optimization of the procedures of CO_2-laser irradiation, an LPFG with strong period deformation is achieved in the 4MF. Based on this LPFG, one can directly convert the fiber fundamental mode (LP_{01}) to the high-order LP core mode (LP_{21}) with an efficiency of 99.7% and then transform the LP_{21} mode into a high-order vortex mode (±2 order).

The main disadvantage of the considered methods of generating OV is their relative complexity, which implies more precise doping and, for example, additional equipment for rotating the fiber when applying Bragg gratings or manufacturing PLs, as well as low modal purity due to the need to use precise equipment. For example, fiber methods such as LPFGs require real-time fiber rotation and other fiber optic tools such as PCs.

In our paper, we will focus on the fiber-optic method for OAM modes generation, based on the application of a chiral (vortex) fiber Bragg grating (ChFBG). The proposed method allows the creation of complete passive fiber-optic devices that can be used in telecommunication systems as OAM mode generators aiming to increase the throughput of communication channels by OAM modes multiplexing. For example, such modes can be applied for radio-optical systems implementation using both optical and radio vortexes [17–19] or in quantum computing [20].

2. The Theoretical Model for a Chiral (Vortex) Fiber Bragg Grating (ChFBG)

As mentioned above, the OAM signal generation in an optical fiber is currently an underinvestigated and urgent scientific and technical problem. One of the main goals in this field of research is to develop a full fiber-optic passive device that does not require any additional optical components such as polarization controllers (PC) in the case of LPFGs [16], mirrors, lenses, etc. In other words, the proposed device should be a complete fiber optic passive element. In this paper, we propose a method for generating a first-order OAM mode based on ChFBG, which in the general case is a continuous diffractive fiber structure with a spiral (helical) shape. Conventional fiber Bragg gratings, represented by a discrete set of fingers [21], can be defined, for example, using an approach similar to the tunneling effect description in quantum mechanics [22]. However, the most common approach to describe FBGs is the coupled modes theory. The OAM mode generation in this context means the conversion of the fundamental mode into a higher-order mode: in terms of vortex mode generation, it is necessary to achieve the highest possible conversion of the OAM_0 mode into OAM_1; therefore, the theoretical model of the ChFBG is based on the aforementioned coupled modes theory, considering ChFBG as a type of a mode-coupling device. It is known that the existence of vortex modes requires a few-mode fiber-optic regime since any non-zero order OAM mode is represented by a superposition of the degenerated TE (transverse electric) and TM (transverse magnetic) modes [23]. Thus, since G.652 fiber is multimode in the first spectral window only, the operation in the optical C-band (1530–1565 nm) requires special few-mode fibers, e.g., the step-index FMF by the Optical Fiber Solutions (OFS) company. For the fiber considered, the approximation of a weakly guiding fiber is applicable, i.e., a fiber in which the difference between the refractive indices of the core and the cladding is less than 1% (note that the most commercially available fibers are considered as weakly guiding). In this case, the apparatus of Bessel functions and LP-modes, mentioned in previous section, can be used to describe the mode composition (including OAM modes) of the optical field propagating through the optical fiber [23]. Hence, using the weakly guiding fiber approximation and Bessel-function formalism, the full-fiber OAM-generation problem statement can be formulated as follows: it is necessary to develop a grating that transforms the zero-order OAM mode (with flat wavefront—LP_{01}):

$$E_{01}(r,\varphi,z) = e^{-i\beta_{01}z}\frac{J_{01}\left(u_{01}\frac{r}{a}\right)}{J_{01}(u_{01})}, 0 \le r \le r_{co}$$

into the first-order OAM mode (superposition of two spatially shifted LP-modes: $LP^e_{11} + iLP^o_{11}$):

$$E_{11}(r,\varphi,z) = e^{i\beta_{11}z}\frac{J_{11}\left(u_{11}\frac{r}{a}\right)}{J_{11}(u_{11})}e^{i\varphi}, 0 \le r \le r_{co}$$

where r_{co} is the core radius (since the grating only exists in the fiber core), J_{01} and J_{11} are Bessel functions of 01 and 11 orders, respectively. The parameters u_{01} and u_{11} will be described further in the text.

From a functional point of view, the ChFBG can be considered as a four-pole circuit with a length from 0 to L (Figure 1a), described by a transfer matrix T; where a_0 is the incident radiation—Gaussian-like incoming field—LP_{01} (OAM = 0); b_0 is the reflected radiation—the aforementioned superposition $LP^e_{11} + iLP^o_{11}$ (OAM = 1); and a_1 is radiation passed through the grating; b_1 is the radiation incident on the grating from the receiver side; in the general case, radiation b_1 arises due to inevitable Rayleigh scattering or in case of duplex communication. Such radiation provides the generation of a reflected signal in the b_1 direction, which is an undesirable circumstance, since additional low-power radiation will appear in the grating, i.e., performing additional noise and re-reflections. One can easily remove undesirable b_1 by placing an optical isolator behind the grating (Figure 1b).

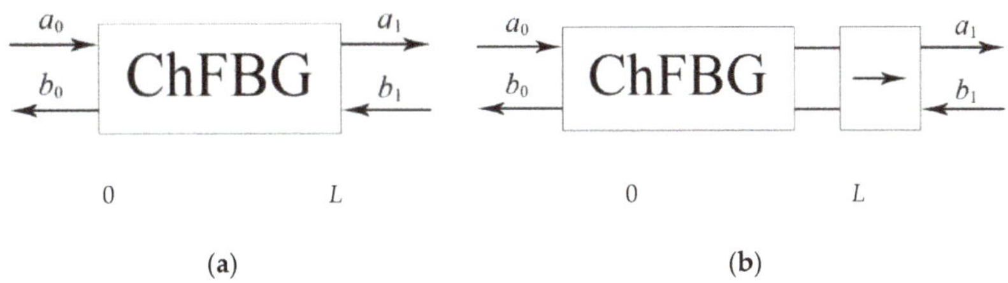

Figure 1. The schematic representation of the chiral (vortex) fiber Bragg grating (ChFBG): (**a**) in the form of a four-pole circuit; (**b**) with isolator for reflection mitigation.

The shape of the ChFBG is presented in Figure 2.

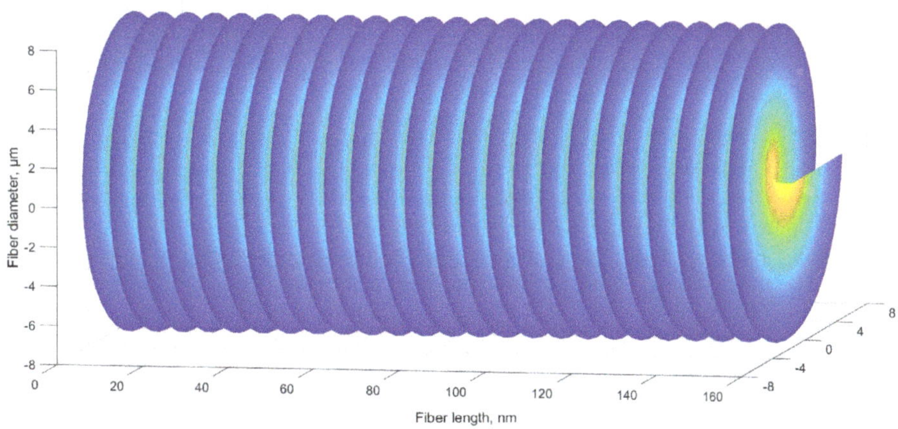

Figure 2. The schematic shape of the regular ChFBG.

Thus, expressing the radiation at the end of the grating ($z = L$) through the radiation at the beginning of the grating ($z = 0$), it can be defined as column vectors:

$$\begin{pmatrix} a_1 \\ b_1 \end{pmatrix} = T\begin{pmatrix} a_0 \\ b_0 \end{pmatrix} \tag{1}$$

The transmission matrix T is represented as follows:

$$T = \begin{pmatrix} T_{11} & T_{12} \\ T_{21} & T_{22} \end{pmatrix} = \begin{pmatrix} w & z \\ z* & w* \end{pmatrix}$$

In order to find the elements of the T matrix, we use the well-known approach described in [24,25], namely, the coupled modes theory [26,27], together with the model of field coupling under arbitrary perturbations of the refractive index of the fiber, considered in [28]. According to these approaches, field reflection and transmission coefficients are complex numbers determined through the matrix F, used in [27] to describe radiation of arbitrary diffraction structures, and associating radiation at the beginning of the grating $E(0)$ with radiation at the end of the grating $E(L)$: $E(0) = F \cdot E(L)$. In contrast to the approach proposed in [27], matrix T is defined by connecting the reflected and transmitted radiation through the incident radiation. Therefore, the matrix T is the inverse of the matrix F:$T = F^{-1}$. According to the matrix-converting rule and taking into account that determinant of the F matrix is unitary, we obtain:

$$T = \begin{pmatrix} T_{11} & T_{12} \\ T_{21} & T_{22} \end{pmatrix} = \begin{pmatrix} F_{22} & -F_{12} \\ -F_{21} & F_{11} \end{pmatrix} \quad (2)$$

Therefore, taking into account (1) and (2), we can state the relation between refracted and incoming light in the general case:

$$\begin{aligned} a_1 &= T_{11}a_0 + T_{12}b_0 = F_{22}a_0 - F_{12}b_0, \\ b_1 &= T_{21}a_0 + T_{22}b_0 = -F_{21}a_0 + F_{11}b_0 \end{aligned} \quad (3)$$

Assuming that $b_1 = 0$ (no backward radiation), it can be found from (3) that the equations for coupling modes in the case of a ChFBG remain the same as for a regular FBG, differing in the coupling coefficient only, therefore, with respect to [27] we can obtain:

$$\begin{aligned} b_0 &= wa_0 = a_0\frac{F_{22}}{F_{11}} = a_0\frac{-k_{ab}\sinh(\gamma L)}{\gamma\cosh(\gamma L)+i\Delta\beta\sinh(\gamma L)}, \\ a_1 &= z * a_0 = \frac{1}{F_{11}}a_0 = a_0\frac{\gamma e^{-i\beta_0 L}}{\gamma\cosh(\gamma L)+i\Delta\beta\sinh(\gamma L)} \end{aligned} \quad (4)$$

where the phase coefficient difference $\Delta\beta$ for counter-propagating modes is equal to [25]:

$$\Delta\beta = \beta_{01} + \beta_{11} - 2\beta_0$$

In relation (4), the complex coefficient w refers to the reflection coefficient, z^* is the transmission coefficient, β_0 is defined below and depending on Λ—the grating period, defined as $\Lambda = \lambda_B/2n_0$, where λ_B is the Bragg reflection wavelength; and β_{01} and β_{11} are the phase coefficients of the incident (OAM = 0) and reflected modes (OAM = 1), respectively. The parameter γ is defined as $\gamma^2 = k_{ab}^2 - \Delta\beta^2$, while k_{ab} is the complex overlap integral (which specifies the coupling coefficient between modes) determined by the inhomogeneity of the refractive index δn [28]:

$$k_{ab} = \frac{i\omega\varepsilon_0}{2}\int_{-\infty}^{\infty}\int_{-\infty}^{\infty}\delta n^2(r,\varphi)\vec{E}_a^*(r,\varphi)\cdot\vec{E}_b(r,\varphi)drd\varphi \quad (5)$$

or, in the case of LP modes:

$$k_{ab} = \frac{i\omega\varepsilon_0}{2}\int_{-\infty}^{\infty}\int_{-\infty}^{\infty}\delta n^2(r,\varphi)\vec{E}_{11}^*(r,\varphi)\cdot\vec{E}_{01}(r,\varphi)drd\varphi \quad (6)$$

It is easy to show that $|w|^2 + |z|^2 = 1$, i.e., the sum of the squares of the absolute values of the reflection and transmission coefficients, is equal to 1. Note that, in the absence of field coupling caused by the transverse refractive index perturbation, the ChFBG acts as a regular FBG without changing the structure of the reflected field.

Finally, the refractive index of the ChFBG is defined as follows:

$$n(r,\varphi,z) = n_0 + \Delta n\left[e^{-i\beta_0 z}\cdot g(z)\cdot f(r)\cdot e^{im\varphi} + \text{c.c.}\right] \quad (7)$$

where Δn is the modulation amplitude of the refractive index (which must be lower than in regular FBG, see below), m is vortex order, $\beta_0 = 2\pi/y(z)$, $y(z)$ is the chirp function (Figure 3), used to make the ChFBG broadband and which in the general case has an arbitrary form.

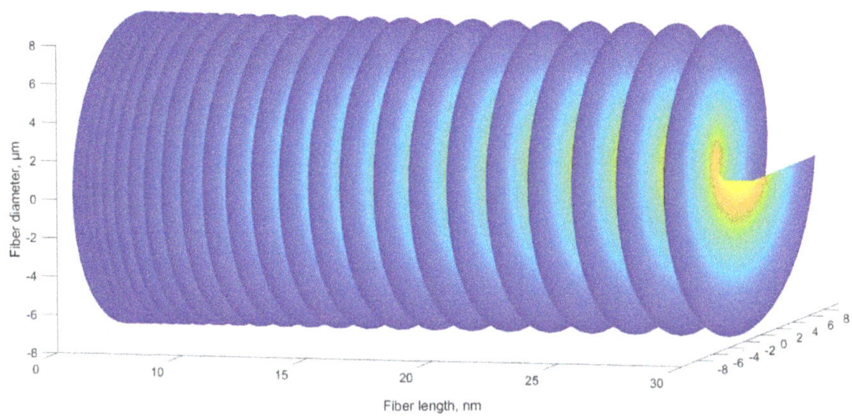

Figure 3. Chirped ChFBG with a linear increase in the lattice period.

Note that the azimuthal position $\Delta\varphi$ of the grating finger is defined by $\Delta\varphi = (2\pi/\Lambda)\cdot\Delta z$. Thus, for linear chirp we can write $y(z) = \Lambda + \Lambda_0 z$ (without chirp—$y(z) = \Lambda$); $g(z)$ is the apodization function, used to narrow down the reflection spectrum and change its shape; in our case $g(z)$ is equal to 1 (there is no apodization), $f(r)$ is the radial function, determining the radial refractive index perturbation (because the grating is uniform in the azimuth), since the degree of mode coupling depends on k_{ab} according to (5).

According to the coupled modes theory, mode coupling occurs at a non-zero functional transverse perturbation of the refractive index (according to [28]). Since in a classical (non-vortex) FBG, there is no functional transverse perturbation (as a function of the fiber radius or azimuth angle), but only a quantitative increment of the refractive index Δn exists, the overlap integral (5) is not zero only in the case when $a = b$, i.e., when the incident and the reflected modes are the same, and equals to zero otherwise. Hence, this follows the conclusion that was theoretically substantiated in [25]: the conventional FBG does not lead to mode coupling. However, it is obvious that in order to convert the OAM_0 mode to the OAM_1 mode, the overlap integral between these modes must be non-zero, and in the case of orthonormal signals, it should be close to 1 (which means 100% energy transfer from the OAM_0 mode to the OAM_1 mode). In other words, one needs to make orthogonal modes non-orthogonal to increase power coupling between them. Based on these considerations, it can be shown that the radial function $f(r)$ can be defined as follows:

$$f(r) = \sigma \frac{J_{11}\left(u_{11}\frac{r}{r_{co}}\right)}{J_{01}\left(u_{01}\frac{r}{r_{co}}\right)} \tag{8}$$

where σ is the normalizing coefficient, and the factors u_{01} and u_{11} are the roots of the characteristic equation for a particular type of fiber in which the grating is written; these factors can be calculated easily for any step-index fiber. Figure 4 shows the shape of the function (8), which describes one period of the considered grating; the corresponding refractive index profile of one finger of the ChFBG is shown in Figure 5. This function, according to (6), theoretically provides an absolute magnitude of the overlap integral equal to 1.

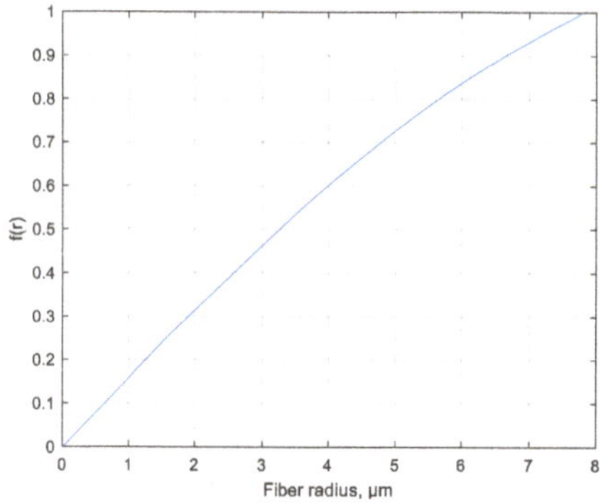

(a)

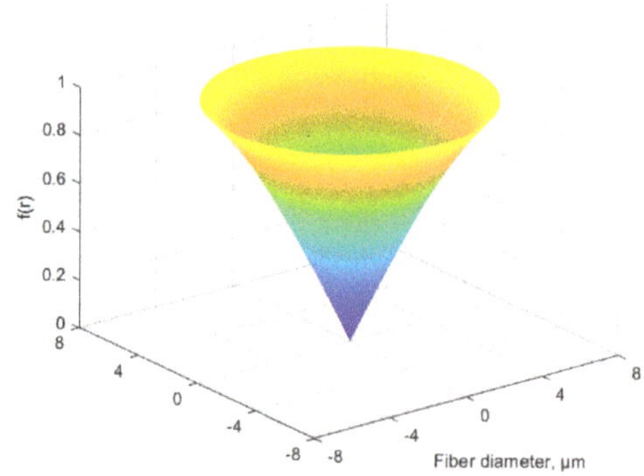

(b)

Figure 4. (**a**) The shape of the radial function $f(r)$ according to (8), describing the transverse profile of the refractive index of the grating finger; (**b**) 3D image of a homogeneous finger with a given profile (in case of non-chiral regular grating, normalized units).

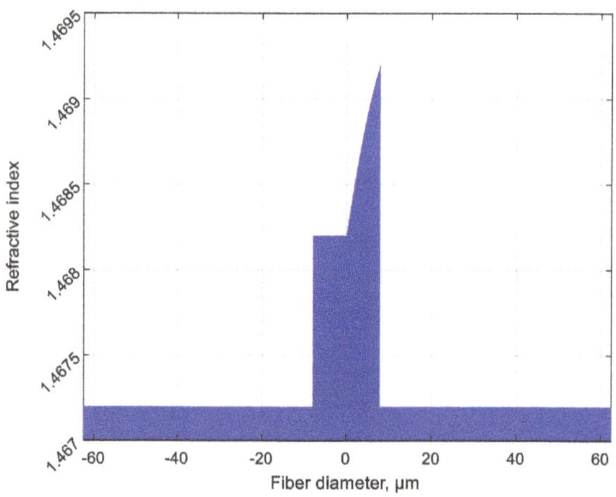

Figure 5. The refractive index profile of the ChFBG's cross-section.

Considering that the magnitude of Δn should be around 10^{-3} (which will be explained below), it can be defined as:

$$n^2 \sim n_0^2 + 2n_0\Delta n \cdot e^{-i\beta_0 z} \cdot f(r) \cdot e^{im\varphi} \tag{9}$$

where $\delta n^2(r,\varphi) = 2n_0\Delta n \cdot f(r) \cdot e^{im\varphi}$. Then expression (6) can be overwritten as follows:

$$k_{ab} = i\omega\varepsilon_0 n_0 \Delta n \cdot \int_0^{2\pi}\int_0^{r_{co}} e^{im\varphi} f(r) \cdot \vec{E}_{11}^{*}(r,\varphi) \cdot \vec{E}_{01}(r,\varphi) dr d\varphi. \tag{10}$$

The factor $e^{im\varphi}$ describing the chirality of the grating provides the integral (10) to be non-zero which means the non-zero probability of conversion of LP_{01} mode to OAM_1 mode. In this case, m should be equal to 1. Note that the reflection coefficient w, according to (4), upon apodization has a dependence on z, i.e., $w = w(z)$. The amplitude of the reflected mode in the case of grating apodization can be defined as [28]:

$$b_0 \approx e^{i\beta_{11}L} \int_0^L w(z) e^{-i(\beta_{01}-\beta_{11})z} dz \tag{11}$$

where the reflection coefficient $w(z)$ is determined by (4)–(9). Otherwise, the amplitude of the reflected filed has a longitudinal phase shift only.

The matrix approach considered in this work is convenient for the analysis of irregular and/or cascade gratings, that can be used, e.g., for broadband operation; in this case, it is possible to split a complex grating into N regular sections (Figure 6), each of which is described by its own matrix T_i, and the resulting field can be expressed as a product of the matrices:

$$\begin{pmatrix} a_1 \\ b_1 \end{pmatrix} = T_N \ldots \cdot T_2 \cdot T_1 \begin{pmatrix} a_0 \\ b_0 \end{pmatrix}.$$

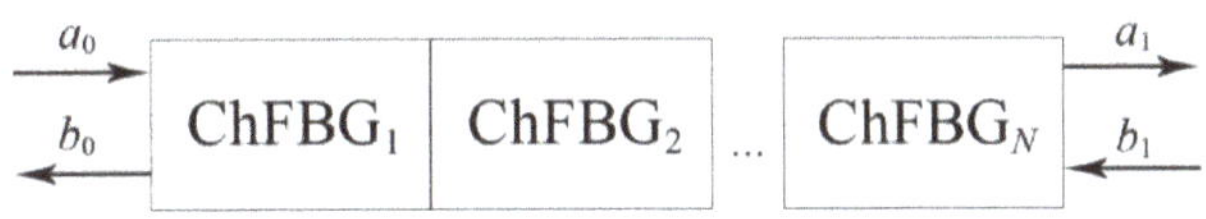

Figure 6. Matrix model of irregular ChFBG.

The reflectivity of the ChFBG depends on several factors, for example, the magnitude of the refractive index modulation Δn, number of periods N, etc. However, mode coupling depends mostly on transverse perturbation of the refractive index. In the following section, ChFBG properties are investigated by numerical simulation.

3. Numerical ChFBG Analysis and Fiber Design

The proposed ChFBG is designed to transform an incident field with a plane wavefront into a field that carries the OAM. Figure 7 shows the incident and reflected fields of the considered grating.

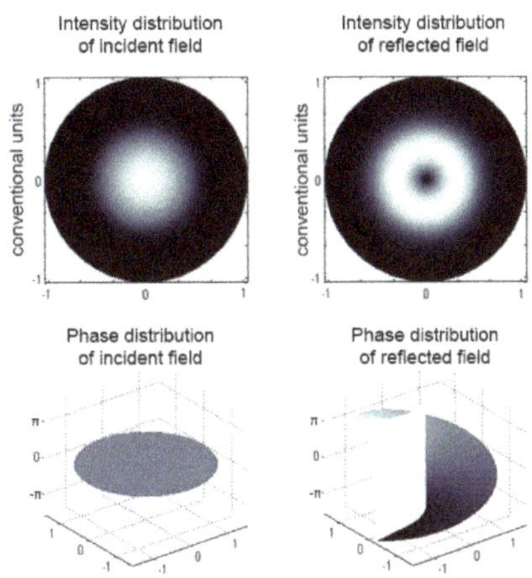

Figure 7. Reflected and incident grating field.

Analyzing Figure 7 we can assert that the vortex structure of the field is formed in the reflected radiation. It can be shown that counter-propagating modes—incident plane wave and reflected OAM wave—are still orthogonal and can be separated. One can separate the reflected field from the incident one, e.g., using a mode splitter [29].

According to expression (4), it is possible to construct the reflection spectrum of the ChFBG, shown in Figure 8.

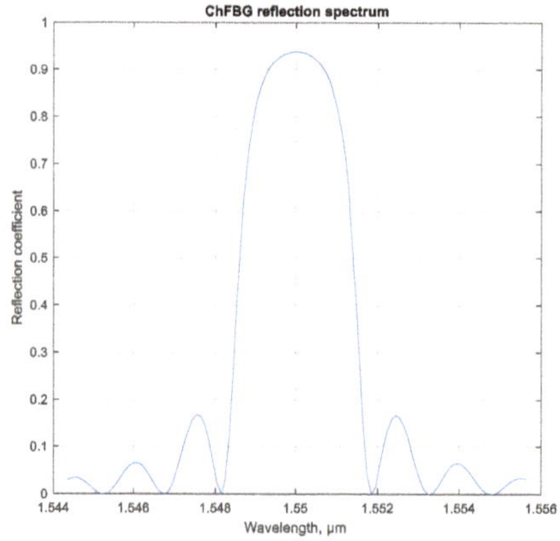

Figure 8. Spectral dependence of the reflection coefficient of the ChFBG.

As is clear from Figure 8, the considered grating has a reflection coefficient around $R = 0.95$ at 1550 nm, however, also significant side lobes occur. If necessary, the spectral characteristics of the ChFBG can be changed by chirping or using the apodization function, as mentioned above. Note that in this calculation we used $\Delta n = 0.003$ and 1000 grating periods.

Since in order to apply the mathematical apparatus of Bessel functions to define OAM modes, it is necessary to remain within the weakly guiding fiber approximation ($\Delta n < 1\%$), the amplitude of the induced grating modulation should be relatively low compared to classical FBGs (about 0.005 for the ChFBG versus 0.01 for classical FBG). In this regard, the reflection coefficient, according to (4), can be improved by increasing the number of grating periods N (Figure 9).

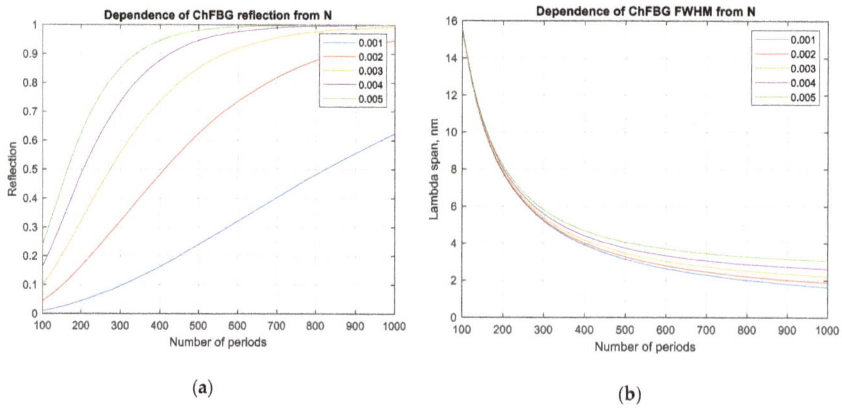

Figure 9. Relation of the reflection coefficient and spectrum of the ChFBG: (**a**) refractive index versus the number of periods (fingers), N; (**b**) width of the reflection spectrum vs. N of the grating at different values of the amplitude of the refractive index induced modulation.

According to Figure 9, the grating parameters can be selected according to the initial requirements or purposes, for example, one can easily find Δn and N values for a given reflection coefficient, or choose Δn and R for a specific N, etc. It should be noted that the overlap integral (10) has only a phase incursion $e^{-i\beta_0 z}$ along the z axis, and the integral's absolute value depends only on the transverse coordinates. This means that the efficiency of the first-order OAM mode generation from the fundamental mode does not depend on the number of grating fingers, but depends on the transverse index perturbation. This statement is also confirmed, for example, by the fact that only one spiral phase plate is sufficient for the vortex radio beam generation [30], and for the free-space optical vortex beam—a single diffractive optical element is used to generate the OAM signal. In this regard, the number of grating periods determines not the efficiency of the OAM mode generation, but the efficiency of its reflection.

One of the key practical issues is the method of manufacturing the grating under consideration. As is known, there are many ways to create fiber Bragg gratings: using phase masks [31] and Talbot interferometers [32], using a Lloyd prism [33], etc. The most promising technology for writing ChFBG seems to be a step-by-step method with simultaneous fiber rotation, but this approach requires a high precision of installation and control of both the grating pitch and the degree of its twist. The required low amplitude of the induced modulation of the refractive index makes the task easier in terms of the laser intensity and the required fiber photosensitivity, but it requires a relatively large number of grating periods. Moreover, the question of the temperature dependence of such a grating and changes in the structure of the reflected field under its influence remains unexplored.

It is obvious that after the considered mode generator, the optical signal must be coupled to the corresponding guiding medium. It was mentioned above that a standard stepped optical fiber is not multimode in the operation wavelength range. Moreover, in the Introduction some works devoted to the development of fibers that support the propagation of vortex optical radiation were considered. It was previously mentioned that the generated OAM optical signal can be branched off using an LP-mode splitter. The authors of this work have previously proposed an optical fiber that supports the propagation of optical vortexes [34]. The proposed fiber design is based on microstructured fibers. It is assumed that the induced chirality on the optical fiber with a special design—microstructured core—will allow us to form the desired order vortex (OAM mode), while the chirality on the standard fiber—the core surrounded by one outer continuous shell—can be used as a "translator", i.e., vortex-maintaining fiber for relatively long distances (note that ChFBG means not twisted fiber with a chiral diffraction structure in it; meanwhile, in [34] we consider the whole twisted fiber). For this purpose, a modification of the rotor of the exhaust tower was carried out: engine rotation speed before modernization was 3 rpm. After refinement, it can now be accelerated to 66 rpm with a drawing speed of 1 m/min. As a result, 2 regimes of optical fiber twisting during the drawing from the workpiece are provided: a step with 10 rpm and a step with 66 rpm. In any case, the engine rotates the workpiece slowly to compensate environmental side effects, so this controllable rotating modification should provide chiral fiber with precise twisting step. Thus, the following scenarios for the implementation of specialized optical fibers for the generation and transmission of OAM are considered.

The first one is an FMF with a step-index refractive index profile and strong induced chirality. We start with the profile of a conventional single-mode fiber (Rec. ITU-T G.652). Next, a theoretical calculation and simulation of fiber properties are carried out in COMSOL Multiphysics in such a way that, with a typical core diameter of 8.3 μm and a correspondingly increased Δn—the difference between core and cladding refractive index (thus increased numeric aperture, NA), the desired modes LP_{01} and LP_{11} satisfy the cut-off condition. Practically, the cut-off wavelength can be shifted by doping GeO_2 core of the fiber. An alternative approach is to slightly increase both the diameter of the fiber core and Δn. In turn, strongly induced chirality will allow these three modes to be "folded" into the desired 1st order OAM. The induced chirality by the aforementioned drawing process modification adds to the desired mode coupling.

The second approach is based on microstructured optical fibers with geometry that mimics the structure of a ring-cored fiber and also has an induced chirality (Figure 10a,b). It is known that fiber with 5.6 μm ("thickness" 1.8 μm) ring core is an applicable solution for the OAM generation. The proposed micro-structured optical fiber is a replica of this design, where the ring is reproduced by the corresponding positioning of the fiber capillaries. In turn, strongly induced chirality should only enhance the effect of the generation and/or maintenance of an OV.

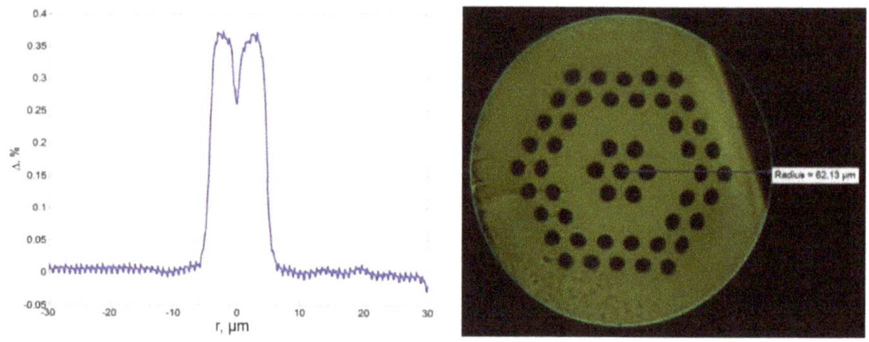

(a) (b)

Figure 10. (**a**) The relative Δn between core and cladding for the proposed fiber. One can obtain a ring-structured profile of refractive index; (**b**) the proposed micro-structured fiber with ring-shape hexagonal geometry [34].

Thus, the considered method of optical OAM modes generation should be used in combination with guiding structures that support the propagation of vortex signals. This full-fiber optic segment can be used in future OAM-based mode-division multiplexing (MDM) systems [35] and radio-over-fiber [36] networks as well.

4. Sensor Application

It is well known that FBGs can be used as sensors of many physical parameters. It can be shown that the proposed grating is also sensitive to temperature and strain, which manifests itself in a corresponding change in the reflection wavelength:

$$\frac{\Delta\lambda_B}{\lambda_B} = (1-p_e)\varepsilon + (\alpha_\Lambda + \alpha_n)\Delta T \tag{12}$$

where p_e is the strain optic coefficient, α_Λ is the thermal expansion coefficient (which is 0.55×10^{-6}/°C for silica), α_n is the thermo-optic coefficient (1.05×10^{-5}/°C for silica) [37], $\varepsilon = \Delta L/L$ is strain, L is ChFBG length, ΔL and ΔT are length and temperature increments, respectively. It should be noted that, in this case, the chirality of the grating does not depend on the wavelength in terms of the topological charge—OAM order (i.e., when the wavelength changes due to heating or stretching of the ChFBG, its chiral nature will be conserved; therefore, the OAM order does not depend on external factors). Since the distance between any two adjacent point of grating fingers with the same spatial phase $\Delta\varphi$ remains unchanged and equal to Λ, environment changes will lead to proportional changes in grating period along the whole ChFBG. The proposed ChFBG, as well as chiral fiber gratings [38], can be used as sensors for temperature, pressure, etc. In [38] it is proven experimentally that chiral fiber gratings can act like sensors. However, in contrast to reference [38], in which authors use high-temperature silica fiber, the proposed ChFBG is supposed to be written in a conventional silica fiber with the temperature parameters listed under expression (12). Thus, we obtain the following dependences for the reflection wavelength on temperature (Figure 11). The results obtained are in good agreement with classical FBG sensors [24]. In contrast to classical FBGs, the proposed grating is not only an effective sensor of physical parameters, but also has an important advantage—invariance of the OAM order (chirality) of the reflected mode to changes in the physical parameters, since as was mentioned above grating chirality remains constant with changes in grating period caused by temperature or/and strain. Thus the main advantage of the proposed solution over the device presented in [38] or in [24] is that it can also replace (or supplement) the complex arrays of fiber Bragg gratings in so-called addressed sensor systems.

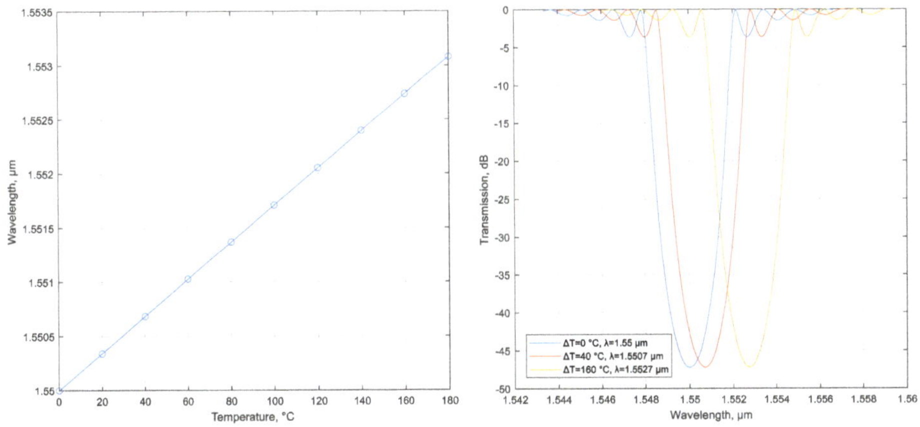

(**a**) (**b**)

Figure 11. (**a**) ChFBG wavelength as a function of temperature change; (**b**) ChFBG transmission as a function of temperature change for three temperature increments: 0 °C, 40 °C, and 160 °C.

One of the main problems of such systems is addressing, i.e., discrimination between signals received from different gratings operating at the same central wavelength or in the case when a change in the measured physical parameters leads to overlapping of the reflection spectra of different gratings. To solve this problem, the authors of [21] proposed a method for addressed sensors based on the use of two-frequency gratings, the difference between the main reflection frequencies of which remains constant for any change in the measured physical field and is thus a unique sensor address in complex systems. The ChFBG, proposed in this article, can be used as both an alternative and an addition to this system: using a ChFBG that generates OAM signals of +1 and −1 orders (the zero order of OAM is generated by conventional FBG), it is possible to construct sensor systems using only one wavelength, but provide addressing due to the spatial distribution of the optical field. When polling such sensors, the signals will (or can) have the same frequency, but be spatially orthogonal and, therefore, such signals can be separated using, for example, mode splitters or other passive fiber-optic components (Figure 12).

Thus, the use of gratings for three OAM orders (−1, 0, +1) will increase the capacity of sensor systems threefold, but, however, requires the use of additional passive optical components for mode separation. The manufacture of apodized or chirped ChFBGs is a complex technical problem, but it will make it possible to develop sensor systems based on narrow-band (or broadband) gratings that change the spatial distribution of the reflected field. Thus, the use of ChFBG in sensor systems allows us to develop addressed sensor arrays with an address in both the frequency and spatial domains, which makes it possible to increase the number of sensors without using additional active equipment.

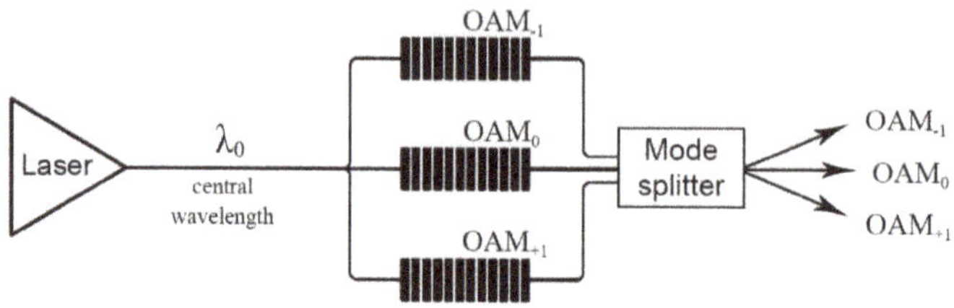

Figure 12. Schematic example of an addressed sensor system based on an array of single-frequency ChFBGs: A coherent optical signal from a laser at a wavelength λ_0, which is a central wavelength for all gratings, enters an array of three ChFBGs, which form signals with orbital angular momentums OAM_{-1}, OAM_{+1} and OAM_0. The signals from the sensors can be separated using a mode splitter and directed to further processing (the photodetector is not shown in the figure).

5. Conclusions

In this paper, we present a theoretical model of a chiral fiber Bragg grating based on the coupled modes theory modes and scattering matrices. The proposed model allows us to study the characteristics of a grating with arbitrary apodization and chirping functions, obtaining a ChFBG with the required spectral characteristics; additionally, it can define complex (cascade) gratings. The radial perturbation function of the refractive index provides up to 100% of the theoretical efficiency in the OAM_1 to OAM_0 mode conversion. Moreover, based on the analysis of the disadvantages of existing methods for optical vortex signal generation, a new optical fiber is proposed, which is a chiral microstructured optical fiber that mimics a ring-core structure and allows the formation of first-order OAM modes. Since the cross-section of the fiber is a periodic structure (formed by capillaries), it can be considered as a photonic crystal and, therefore, as a spatial-phase filter for the generation of vortices with the desired order. As a vortex-maintaining fiber, we propose a strongly chiral multimode fiber of size 100/125 μm with low modal dispersion that allows the vortex structure of the optical field propagating over relatively large distances to be maintained. These novel fibers can be applied in future communication systems, e.g., in the radio-over-fiber backbone infrastructure. Thus, in this paper, a full-fiber vortex generation method and vortex maintaining fiber have been proposed. New application of the results obtained in this work as address sensors with an address in the form of a constant topological charge in multi-sensor systems for monitoring various physical fields is also proposed.

Author Contributions: Conceptualization, V.B. and I.V.; methodology and funding acquisition, I.M.; software, visualization and formal analysis, A.G.; writing—original draft preparation, review and editing, E.G.; supervision, V.B. and I.V.; project administration, A.S. All authors have read and agreed to the published version of the manuscript.

Funding: This work was partially supported by a grant from the Russian Science Foundation, project No. 18-19-00123, ("Numerical ChFBG analysis and fiber design" paragraph), and by the Ministry of Science and Higher Education of the Russian Federation for research under the State Assignment of FSBEI HE USATU No. FEUE-2020-0007 on the topic "Theoretical foundations of modeling and semantic analysis of the processes of transformation of vortex electromagnetic fields in infocommunication systems" ("The theoretical model for ChFBG" paragraph).

Conflicts of Interest: The authors declare no conflict of interest.

References

1. Saito, Y.; Kishiyama, Y.; Benjebbour, A.; Nakamura, T.; Li, A.; Higuchi, K. Non-orthogonal multiple access (NOMA) for cellular future radio access. In Proceedings of the 2013 IEEE 77th Vehicular Technology Conference, Dresden, Germany, 2–5 June 2013; pp. 1–5.
2. Basnayake, V.; Jayakody, D.N.K.; Sharma, V.; Sharma, N.; Muthuchidambaranathan, P.; Mabed, H. A New Green Prospective of Non-orthogonal Multiple Access (NOMA) for 5G. *Information* **2020**, *11*, 89. [CrossRef]
3. Liang, L.; Cheng, W.; Zhang, W.; Zhang, H. Mode Hopping for Anti-Jamming in Radio Vortex Wireless Communications. *IEEE Trans. Veh. Technol.* **2018**, *67*, 7018–7032. [CrossRef]

4. Zin, A.M.; Bongsu, M.S.; Idrus, S.M.; Zulkifli, N. An overview of radio-over-fiber network technology. In Proceedings of the International Conference on Photonics, Langkawi, Malaysia, 5–7 July 2010; pp. 1–3.
5. Yue, Y.; Huang, H.; Ren, Y.; Pan, Z.; Willner, A.E. Special issue on novel insights into orbital angular momentum beams: From fundamentals, devices to applications. *Appl. Sci.* **2019**, *9*, 2600. [CrossRef]
6. Lin, Z.; Wang, A.; Xu, L.; Zhang, X.; Sun, B.; Gu, C.; Ming, H. Generation of optical vortices using a helical fiber Bragg grating. *J. Lightwave Technol.* **2014**, *32*, 2152–2156. [CrossRef]
7. Zhang, X.; Wang, A.; Chen, R.; Zhou, Y.; Ming, H.; Zhan, Q. Generation and conversion of higher order optical vortices in optical fiber with helical fiber Bragg gratings. *J. Lightwave Technol.* **2016**, *34*, 2413–2418. [CrossRef]
8. Mao, B.; Liu, Y.; Zhang, H.; Yang, K.; Han, Y.; Wang, Z.; Li, Z. Complex analysis between CV modes and OAM modes in fiber systems. *Nanophotonics* **2018**, *8*, 271–285. [CrossRef]
9. Huang, W.; Liu, Y.G.; Wang, Z.; Zhang, W.; Luo, M.; Liu, X.; Guo, J.; Liu, B.; Lin, L. Generation and excitation of different orbital angular momentum states in a tunable microstructure optical fiber. *Opt. Express* **2015**, *23*, 33741–33752. [CrossRef]
10. Seghilani, M.; Azaña, J. All-Fiber OAM generation/conversion using helically patterned photonic crystal fiber. *IEEE Photonics Technol. Lett.* **2018**, *30*, 347–350. [CrossRef]
11. Heng, X.; Gan, J.; Zhang, Z.; Qian, Q.; Xu, S.; Yang, Z. All-fiber orbital angular momentum mode generation and transmission system. *Opt. Commun.* **2017**, *403*, 180–184. [CrossRef]
12. Eznaveh, Z.S.; Zacarias, J.C.A.; Lopez, J.E.A.; Shi, K.; Milione, G.; Jung, Y.; Thomsen, B.C.; Richardson, D.; Leon-Saval, S.; Correa, R.A. Photonic lantern broadband orbital angular momentum mode multiplexer. *Opt. Express* **2018**, *26*, 30042–30051. [CrossRef]
13. Eznaveh, Z.S.; Zacarias, J.A.; Lopez, J.A.; Jung, Y.; Shi, K.; Thomsen, B.C.; Richardson, D.J.; Leon-Saval, S.; Correa, R.A. Annular core photonic lantern OAM mode multiplexer. In Proceedings of the Optical Fiber Communication Conference of Optical Society of America, Los Angeles, CA, USA, 19–23 March 2017.
14. Ramachandran, S.; Kristensen, P.; Yan, M.F. Generation and propagation of radially polarized beams in optical fibers. *Opt. Lett.* **2009**, *34*, 2525–2527. [CrossRef] [PubMed]
15. Ung, B.; Vaity, P.; Wang, L.; Messaddeq, Y.; Rusch, L.A.; LaRochelle, S. Few-mode fiber with inverse-parabolic graded-index profile for transmission of OAM-carrying modes. *Opt. Express* **2014**, *22*, 18044–18055. [CrossRef] [PubMed]
16. Wu, H.; Gao, S.; Huang, B.; Feng, Y.; Huang, X.; Liu, W.; Li, Z. All-fiber second-order optical vortex generation based on strong modulated long-period grating in a four-mode fiber. *Opt. Lett.* **2017**, *42*, 5210. [CrossRef] [PubMed]
17. Vinogradova, I.L.; Meshkov, I.K.; Grakhova, E.P.; Sultanov, A.K.; Bagmanov, V.K.; Voronkova, A.V.; Gizatulin, A.R. Secured RoF segment in subterahertz range providing independent optical modulation of radiochannel frequency characteristics and phased antenna array beamsteering parameter. *Comput. Opt.* **2018**, *42*, 786–799. [CrossRef]
18. Bagmanov, V.K.; Sultanov, A.K.; Gizatulin, A.R.; Meshkov, I.K.; Kuk, I.A.; Grakhova, E.P.; Abdrakhmanova, G.I.; Vinogradova, I.L. Optics-to-THz-conversion of vortex beams using nonlinear difference frequency generation. *Comput. Opt.* **2019**, *43*, 983–991. [CrossRef]
19. Bagmanov, V.K.; Sultanov, A.K.; Gizatulin, A.R.; Meshkov, I.K.; Kuk, I.A.; Grakhova, E.P.; Abdrakhmanova, G.I.; Vinogradova, I.L. The vortex beams conversion from the optical range into the radio domain based on the nonlinear generation of the difference frequency. In Proceedings of the 2019 27th Telecommunications Forum (TELFOR), Belgrade, Serbia, 26–27 November 2019; pp. 1–4.
20. Vashukevich, E.A.; Golubeva, T.Y.; Golubev, Y.M. Conversion and storage of modes with orbital angular momentum in a quantum memory scheme. *Phys. Rev.* **2020**, *101*, 033830. [CrossRef]
21. Morozov, O.G.; Sakhabutdinov, A.J. Addressed fiber Bragg structures in quasidistributed microwave-photonic sensor systems. *Comput. Opt.* **2019**, *43*, 535–543. [CrossRef]
22. Tai, H. Theory of fiber optical Bragg grating: Revisited. In Proceedings of the SPIE Optical Modeling and Performance Predictions, San Diego, CA, USA, 3–8 August 2003; Volume 5178.
23. Soifer, V.A. *Computer Design of Diffractive Optics*; Woodhead Publishing: Sawston, UK, 2012; p. 896.
24. Othonos, A. Fiber Bragg gratings. *Rev. Sci. Instrum.* **1997**, *68*, 4309. [CrossRef]
25. Kashyap, R. *Fiber Bragg Gratings*; Academic Press: London, UK, 1999; p. 478.
26. Yariv, A.; Nakamura, M. Periodic Structures for Integrated Optics. *IEEE J. Quantum Electron.* **1977**, *Qe-13*, 133–153. [CrossRef]

27. Yamada, M.; Sakuda, K. Analysis of almost-periodic distributed feedback slab waveguides via a fundamental matrix approach. *Appl. Opt.* **1987**, *16*, 3474–3478. [CrossRef]
28. Ho, K.-P.; Kahn, M. Linear propagation effects in mode-division multiplexing systems. *J. Lightwave Technol.* **2014**, *32*, 614–628. [CrossRef]
29. Wang, Z.; Wang, Y.; Li, Y.; Wu, C. Mode splitter based on triple-core waveguide. *Opt. Express* **2006**, *14*, 10324–10331. [CrossRef] [PubMed]
30. Shen, Y.; Wang, X.; Xie, Z.; Min, C.; Fu, X.; Liu, Q.; Gong, M.; Yuan, X. Optical vortices 30 years on: OAM manipulation from topological charge to multiple singularities. *Light Sci. Appl.* **2019**, *8*, 90. [CrossRef] [PubMed]
31. Singh, N.; Jain, S.; Aggarwal, A.; Bajpai, R. Fiber Bragg Grating Writing Using Phase Mask Technology. *J. Sci. Ind. Res.* **2005**, *64*, 108–115.
32. Zhang, Z.; Xu, B.; He, J.; Hou, M.; Bao, W.; Wang, Y. High-Efficiency Inscription of Fiber Bragg Grating Array with High-Energy Nanosecond-Pulsed Laser Talbot Interferometer. *Sensors* **2020**, *20*, 4307. [CrossRef]
33. Laffont, G.; Cotillard, R.; Ferdinand, P. Multiplexed regenerated Fiber Bragg Gratings for high temperature measurement. *Meas. Sci. Technol.* **2013**, *24*, 094010. [CrossRef]
34. Gizatulin, A.R.; Meshkov, I.K.; Grakhova, E.P.; Ishmiyarov, A.A.; Kuk, I.A.; Sultanov, A.K.; Bagmanov, V.K.; Abdrakhmanova, G.I.; Vinogradova, I.L.; Bourdine, A.V.; et al. Design of vortex optical fibers for RoF systems: Part I: Overview and alternative solutions. In Proceedings of the SPIE, Optical Technologies for Telecommunications, Kazan, Russia, 19–1 November 2019; Volume 11516, p. 115161S.
35. Bozinovic, N.; Yue, Y.; Ren, Y.; Tur, M.; Kristensen, P.; Willner, A.; Ramachandran, S. Orbital Angular Momentum (OAM) Based Mode Division Multiplexing (MDM) over a Km-length Fiber. In Proceedings of the European Conference and Exhibition on Optical Communication, Amsterdam, The Netherlands, 16–20 September 2012.
36. Tatarczak, A.; Lu, X.; Rommel, S.; Rodriguez, S.; Olmos, J.J.V.; Monroy, I.T. Radio-over-fiber transmission using vortex modes. In Proceedings of the International Topical Meeting on Microwave Photonics (MWP), Paphos, Cyprus, 26–29 October 2015; pp. 1–3.
37. Doyle, C. Fibre Bragg Grating Sensors: An Introduction to Bragg Gratings and Interrogation Techniques. Available online: https://www.researchgate.net/publication/265307606_Fibre_Bragg_Grating_Sensors_An_Introduction_to_Bragg_Gratings_and_Interrogation_Techniques (accessed on 2 June 2020).
38. Kopp, V.; Churikov, V.; Zhang, G.; Singer, J.; Draper, C.W.; Chao, N.; Neugroschl, D.; Genack, A.Z. Chiral fiber gratings: Perspectives and challenges for sensing applications. In Proceedings of the SPIE 6619, Third European Workshop on Optical Fibre Sensors, Napoli, Italy, 4–6 July 2007; pp. 66190B1–66190B19.

sensors

MDPI

Letter

Optical Designs with Curved Detectors for Fiber Bragg Grating Interrogation Monitors

Eduard Muslimov [1,*], Nadezhda Pavlycheva [1], Emmanuel Hugot [2,3], Simona Lombardo [2], Ilnur Nureev [1] and Oleg Morozov [1]

1 Kazan National Research Technical University Named after A.N. Tupolev-KAI, 420111 Kazan, Russia; nkpavlycheva@kai.ru (N.P.); iinureev@kai.ru (I.N.); microoil@mail.ru (O.M.)
2 Aix Marseille University, CNRS, CNES, LAM, Groupe R&D Optique & Instrumentation, 13388 Marseille, France; emmanuel.hugot@lam.fr (E.H.); simona.lombardo@lam.fr (S.L.)
3 CURVE-S.A.S., 13388 Marseille, France
* Correspondence: ermuslimov@kai.ru

Abstract: In this paper, we evaluate the application of curved detectors and freeform optics technologies for fiber Bragg gratings (FBGs) interrogation monitors design. It is shown that, in a high-dispersion spectrograph scheme, the camera part operates in special conditions, which result in a field curvature change. This field curvature can be compensated by the use of a curved detector. When used together with freeform optics, the curved detectors allow for reduction of the number of optical components to two or even one element by merging their functions. Three design examples for the range of 810–860 nm reaching the spectral resolution limit of 89–139 pm at $NA = 0.14$ are presented to demonstrate the achieved performance and the technological trade-offs.

Keywords: fiber Bragg grating sensors; spectrograph; high angular dispersion; curved detectors; freeform optics

check for updates

Citation: Muslimov, E.; Pavlycheva, N.; Hugot, E.; Lombardo, S.; Nureev, I.; Morozov, O. Optical Designs with Curved Detectors for Fiber Bragg Grating Interrogation Monitors. *Sensors* **2021**, *21*, 34. https://dx.doi.org/10.3390/s21010034

Received: 30 September 2020
Accepted: 18 December 2020
Published: 23 December 2020

Publisher's Note: MDPI stays neutral with regard to jurisdictional claims in published maps and institutional affiliations.

1. Introduction

The sensors based on fiber Bragg gratings (FBGs) have a number of advantages, which have been demonstrated for different applications [1]. They can be used for temperature and pressure measurements, while the sensors of both types can be combined in a single fiber. Their outstanding thermal range and electromagnetic stability makes them a primary choice for operations under harsh environments, like the aerospace [2] or the oil and gas [3] industries. The possibility to build a distributed sensing network of FBGs [4] encourages their application for structural health monitoring. The number of other application examples is impressively large, and it keeps growing.

The readout of FBG sensors requires measurement of their spectral response change with a very high accuracy. It can be done with interrogation monitors, which essentially represent high-resolution diffractive spectrographs with a photosensitive arrays. Such a device allows us to perform a simultaneous readout from tens of FBGs at high sensitivity without mechanical movements. But reaching a high spectral resolution in the dedicated waveband with limited size of the detector and instrument represents a separate engineering challenge.

In recent years, some technologies driving the field of imaging optics have experienced a fast growth. Namely, curved detectors technology has evolved from early concepts [5–7] to fully-functional instruments [8]. It has been proven that a complementary metal-oxide-semiconductor (CMOS) detector equivalent to a commercial flat one in terms of sensitivity and electronic performance can be curved to a desired shape with a high precision. Use of such a detector allows us to exclude the field curvature correction measures from the optical design, increase the resolution, and achieve a lower distortion and higher image illumination, as well as decrease the overall system dimension. In particular, it was demonstrated that the use of curved detectors makes it possible to significantly increase the performance of astronomical instrumentation and receive images of previously unresolvable objects [9,10]. Similar advantages apply to the technology of freeform optics [11], which has emerged during the last decade and allows to create optical systems notable for their outstanding resolution, high aperture, large field of view, and compactness.

The main goal of the present study was to explore a potential gain in performance and packaging achievable by applying freeform optics and curved detectors technologies in the optical designs of high-resolution spectrographs for FBG interrogation.

There are a few requirements intrinsic to the design of FBG interrogation monitors, which urged us to consider these technologies. First, there is a clear trend to decrease the number of optical components, thus reducing the size and increasing the environmental stability of the device. Use of freeform optics allows us to remove some optical components and merge a few functions in one element, while maintaining or increasing the optical performance. Second, reaching a high spectral resolution, together with a limited size, requires a high angular dispersion in a narrow spectral region. In some cases, such a high angular dispersion is achieved by using a complex dispersive unit representing a combination of gratings and/or prisms. Correction of the field curvature in such a design becomes an unusual task. The camera part is operating in a dispersed beam. To estimate the effect of dispersion on the field curvature, we consider an equivalent pupil offset in a dispersive unit with two transmission grating similar to that used in Reference [12]. It consists of two transmission gratings set up one after another to increase the angular dispersion in sequence (see Figure 1). In this figure, ϕ is the incidence angle on the first grating, γ is the angle between the gratings normals, r is the distance between the grating centers, and s and s' are the distances from the second grating center to the camera mirror and the equivalent pupil center, respectively. If we trace a chief ray at two different wavelengths through this unit, we can show that they have a virtual intersection point behind the second grating.

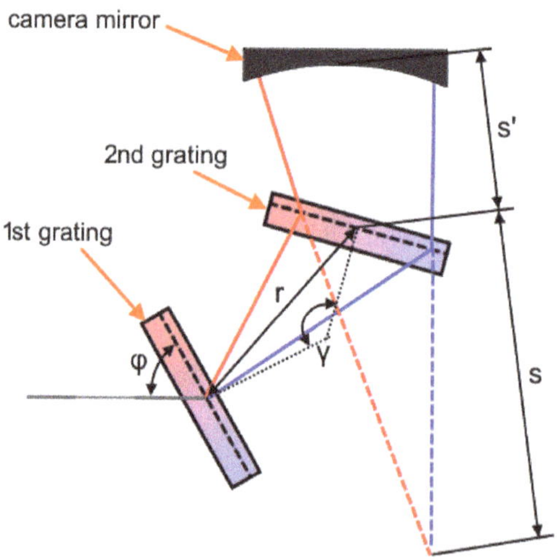

Figure 1. Chief ray tracing through a double dispersive unit. The $s + s'$ distance corresponds to the equivalent pupil shift.

If the spectral image is focused by a single camera mirror, this setup is equivalent to a mirror working with a shifted pupil. Then, the field curvature R_m can be approximately found from the following equation [13]:

$$\frac{1}{R_m} = \frac{2}{R} - \frac{4\left[k\sigma^2 + (1-\sigma)^2\right]}{R}, \tag{1}$$

where R is the mirror vertex radius, k is the conic constant, and σ is

$$\sigma = \frac{s + s'}{R}. \tag{2}$$

We apply this procedure for the following typical set of parameters: wavelengths λ = 810–860 nm; angle between the gratings $\gamma = 90°$; incidence on the first grating $\phi = 45°$; distance between the gratings $r = 60$ mm; distance to the mirror $s' = 50$ mm; mirror's vertex radius $R = -200$ mm. The resultant equivalent pupil shift and the field curvature radius as functions of the gratings spatial frequency N are shown in Figure 2.

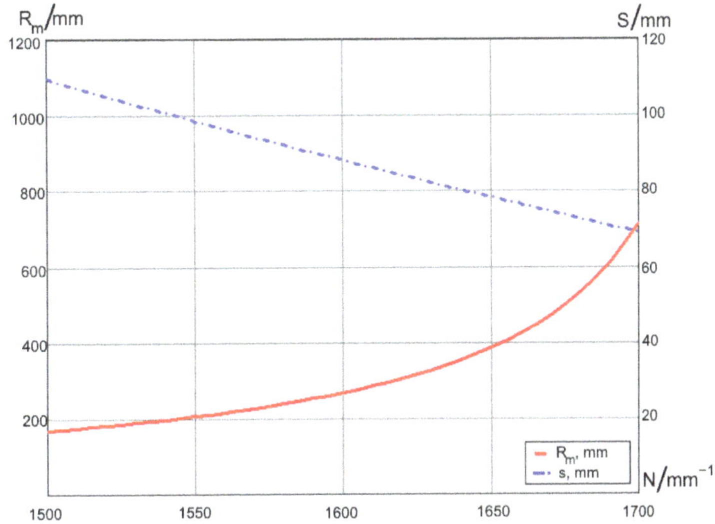

Figure 2. Equivalent pupil shift and the field curvature as a function of the grooves frequency.

The plot shows that the ray's intersection position and the curvature changes significantly with the dispersion. This effect must be taken into account in the spectrograph optical design.

Further, we consider a number of optical designs taking advantage of curved detectors and freeform optics. We gradually increase the level of design integration, merging a few functions in one optical element. In each case, we compare design versions with flat and curved detectors to show their impact. We use a design with a pair of transmission grating similar to that described in Reference [12] as the starting point. For all of the designs, we use the same set of basic parameters inspired by a commercial FBG interrogator [14] and accounting for parameters of an existing linear CMOS sensor [15] with 4096 pixels and a high sensitivity of 650 $\frac{V}{Lx \cdot s}$. The parameters are summarized in Table 1.

Table 1. Initial values for the spectrographs designs.

Parameter	Value
Working spectral range/nm	810–860
Input NA	0.14
Detector length/mm	28.7
Pixel size/μm	7 × 200
Collimator and camera focal lengths/mm	100

2. Optical Designs

The first optical design under consideration consists of a doublet lens collimator, two transmission gratings, and a freeform mirror acting as a camera objective. The gratings have straight equally spaced grooves and are imposed on flat surfaces. For simplicity, we assume that they have the same grooves frequency, equal to 1639 mm^{-1}. The freeform mirror shape is described by Zernike polynomials. The general view of the optical design is shown in Figure 3. Hereafter, this design is denoted as "Design A". For this case, Equations (1)–(2) can be applied directly. We applied them to obtain the first estimation of the field curvature and then repeated optimization with curved and flat detectors. It appears that the curvature can be efficiently compensated by the freeform, since the image surface radius in the resultant design is –5134 mm.

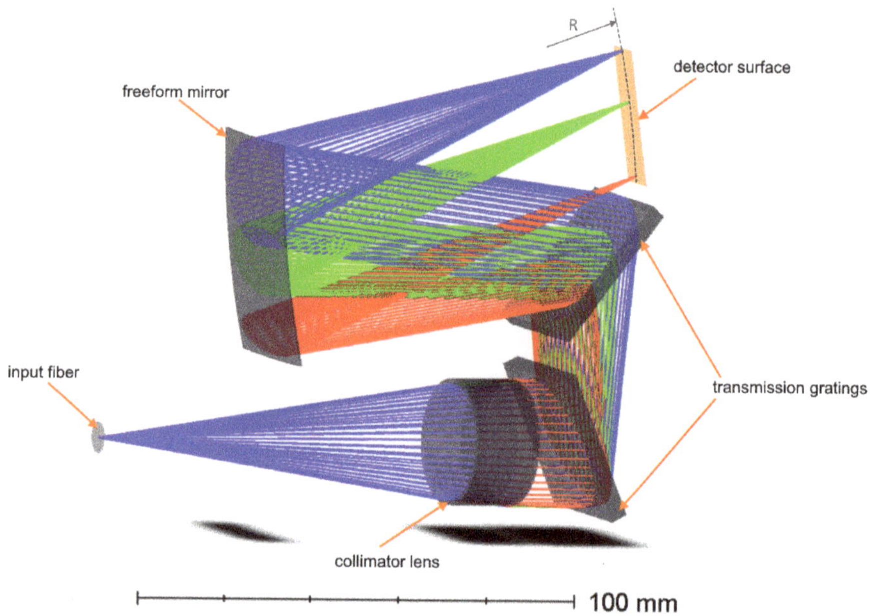

Figure 3. Optical design of the high-dispersion spectrograph with a pair of transmission gratings (Design A).

In the second design ("Design B"), some functions are performed by the same optical element. Thus, the dispersive unit has only one grating working in a double-pass mode to increase the dispersion. In order to create this double-pass geometry a transmission grating with spatial frequency of 1900 mm^{-1} is imposed on the first surface of the prism. The prism is made of Schott N-LASF41 (glass catalogue of Schott AG, Mainz, Germany) glass and has an apex angle of 59.3°. The prism rear surface has a reflective coating. This solution is to some extent similar to immersed gratings proposed in Reference [16,17]. The freeform mirror in Design B serves as both the collimator and camera. To avoid geometrical conflicts between the beams in forward and backwards propagation, a sagittal shift of the source is used. The fiber input has the sagittal coordinate $x = -5$ mm, so the image coordinate is $x' = 5$ mm. The layout of Design B is shown in Figure 4. To find the image curvature, we found the prism geometrical development, found the initial value of the field curvature, and performed numerical optimization for designs with flat and curved detectors, as it was done for the "Design A". The optimal image radius found after optimization is 186.2 mm.

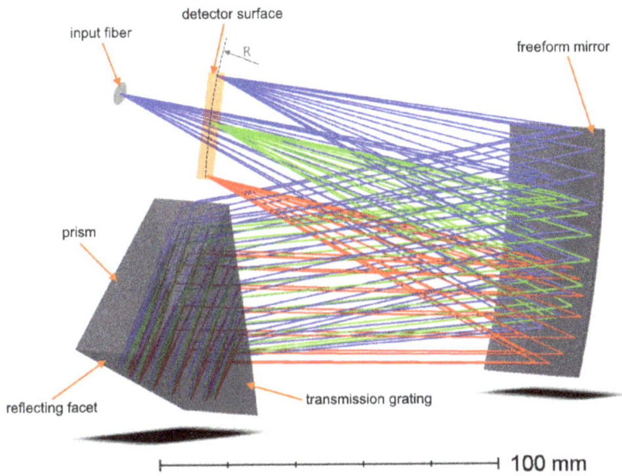

Figure 4. Optical design of the high-dispersion spectrograph with a double-pass immersed grating (Design B).

Finally, we consider combining all the functions in one optical element: a grating with non-equally spaced grooves imposed on a concave XY-polynomial freeform surface working as the collimator, dispersive unit, and camera at the same time. It becomes practically impossible to use a double-dispersion geometry, so the grating has spatial frequency of 1623 mm^{-1} in its vertex and works with high dispersion angles. This design ("Design C") is presented in Figure 5. This design follows the model of spectrographs on Rowland circle [18], although the number of correction parameters is much higher. In this case, the pupil coincides with the grating and the spectral image lies on a cylindrical surface with radius of $R/2$. The radius found after optimization is 107.9 mm.

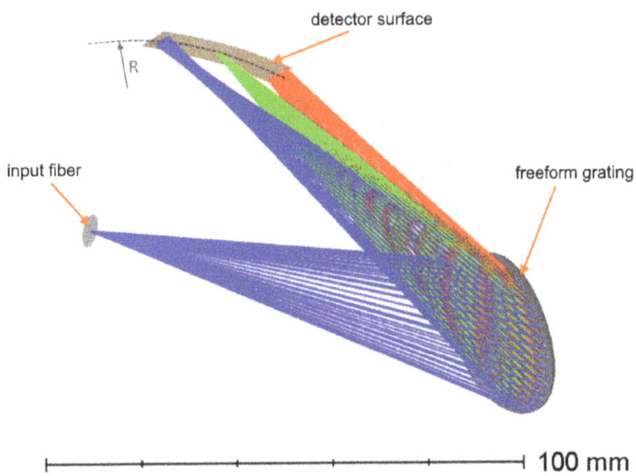

Figure 5. Optical design of the high-dispersion spectrograph with a freeform grating (Design C).

3. Imaging Performance Assessment

The imaging performance of each design is estimated through the spot diagrams and instrument functions at three reference wavelengths, corresponding to the center and edges of the working range. The spot diagrams for Design A are shown in Figure 6.

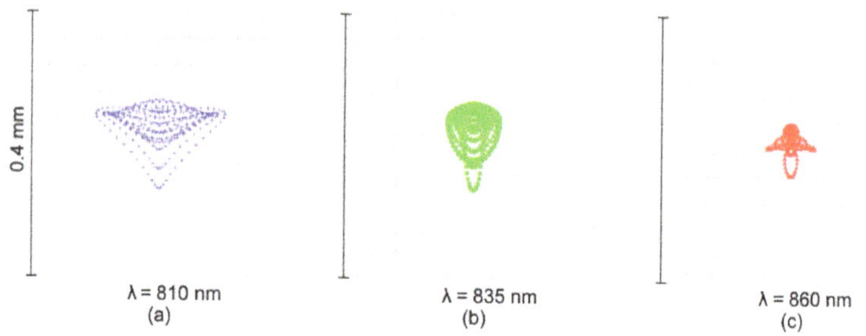

Figure 6. Spot diagrams of the high-dispersion spectrograph Design A.

The diagrams show that the aberrations are corrected in both tangential and sagittal plane. The image quality is high over the entire range (Figure 6a–c), although the longwave edge (Figure 6c) is slightly over-corrected. The root mean square (RMS) spot radius varies from 25 to 63 μm, while the maximum radii are 38–111 μm.

The instrument function (IF) is computed for an input fiber width of 50 μm. The IF graphs, i.e., the relative illumination in the monochromatic image of fiber end at each of the three reference wavelengths, are shown in Figure 7a–c. Here, the results obtained with a flat detector are shown in grey for comparison. It is clear that, in this particular case, the freeform camera provides a sufficient aberration correction, and the curved sensor contribution is as low as 2.6%.

For the given values (see Table 1), the average reciprocal linear dispersion is 1.74 nm/mm. This value, together with the pixel size, defines the device spectral sensitivity. For instance, a pixel size of 7 μm is equivalent to the spectral shift of 12 pm or 0.012 nm. The product of the IF full-width-at-the-half-maximum (FWHM) by the reciprocal linear dispersion represents the spectral resolution limit. It affects both the sensitivity to the spectral shift and also the number of FBG sensors, which can be measured simultaneously. The spectral resolution limit values for all the designs are summarized in Table 2.

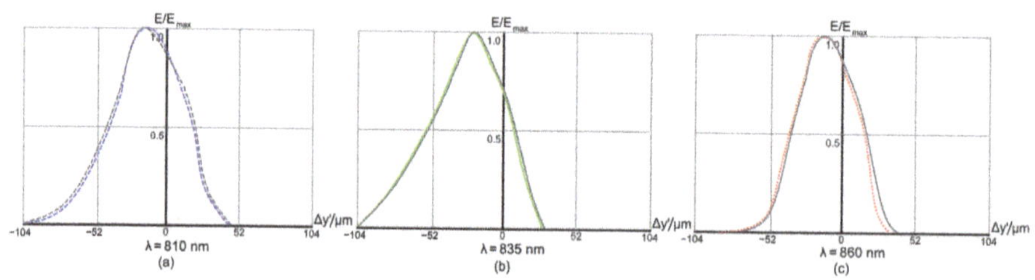

Figure 7. Instrument functions of the high-dispersion spectrograph Design A for a 50 μm input fiber. The grey lines correspond to the design with flat detector.

Similarly, the spot diagrams for Design B are given in Figure 8. The plots show that the aberration correction is good and uniform, though the spots are more affected by high-order aberrations. The RMS radii are 43–84 μm, and the maximum radii are 71–144 μm. However, the spot at central wavelenght is elongated in the sagittal direction (Figure 8b) and the spots blurring in tangential direction at the edges (Figure 8a,c) is negligible. So the spectral resolution remains high.

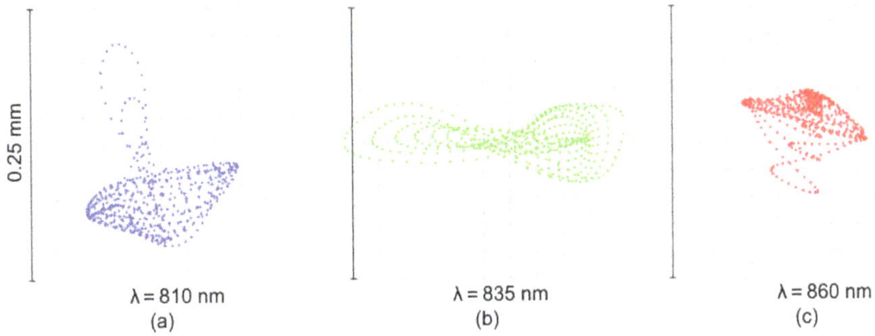

Figure 8. Spot diagrams of the high-dispersion spectrograph Design B.

The Design B IF graphs are shown in Figure 9. The resolution is compatible to that achieved in Figure 7, but the effect of the curved detector is more significant. The use of a curved image plane allows to gain up to 23% in terms of resolution (see Figure 9c).

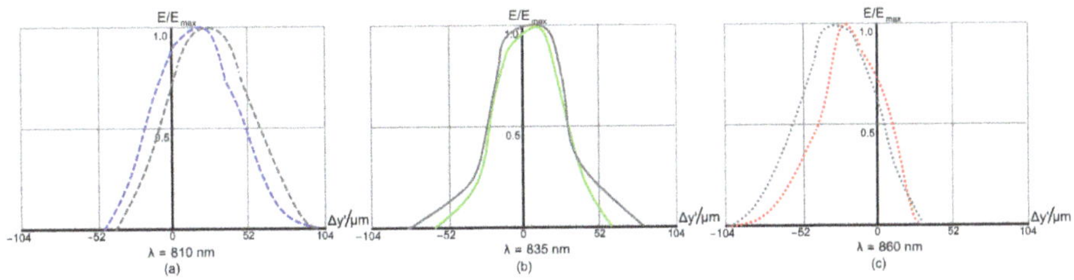

Figure 9. Instrument functions of the high-dispersion spectrograph Design B for a 50 μm input fiber. The grey lines correspond to the design with flat detector.

Finally, the spot diagrams of Design C are presented in Figure 10. It is clear that the design is driven by a high astigmatism, intrinsic for the parent Rowland circle-based solution. It is corrected only for the central wavelength (Figure 10b). The astigmatic elongation of the spot image at the edges of spectral range reaches ±3.8 mm (Figure 10c). However, in the spot dimension, the tangential section remains relatively small and varies from ±43 to ±290 μm.

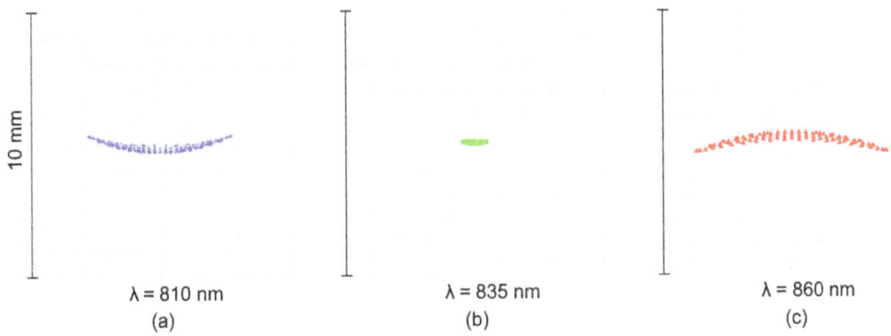

Figure 10. Spot diagrams of the high-dispersion spectrograph Design C.

The corresponding IFs are shown in Figure 11. In this case, the gain in resolution obtained by using a curved detector is as high as 92.3% (see Figure 11b).

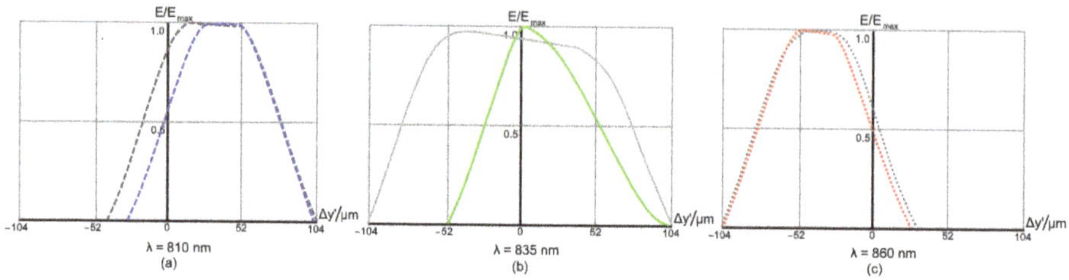

Figure 11. Instrument functions of the high-dispersion spectrograph Design C for a 50 μm input fiber. The grey lines correspond to the design with flat detector.

Table 2 summarizes all of the computed values of the spectral resolution limit. For each design and each reference wavelength, the values with curved/flat detectors are given.

Table 2. Spectral resolution limit with curved/flat detector (in pm).

	Design A	Design B	Design C
810 nm	103/106	109/109	89/89
835 nm	120/120	92/98	89/109
860 nm	145/170	172/273	139/148

4. Components Analysis

In the present section, we investigate the parameters of the freeform mirrors and curved detectors found during the optimization in more detail. The main purpose of this analysis is to assess their complexity and technological feasibility.

4.1. Freeform Elements

In designs A and B, the freeform surface shape was described by the standard Zernike polynomials:

$$z = \frac{cr^2}{1+\sqrt{1-(1+k)c^2r^2}} + \sum_{i=1}^{n} A_i Z_i(\rho,\phi), \tag{3}$$

where c is the vertex curvature, r is the radial coordinate on the surface, k is the conic constant, Z_i are the Zernike polynomials, and A_i are the corresponding coefficients. In both cases, the number of polynomials was limited by the 5th order. Since the design A is symmetric with respect to the YZ plane, only 9 Zernike symmetrical modes were used. Design B brakes the plane symmetry to separate the incoming and outcoming beams, so additional modes, namely Z_5 and Z_8, were introduced. The residual surface sags after subtracting the best fit sphere for the freeform surfaces in Designs A and B are shown in Figure 12a,b, respectively. The corresponding numerical estimates are given in Table 3.

In design C, the surface of the freeform grating is described by the following equation:

$$z = \frac{c(a^2x^2+b^2y^2)}{1+\sqrt{1-a^2x^2-b^2y^2}} + \sum_{i=1}^{n} A_i E_i(x,y). \tag{4}$$

Here, a, b, and c are semi-axes of the ellipsoid, E_i are non-normalized XY-polynomials, and A_i are the corresponding coefficients. Although use of this equation may decrease the optimization efficiency, it allows to apply standard modeling features. Eight XY polynomials, including all the YZ-symmetric modes up to the 4th order, were used. The freeform shape is also presented in Figure 12c and Table 3. The surface shape is driven by the large astigmatism of Rowland-type mounting. It causes a high difference between the curvatures in tangential and sagittal directions, so it becomes impossible to find the BFS, and, as a result, a plane is used as the reference shape.

The grooves pattern of the grating in Design C has a varying period, which follows the law:

$$\frac{1}{N} = \frac{1}{N_0} + \alpha y + \beta y^2 + \Gamma y^3 + \Delta y^4 + \epsilon y^5, \quad (5)$$

where N_0 is the grooves frequency in the vertex in μm^{-1}, $\alpha - \epsilon$ are the non-uniformity coefficients and y is the coordinate of point on the grating surface in mm. Such a pattern can be formed by holographic recording or etched through a mask. The values found by optimization are: $\alpha = 6.518 \times 10^{-4}\ \frac{\mu m}{mm}$, $\beta = 1.718 \times 10^{-5}\ \frac{\mu m}{mm^2}$,$\gamma = 5.184 \times 10^{-7}\ \frac{\mu m}{mm^3}$, $\delta = 3.314 \times 10^{-9}\ \frac{\mu m}{mm^4}$, $\epsilon = 2.990 \times 10^{-11}\ \frac{\mu m}{mm^5}$. Applying these coefficients in Equation (5) results in varying of the grating period by 0.042 µm, or 6.4%, which is technologically achievable [19].

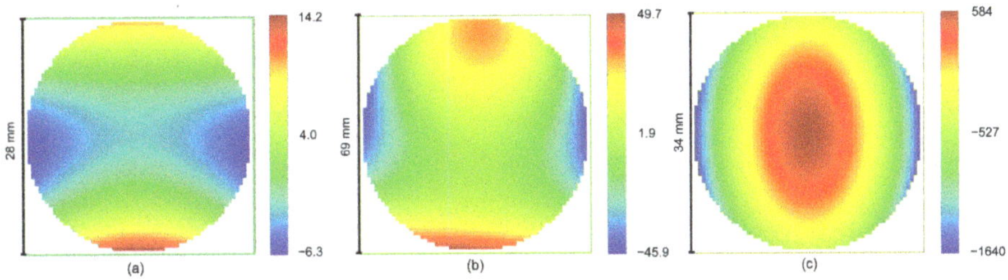

Figure 12. Deviation from the best fit sphere (in microns) of the freeform mirrors surfaces.

Table 3. Freefrom mirros asphericity data.

	Design A	Design B	Design B
Type	Zernike sag	Zernike sag	Ellipsoid + XY pol.
Diameter/mm	28.3	68.8	34.0
BFS radius/mm	195.9	282.5	flat
Max. BFS deviation/µm	14.2	29.7	543.6
RMS BFS deviation/µm	3.6	9.8	1636.7

The data indicate that the freeforms designs A and B are both driven by the primary coma and astigmatism and have slowly varying sag with a moderate peak deviation. Both of them should be feasible with the current level of the freeform polishing technology [20]. One may also note that the freeform in the design B is slightly asymmetrical to compensate the beam deviation from the YZ plane. The freeform shape obtained in the design C appears to be more challenging. However, the surface is relatively smooth; the higher XY orders contribute to the total sag only by 23.7 µm or 1.1%.

4.2. Curved Detectors

The feasibility of the found curved detectors shapes can be estimated through the required thickness of the silicon dye. The curving process was modeled in Reference [21], both analytically and numerically by finite-element analysis. As a result, a set of calibration curves was produced. They define the minimum radius of curvature, which can be achieved for a sample of given length and thickness without risk of breakage. The calibration curves for a 28-mm length sample are shown in Figure 13. The optimal detector radius in design A is extremely large, so it can be considered as flat. Applying the calibration curve for the detector radius in design B, which equals to $R_B = 186.2$ mm, we find the required thickness of $t_b = 306.4$ µm. This value is technologically achievable. Repeating the same procedure for the design C radius of $R_C = 107.9$ mm, we obtain the thickness equal to 21.9 µm. This value is practically unreachable with the current back-thinning process. This means that the existing curving process, which was developed to produce spherically shaped image detectors, cannot be applied in this case directly. However, typical spectrographs work with linear detector arrays. This implies that the size of detector in sagittal direction can be smaller than that in tangential direction by two orders of magnitude. In this case, one can neglect the curvature in the sagittal plane and consider curving the detector into a cylindrical shape instead of a spherical one. A cylindrical curving generates less mechanical stress, and a simplified linear model can be used to compute the required thickness. Referring to the linearized model from the same source [21] (see the dashed line in Figure 13), we find the dye thickness of $t'_C = 70.6$ μm. This value is close enough to the thicknesses obtained with the existing technology. This implies that the computed detector shape is technically feasible but requires some revision of the curving process to generate a cylindrical shape instead of a spherical one.

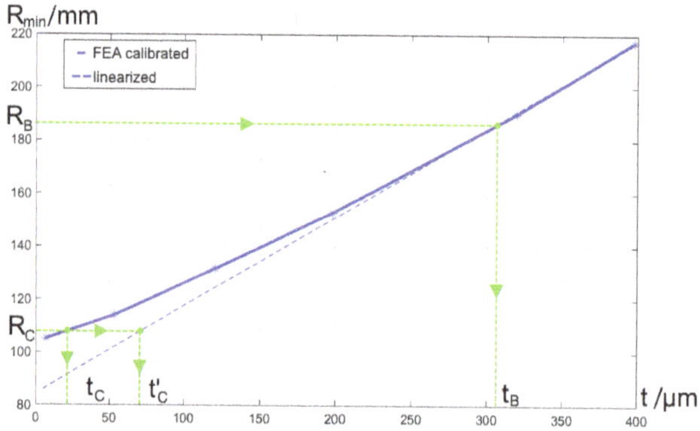

Figure 13. Dependence of the achievable detector radius of curvature on the dye thickness.

5. Results and Discussion

In the previous sections, it was shown that field curvature aberration is intrinsic for high dispersion spectrographs and that they can be successfully eliminated by the use of curved detectors. In addition, it has been demonstrated that application of the freeform optics technologies allows us to create new optical designs with merged functionality of the components. The imaging performance analysis indicates that the new designs provide high image quality, while decreasing the number of components and the overall dimensions. The research stage presented in this paper is focused on comparison of the optical design concepts and estimation of the performance gain reachable with the freeforms and curved detectors. But, it is also important to assess the performance difference for a certain target application, i.e., for FBG response sensing. It would be practically difficult and even excessive to make an experimental proof for each of the proposed designs, so we have to rely on simulations.

We consider an FBG, with a Gaussian profile of 0.05 nm FWHM centered around 835 nm [22]. We take the narrowest profile for simulations to make the difference more visible. It is recomputed to an ideal image profile with the scale of spectrograph linear dispersion equal to 1.74 nm/mm. Then, it is convolved with the 50 μm-width rectangular function representing the input fiber acting as a spatial filter. Finally, the result is convolved with the line spread functions (LSF) computed for each design. The computation is repeated for a centered FBG signal and a signal spectrally shifted by 30 pm. The results of the image simulations are shown in Figure 14.

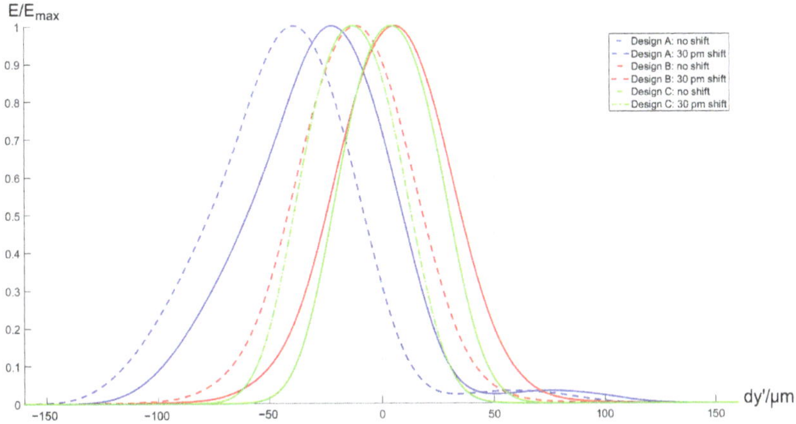

Figure 14. Simulation of fiber Bragg grating (FBG) profiles obtained with different designs.

The plot shows that the curve displacement corresponding to the signal spectral shift is identical for all the designs. The output curve for Design A has a notable asymmetrical broadening with FWHM of 69.1 μm, 9.9% of FWHM asymmetry, and lateral offset of 21.7 μm due to the coma-type aberration. In addition, it has a 3.3% pedestal-type artifact at the bottom part. All of these imperfections are signatures of the optical design. The offset effect may require an additional calibration, while the curve asymmetry and broadening will decrease the precision of the line position measurement, and the artifact will decrease the maximum number of sensors, which can be readout at the same time.

The simulated output for Design B is almost free of the aberrations signatures. It has a minor offset of 1.0 μm and a residual symmetrical broadening with 60.04 μm FWHM and 1.03% asymmetry.

The signal for Design C is even closer to the initial convolution of Gaussian profile and rectangular function. The offset is 0.02 μm, the broadened FWHM is 51.9 μm, and the asymmetry is 0.02%, which is negligible.

Thus, all of the developed designs can be applied for the FBG sensing tasks. Design A is notable for using simpler optics and low field curvature. However, it uses more optical components, has larger dimensions and demonstrates decentering and broadening in the output due to the residual aberrations. Design B uses less components and has a reduced volume together with a better imaging performance in an FBG sensing application. This solution may be more sensitive to manufacturing errors because of its double-pass geometry. In addition, such an off-plane design is challenging for development of the mechanical design and performing the alignment. Finally, Design C shows a superior imaging performance and consists of only two components. This result is achieved by the use of technologically challenging freeform grating and cylindrical curved detector, as well as by allowing a relatively high astigmatic elongation of the image.

6. Conclusions

In the present paper, we considered the prospects of application of freeform optics and curved detectors technologies in high-resolution spectrographs to be used for FBG sensors readout. We considered a high-dispersion spectrograph design with two transmission gratings for the range of 810–860 nm. The optimization and modeling results have shown that using of a freeform camera mirror in this design allows to correct the aberrations and reach a spectral resolution limit up to 89 pm.

Simultaneous application of the freeform optics and curved detectors technologies makes it possible to merge the optical components functions and create a design of two components: a freeform focusing mirror and a double-passed immersed grating. To the best of our knowledge, the proposed design is new for this type of application. This solution has approximately the same optical performance, and both the freeform and curved detector are feasible with the current technological level.

Finally, it was shown that using a freeform grating with a complex grooves pattern and a curved detector allows us to revise a classical design with a concave grating in a Rowland-circle mounting. This design consists of only one optical element and allows us to reach spectral a resolution limit of 139 pm. The grating, in this case, has an astigmatism-driven freeform shape with low contribution of the high orders, while the detector has a high curvature, which can be achieved for a linear detector curved into a cylindrical shape.

It was demonstrated by means of simulation that all of the designs exhibit a good performance in an FBG sensing application. Each of them has some intrinsic advantages and drawbacks in terms of image quality, size, and technological complexity. Depending of the practical priorities, one of these designs can be chosen for further investigations, including practical implementation and experimental studies.

Author Contributions: Conceptualization, E.M. and N.P.; methodology, E.M.; software, E.M.; validation, N.P. and O.M.; formal analysis, O.M. and I.N.; investigation, I.N.; resources, E.H. and S.L.; data curation, E.M. and S.L.; writing—original draft preparation, E.M.; writing—review and editing, N.P.; visualization, E.M.; supervision, E.H.; project administration, E.H.; funding acquisition, E.H. All authors have read and agreed to the published version of the manuscript.

Funding: This research was funded by the European Research council through the H2020-ERC-STG-2015-678777ICARUS program. O.M. was funded by RFBR, DST, NSFC and NRF according to the research project 19-57-80006 BRICS_t, I.N. was funded by Ministry of Science and Higher Education of RF according to Agreement N° 075-03-2020-051 (fzsu-2020-0020).

Acknowledgments: The authors would like to thank their colleagues Wilfried Jahn and Thibault Behaghel from CURVE-ONE for their help with the curving process model calibration. The authors are grateful to their colleagues, Isabelle Racicot and Felipe Pedreros, from Laboratoire d'Astrophysique de Marseille for proofreading the manuscript.

Conflicts of Interest: The authors declare no conflict of interest.

References

1. Alvarez-Botero, G.; Baron, F.E.; Cano, C.C.; Sosa, O.; Varon, M. Optical sensing using Fiber Bragg Gratings: Fundamentals and applications. *IEEE Instrum. Meas. Mag.* **2017**, *20*, 33–38. [CrossRef]
2. Cheng, L.; Ahlers, B.; Toet, P.; Casarosa, G.; Appolloni, M. Multi-parameter Fibre Bragg Grating sensor-array for thermal vacuum cycling test. *Internat. Conf. Space Opt.* **2008**, *10566*, 105661E.
3. Nureev, I.I.; Morozov, O.G.; Agliyullin, A.F.; Purtov, V.V.; Ovchinnikov, D.L.; Anfinogentov, V.I.; Vinogradov, V.Y. Microwave photonic polyharmonic probing for fiber optical telecommunication structures and measuring systems sensors monitoring. *Proc. SPIE* **2018**, *10774*, 10741J.
4. Kafarova, A.M.; Faskhutdinov, L.M.; Kuznetzov, A.A.; Minaeva, A.Y.; Sevruk, N.L.; Nureev, I.I.; Vasilets, A.A.; Bourdine, A.V.; Morozov, O.G.; Burdin, V.A. Experimental researches of Fiber Bragg Gratings operating in a few-mode regime. *Proc. SPIE* **2016**, *9807*, 98070L.
5. Rim, S.B.; Catrysse, P.B.; Dinyari, R.; Huang, K.; Peumans, P. The optical advantages of curved focal plane arrays. *Opt. Express* **2008**, *16*, 4965–4971. [CrossRef]
6. Dinyari, R.; Rim, S.B.; Huang, K.; Catrysse, P.B.; Peumans, P. Curving monolithic silicon for nonplanar focal plane array applications. *Appl. Phys. Lett.* **2008**, *92*, 091114. [CrossRef]
7. Iwert, O.; Ouellette, D.; Lesser, M.; Delabre, B. First results from a novel curving process for large area scientific imagers. *Proc. SPIE* **2012**, *8453*, 84531W.
8. Lombardo, S.; Behaghel, T.; Chambion, B.; Jahn, W.; Hugot, E.; Muslimov, E.; Roulet, M.; Ferrari, M.; Gaschet, C.; Henry, D.; et al. Curved CMOS sensor: Characterization of the first fully functional prototype. *Proc. SPIE* **2018**, *10679*, 1067910.
9. Lombardo, S.; Behaghel, T.; Chambion, B.; Caplet, S.; Jahn, W.; Hugot, E.; Muslimov, E.; Roulet, M.; Ferrari, M.; Gaschet, C.; et al. Curved detectors for astronomical applications: Characterization results on different samples. *Appl. Opt.* **2019**, *58*, 2174–2182. [CrossRef] [PubMed]
10. Lombardo, S.; Muslimov, E.; Lemaître, G.; Hugot, E. Next-generation telescopes with curved focal surface for ultralow surface brightness surveys. *Mon. Not. R. Astron. Soc.* **2019**, *488*, 5057–5064. [CrossRef]
11. Thompson, K.P.; Rolland, J.P. Freeform Optical Surfaces: A Revolution in Imaging Optical Design. *Opt. Photonics News* **2018**, *23*, 30–35. [CrossRef]
12. Rose, B. Wavelength Division Multiplexed Device. U.S. Patent 6,978,062,B2, 20 December 2005.

13. Smith, W. Image Formation: Geometrical and Physical Optics. In *Handbook of Optics*; Driscoll, W., Ed.; McGraw-Hill: New York, NY, USA, 1978; pp. 40–41.
14. I-MON 835 OEM. Available online: https://ibsen.com (accessed on 30 September 2020).
15. CMOS Linear Image Sensor. S13496. Available online: https://www.hamamatsu.com (accessed on 30 September 2020).
16. van Amerongen, A.; Krol, H.; Grèzes-Besset, C.; Coppens, T.; Bhatti, I.; Lobb, D.; Hardenbol, B.; Hoogeveen, R. State of the art in silicon immersed gratings for space. *Proc. SPIE* **2015**, *105643*, 105642R.
17. Rodenhuis, M.; Tol, P.J.; Coppens, T.H.; Laubert, P.P.; van Amerongen, A.H. Performance of silicon immersed gratings: Measurement, analysis, and modeling. *Proc. SPIE* **2015**, *9626*, 96261M.
18. Palmer, C.A.; Loewen, E.G. *Diffraction Gratings Handbook*; Rochester: Newport Corporation, NY, USA, 2014; p. 271.
19. Voronov D. L. Variable line spacing diffraction grating fabricated by direct write lithography for synchrotron beamline applications. *Proc. SPIE* **2014**, *9207*, 920706.
20. Muslimov, E.; Hugot, E.; Lombardo, S.; Roulet, M.; Ferrari, M. Freeform optics complexity estimation: Comparison of methods. *SPIE Proc.* **2018**, *10679*, 106791M.
21. Jahn, W. Curved sensors for wide field optical systems. In *Innovative Focal Plane Design for High Resolution Imaging and Earth Observation: Freeform Optics and Curved Sensors*; Aix-Marseille University: Marseille, France, 2017; pp. 109–121.
22. Werneck, M.M.; Allil, R.C.S.B.; Ribeiro, B.A.; de Nazaré, F.V.B. *A Guide to Fiber Bragg Grating Sensors: Current Trends in Short and Long-Period Fiber Gratings*; INTECH: Vienna, Austria, 2013; 24p.

sensors

MDPI

Letter

Demonstration of a Filterless, Multi-Point, and Temperature-Independent Fiber Bragg Grating Dynamical Demodulator Using Pulse-Width Modulation

Joao B. Rosolem [1,*], Marcio C. Argentato [1], Fábio R. Bassan [1], Rivael S. Penze [1], Claudio Floridia [1], Artur de A. Silva [1], Deleon Vasconcelos [2] and Marcelo A. Ramos Junior [2]

1 CPQD Research and Development Center in Telecommunications, Campinas, SP 13086-902, Brazil; marcio.colazza@gmail.com (M.C.A.); fbassan@cpqd.com.br (F.R.B.); rpenze@cpqd.com.br (R.S.P.); floridia@cpqd.com.br (C.F.); arturs@cpqd.com.br (A.d.A.S.)
2 Centrais Elétricas da Paraíba, João Pessoa, PB 58.000-000, Brazil; deleon.vasconcelos@utepasa.com.br (D.V.); marcelo.agra@utepasa.com.br (M.A.R.J.)
* Correspondence: rosolem@cpqd.com.br; Tel.: +55-19-3705-6796

Received: 10 September 2020; Accepted: 13 October 2020; Published: 15 October 2020

check for updates

Abstract: We demonstrated in this work a filterless, multi-point and temperature-independent FBG (fiber Bragg grating) dynamical demodulator using pulse-width-modulation (PWM). In this approach, the FBG interrogation system is composed of a tunable laser and a demodulator that is designed to detect the wavelength shift of the FBG sensor without any optical filter making it very suitable to be used in harsh environments. In this work, we applied the proposed method that uses the PWM technique for FBG sensors placed in high pressure and high-temperature environments. The proposed method was characterized in the laboratory using an FBG sensor modulated in a frequency of 6 Hz, with a 1 kHz sweeping frequency in the wavelength range from 1527 to 1534 nm. Also, the method was evaluated in a field test in an engine of a thermoelectric power plant.

Keywords: FBG; PWM; engine; high temperature; ECU; FBG demodulator

---

1. Introduction

For practical reasons, the control electronics for many types of power machinery are usually placed inside or close to the operating environment of the engine. For example, the monitoring modules for dynamic pressure, temperature, and knock are placed directly in contact with the engine [1]. The electronic modules that have been developed for monitoring different types of engines (naval, thermoelectric, planes, military, and automotive) with specific emphasis on durability at high-temperature operation [2–4].

Monitoring the instantaneous combustion chamber pressure data is required for the closed-loop control of the fuel mass fraction burned in the engines [5,6]. The pressure sensors for this control must be durable and accurate. Using closed-loop control improves engine performance and reduces the emission of pollutants. The feedback system contains an intelligent data analysis system working with an ECU (engine control unit) to precisely dose the fuel quantity in each combustion cycle of each of the engine cylinders. In thermoelectric engines, the dynamical pressure has peaks higher than 250 bar, and the temperature in the combustion chamber is higher than 300 °C. Piezoelectric sensors used to measure the pressure of the combustion chamber currently are not durable when used continuously in high temperatures (>300 °C) [5]. Thus, this application needs a robust and trusty pressure sensor.

Optical fiber sensors are a good alternative to electronic sensors in many engineering applications due to some intrinsic advantages, such as high temperature and chemical resistance and potential for long-lifetime operation. A fiber Bragg grating (FBG) sensors have flexible characteristics, low cost and are readily available [7]. Examples of FBG sensor applications include structural health monitoring in civil engineering [8,9], electric power systems [10,11], railways and roadways monitoring [12,13], in oil tanks monitoring and as chemical sensors [14,15], in biomechanics and in medicine [16,17]. FBG sensors can be fabricated using optical fibers made of different materials, such as glass [18], polymers [19], or sapphire [20].

Another issue of sensing system operation in a thermoelectric power plant is regarding the FBG interrogators. Although many commercial FBG interrogators modules can measure parameters running in frequencies higher than the engine's combustion cycles, they cannot be installed close to or inside the engines. The thermoelectric engine is high power machinery that dissipates a lot of heat. The temperature can vary depending on each part of the engine casing. For example, at the pressure monitoring point, the temperature in the case is higher than 300 °C. In the other parts that have water cooling, the temperature is lower. The average temperature in the machine room reaches 55 °C. Considering that the ECU and the FBG demodulator need to be installed close to the engine or its external body, the high temperatures will affect the operation of a supposed interrogation module. In the interrogation module that uses lasers, semiconductor optical amplifiers (SOAs) and Fabry-Perot filters [21], the cooler of the optoelectronics elements of the interrogator, will work in an excessive regime of operation, causing the device to fail in advance [22]. Passive devices such as WDM multiplexers, splitters, circulators also will be affected by the high temperatures. On the other hand, the current technology of electronic devices that can be used in a high-temperature environment is much more available than optoelectronic technology [2]. In addition, standard interrogators are not able to send the signals from all FBG pressure sensors obtained on each sweep to each engine ECU. Many commercial FBG interrogators record the measured data in an external computer in csv or text files. When a single data reading must be taken for control of each cylinder, the receiver of the optical sensing system could use, for example, the optical-edge-filtering technique [23] to detect the dynamic pressure. However, the FBG center wavelength shift due to temperature variation in the combustion chamber depends on the engine load [24]. Thus, the correct positioning of the optical edge filter in real operation is critical to reproduce the exact dynamic pressure behavior of the combustion chamber. Although some techniques were implemented to solve this problem [25], realizing this measurement in a real application is still a serious challenge and the complexity of the optical source and filter control of the edge filtering technique limits its use in this application. An alternative method based on the dispersion delay effect of a dispersion-compensating fiber (DCF) can also be used to convert the FBG wavelengths into the time domain [26], but its demodulator cannot be used near the engine.

In this work, we proposed an innovative filterless, multi-point, and temperature-independent FBG dynamical demodulator using the pulse-width-modulation (PWM) technique. PWM is a modulation technique that generates variable-width pulses to represent the amplitude of an analog input signal [27]. In [28], an interrogation system is presented based on pulse-modulation, that automatically recognized reflection signals of FBGs even when the FBGs are installed in an arbitrary order or at a long distance and affected by delays. This recognition technique was realized using pulse-modulating in the wavelength-swept laser. However, in contrast to [28], our proposed scheme has a tunable laser that sweeps a pre-set wavelength band where the FBG sensors work in continuous-wave (CW) mode. The demodulator using robust electronic devices can be used near the engine, near the pressure sensor and even integrated with the ECU to transform the wavelength variation-based signal, to a PWD signal and finally in an analog intensity signal compatible with the ECU input port. This approach can be used to measure many types of parameters using FBG sensors. In this work, we applied this method for FBG sensors placed in high pressure and high-temperature environments. The system was characterized in the laboratory using an FBG sensor modulated at a frequency of 6 Hz and a tunable laser with a 1 kHz sweeping frequency and wavelength range from 1527 to 1534 nm. A commercial FBG interrogator was used to compare the results obtained in this application. Besides, it was evaluated in a field test in an engine of a thermoelectric power plant.

2. The Proposed FBG-PWM Demodulator

Figure 1 shows the entire interrogation system where the FBG-PWM demodulator is used. The optical source for this system is a tunable laser with an appropriate sweep frequency. The sweep frequency must be higher than the maximum sensor frequency response to have enough sampled points during the measurements. In the diagram of Figure 1, the laser output can be divided for many demodulators in a power plant. Each splitter output is connected to the sensor using port 2 of an optical circulator. The FBG sensor in Figure 1 is used to measure dynamical pressure inside an engine of a thermoelectric plant engine where the temperature is very high (>400 °C) and is not stable. Port 3 of the circulator is used to connect the FBG reflected signal to the demodulator input. In the demodulator unit, the FBG wavelength-shifted signal follows first to the photodetector. In the photodetector, the optical signal is converted to electrical. Next, a transimpedance amplifier amplifies and clips the signal. Next, a flip-flop type D (FFD) digital circuit transforms it in a PWM signal. Next, low pass active filters are used at the output of the FFD to smooth the pulse train into a stable analog voltage. This analog voltage is the recovered FBG wavelength shifting signal. In other words, the variation of pulse width (PWM) is converted to an analog voltage directed related to the FBG wavelength shifting that in turn, is related to the original engine cycles modulation. This signal is sent to an ECU, which analyzes the signal and provides the correct commands to the engine in closed-loop control.

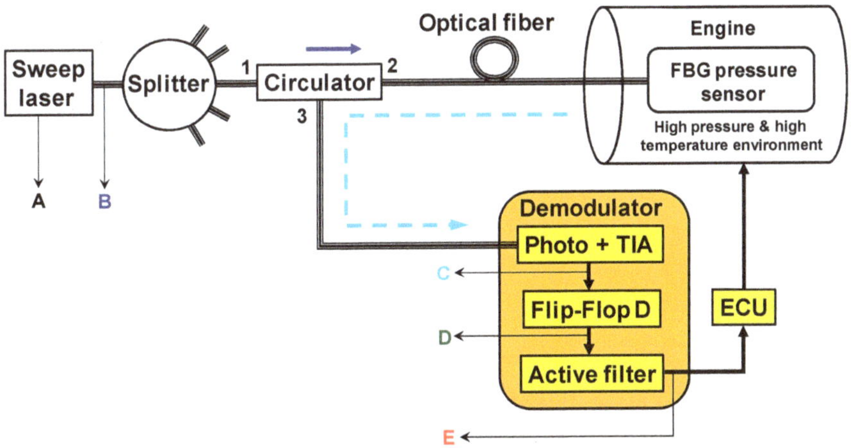

Figure 1. The proposed FBG-PWM demodulator. The letters from A to E represent points with signals waveforms from the sweep laser to the demodulator output.

In this proposed sensing system scheme, only the pressure sensors and the demodulators need to be placed close or on the case of the engine integrated with the ECU. The tunable laser mainly can be installed in a room with controlled temperature.

The key elements to implement the filterless, multi-point, and temperature-independent FBG dynamical demodulator are a tunable ring sweep laser [29] and the FFD digital circuit plus the active filter. The tunable source for this application can be fiber lasers based on semiconductor optical amplifiers or erbium-doped fiber using the Fabry-Perot filter. These lasers have narrow linewidth (<5 pm) and high output power (>10 dBm).

Once the laser output changes in wavelength overtime during the sweep, the optical wavelength variation of FBG in the engine becomes an electrical time variation in the demodulator. When the FFD receives the analog time variation signal of the FBG, it transforms this signal in a digital signal form with a pulse width variation. The pulse width variation has the information of dynamical pressure modulated in the FBG. Low pass active filters (two second-order Butterworth low pass filters) are used at the output of the PWM circuit (D) to smooth the pulse train into a stable analog voltage. The electrical-active-filter removes the digital modulation of the PWM signal recovering the original FBG modulated-signal. Then, this recovered signal is sent to the ECU. No synchronization signal is necessary for this system.

Figure 2 shows the signal waveforms from the tunable laser to the demodulator output. In this figure, the signal E shows a typical engine combustion cycle. The signal-A shows the electrical sweep of the tunable laser. The signal B is the laser output intensity. Although the signal intensity in B is constant in time, the wavelength increases during the positive sweep slope, and it decreases during the negative sweep slope. The signal-C is the FBG electrical signal that is modulated by the dynamical wavelength variation. The signal-D is the digital output of the FFD circuit with pulse width modulation. The signal-E is the active filter output showing an example of a typical engine combustion cycle.

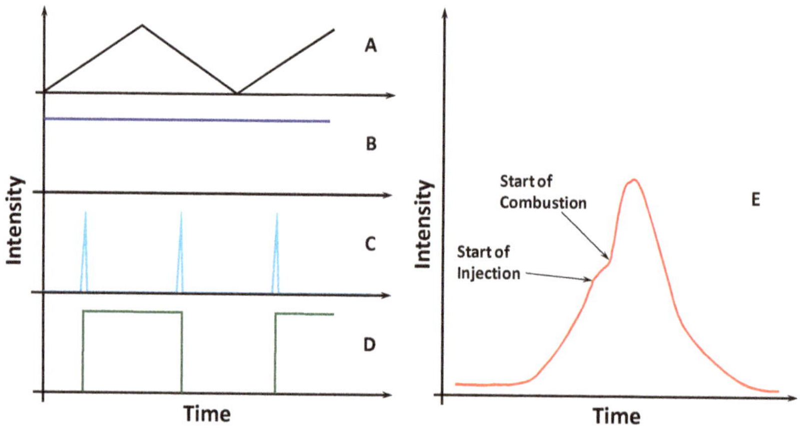

Figure 2. The letters from A to E represent points with signals waveforms from the sweep laser to the demodulator output. In this figure, the signal in E shows a typical engine combustion cycle.

3. Demodulator Evaluation in Laboratory

To demonstrate this technique, we first tested the proposed system in the laboratory. Figure 3a shows the scheme to simulate the temperature changes and dynamic pressures on the FBG. An arbitrary waveform generator (BK4054B, B&K Precision Corporation, Yorba Linda, CA, USA) produced a typical 6 Hz engine combustion frequency. This generator waveform voltage was amplified by one piezoelectric driver connected to a piezoelectric transducer (model PK2FQP2- Figure 3b, (Thorlabs, Newton, NJ, USA) that stressed a polyimide coated FBG coupled into the transducer. The tunable laser used a triangle waveform frequency of 1 kHz to sweep the central wavelength from 1527 to 1534 nm. Figure 3c shows more details of the demodulator-electronic-circuit. The photodetector plus transimpedance amplifier has a bandwidth of 400 kHz and, the active filter was composed of two second-order Butterworth low pass filters with 50 Hz bandwidth. The FFD used was a 74HC74 digital circuit.

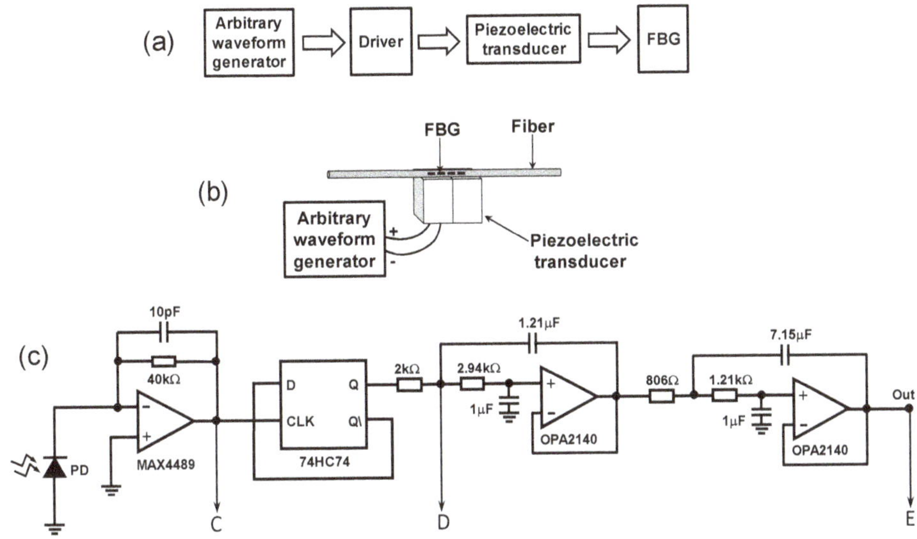

Figure 3. (**a**) Scheme to simulate the temperature changes and dynamical pressures on FBG, (**b**) FBG coupled to a PZT, and (**c**) electronic circuit of the demodulator.

The type D flip flop circuit changes the output logic level in the rising edge of the signal in the clock input port. The FBG electrical signal from the transimpedance amplifier circuit is inserted into the type D flip-flop clock input port and, the inverted flip-flop output port is connected to the input port of the same chip. Thus, when the swept light is reflected by the FBG just the rising edge of the spectrum alters the flip-flop output. When the sweep occurs from shorter to longer wavelengths (positive slope of signal A), it is the left edge that changes the flip-flop output, and, when the sweep occurs from longer to shorter wavelengths (negative slope of signal A), the right edge changes the flip-flop. Since the FBG is varying dynamically from longer to shorter wavelengths and vice versa, variations in rising and falling edge will create a PWM modulation.

The output voltage in E can be express by (1) [30]:

$$V_O = \delta \cdot V_{PWM} \tag{1}$$

where V_O is the averaged output voltage, δ is the duty cycle of the PWM waveform and V_{PWM} is its amplitude. Considering that the tunable laser sweep time determines the total spectral range (BW) and the FBG produces a dynamical time variation signal proportional to dynamical wavelength shift ($\Delta\lambda$), δ can be written as:

$$\delta = \Delta\lambda / BW \tag{2}$$

and V_O can be written as:

$$V_O = (\Delta\lambda / BW) \cdot V_{PWM} \tag{3}$$

Therefore V_O can be increased without reducing the noise-signal ratio reducing the spectral range of the sensing system. Figure 4a,b show the measured signals in C (red) and D (blue) respectively for FBG position in minimum (a) and maximum PZT displacement (b). Figure 5a,b are the PWM signals in D for two distinct PZT displacement amplitudes. These signals were measured in E using an oscilloscope.

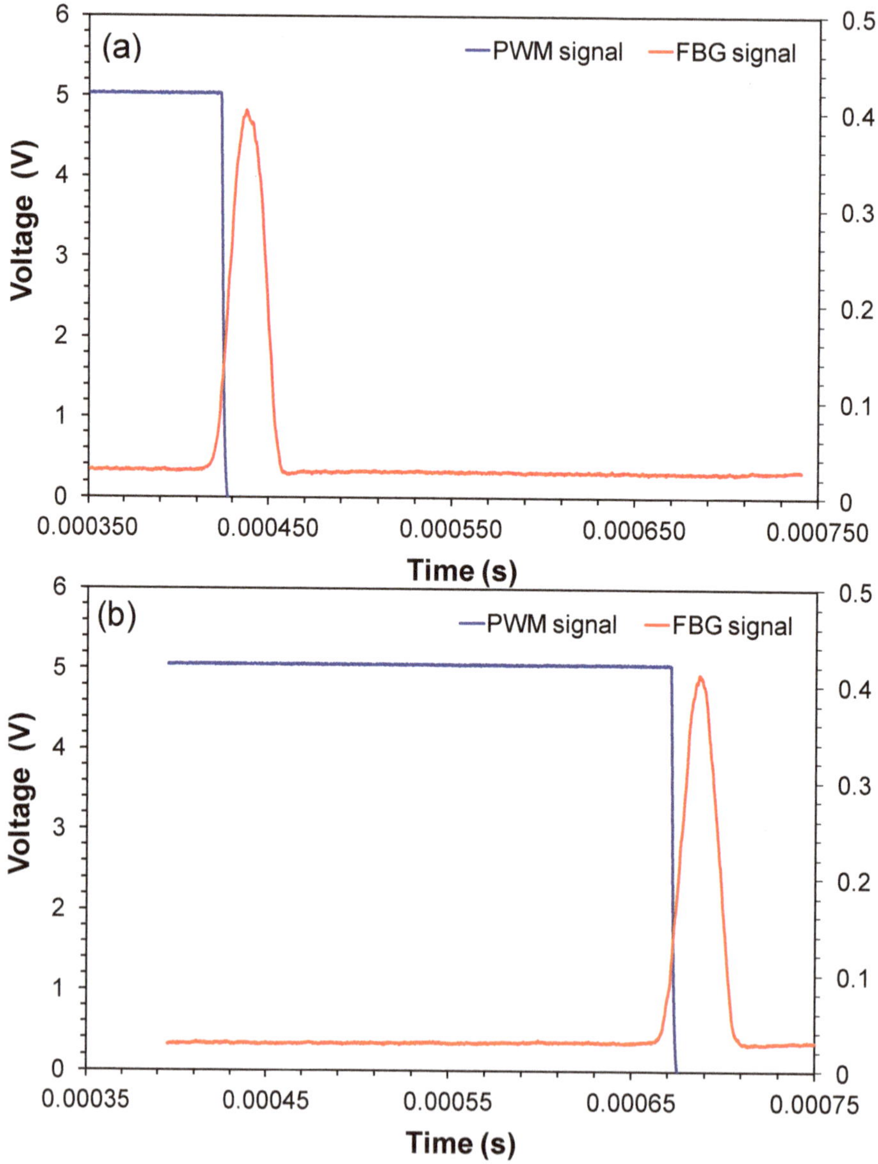

Figure 4. (**a**) and (**b**) Signal in C and D respectively for FBG position in minimum and maximum PZT strain.

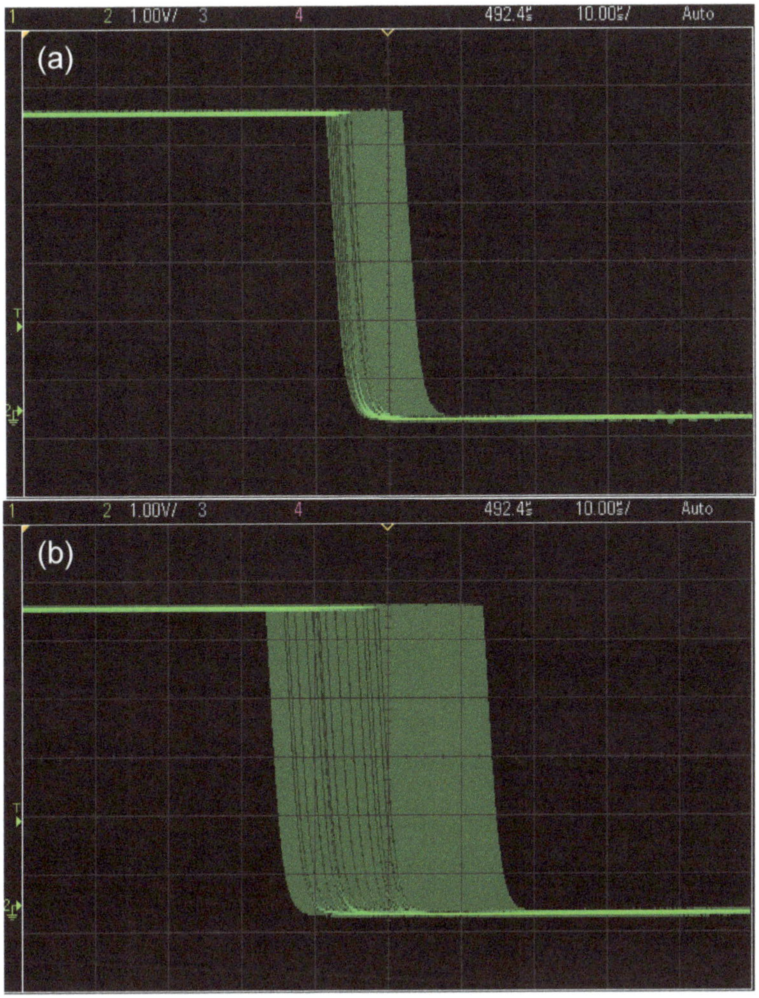

Figure 5. PWM signals in D for two distinct PZT strain amplitudes, (**a**) strain amplitude = 0.2 Vpp and (**b**) strain amplitude = 1.4 Vpp. The vertical axis is voltage (1V/div) and the horizontal axis is time (10 µs/div).

A critical evaluation regarding this proposed system refers to the characteristic of the PWM demodulated signal in terms of trustworthiness to the original FBG modulated signal. We compare in Figure 6 the modulation signal of the arbitrary-waveform-generator, the FBG signal measured in E, and the signal measured of FBG using a commercial FBG interrogator (100 Hz sweep frequency si155 Hyperion from Micron Optics, Atlanta, GA, USA). The commercial interrogator signal was obtained after post-processing. We can observe in Figure 6 that the demodulated signal in E is a good copy of the generator signal waveform; however, some noise can be observed in the signal base. We will comment on the noise source in Section 4. Also, we observed that the signal of the commercial interrogator has not enough sampled points to define all the events in an engine cycle curve. Next, we evaluated quantitatively, the demodulated signal characteristics.

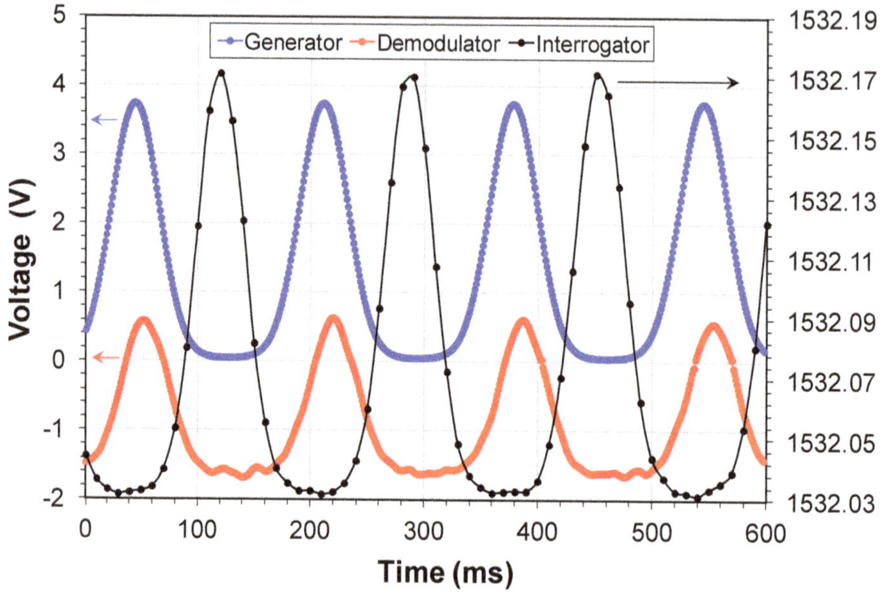

Figure 6. Qualitatively comparison of the signal of the waveform arbitrary generator, the signal in the demodulator output, and the signal measured using a commercial FBG interrogator.

Figure 7a,b show the linearity performance respectively for the demodulator and for the commercial interrogator for four different FBG center wavelengths from 1532.03 to 1532.78 nm. The optical input power in the photodetector was −16 dBm.

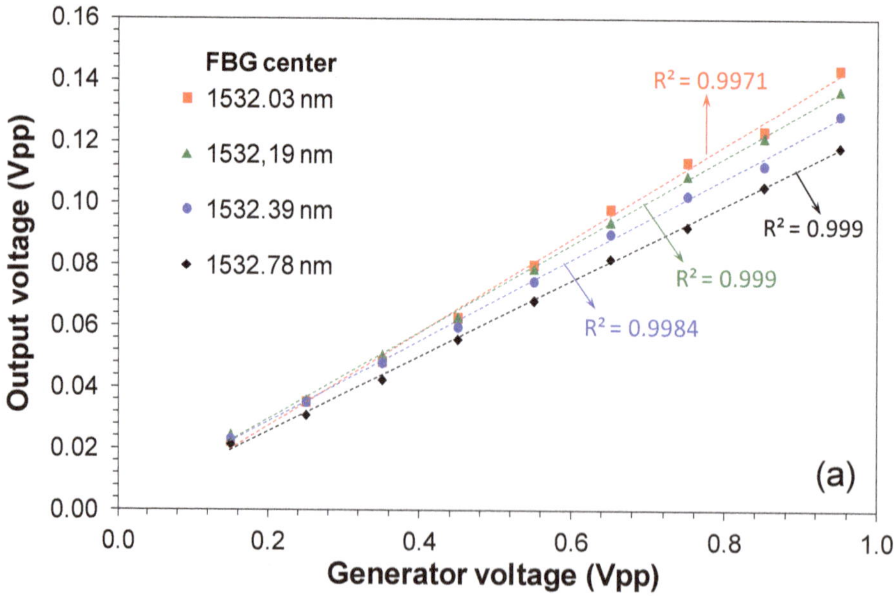

Figure 7. *Cont.*

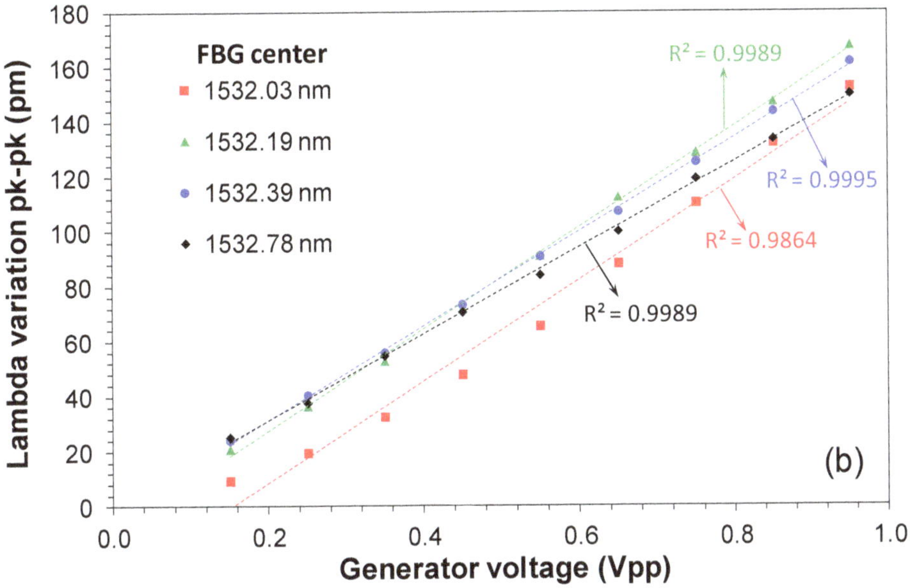

Figure 7. Signal linearity measured in four FBG center wavelengths, (**a**) measured in the demodulator output (E point) and (**b**) measured by a commercial FBG interrogator.

These wavelengths centers represent the FBG at different temperatures inside the engine. This wavelength range (750 pm) corresponds to a temperature variation of 57 °C for an FBG sensitivity of 13 pm/°C [24]. In this measurement procedure, we were limited in the wavelength range due to the PZT voltage limitation. The demodulator could measure the signal in the entire wavelength range determined by the tunable laser sweep (7 nm). We observed that the signals' linearities in the demodulator output are enough to reproduce the original characteristics of the FBG modulated signal with good quality. We attributed the variation in the offset of the curves to the PZT technical characteristics that are not stable with the time.

4. Demodulator Evaluation in a Field Test

The objective of the field test was to verify the performance of the PWM demodulator in terms of signal processing using a dynamical FBG pressure sensor installed in an environment with variable temperature. It was not the test proposal to test the demodulator itself at high temperatures in this project stage. The field tests take place in Centrais Elétricas da Paraíba (EPASA), which is a thermoelectric power plant. This thermoelectric power plant has an installed power of 340 MW, obtained from 40 model 3240 engines (MAN Diesel SE, Augsburg, Bavaria, Germany). The angular speed of each motor is 720 rpm, and heavy fuel oil (OCB1) is used to combustion engines. Each engine has 18 cylinders and uses a mechanical injection pump to control the fuel oil injection. This mechanism reduces the possibility of adjustments in the injected fuel volume and the same proportion limits the better management of the engines. In the field tests, the pressure sensors were connected in a pressure monitoring point available for each engine cylinder. To compare the pressure signals, we used again the commercial FBG interrogator and the data previously obtained from a reference sensor (model HLV 4.0 from Kistler Group, Winterthur, Switzerland), which is a standard sensor used in the thermoelectric power plant.

The thermoelectric power plant has a harsh environment. The internal average temperature in the machine room is around 55 °C, and close to the engines, it can be higher, limiting the continuous uses of standard electronic equipment.

Figure 8a shows the pressure sensor scheme [31]. The FBG was fixed in two points of a stainless steel substrate. A pre-stress was applied in FBG before the fixation. According to Figure 8a, the FBG is placed outside of the engine combustion chamber, and it is stressed by a mechanism composed of one 1-mm thickness membrane and one piston. Only one side of the membrane contacts the high-pressure and high-temperature gas inside the engine's combustion chamber. When the membrane is deformed, by the pressure, it moves a piston that stresses the FBG accordingly. In the field tests, we use FBG pressure sensors connected in a point of pressure monitoring available for each engine cylinder, as we can observe in Figure 8b.

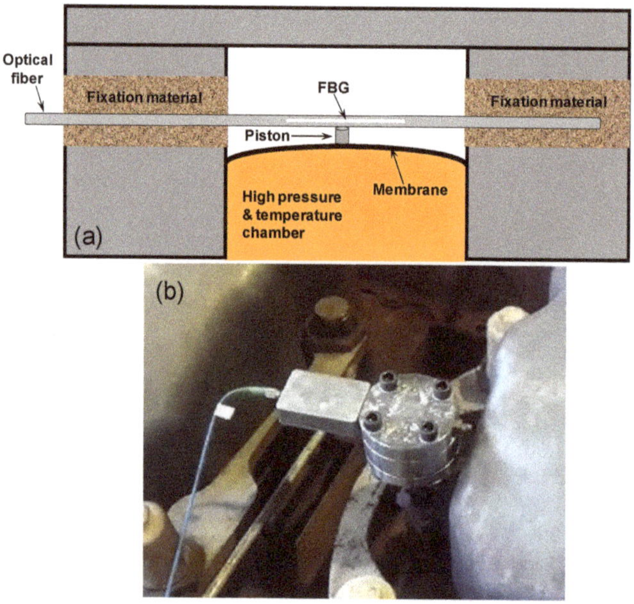

Figure 8. (**a**) Scheme of the FBG pressure sensor and (**b**) sensor connected in the pressure monitoring point of the engine.

Figure 9a shows the point in the engine where the FBG pressure sensor was installed. Figure 9b shows the demodulator, the tunable laser kit, and the interrogator installed in a control room, 50 m from the engine under test.

Figure 9. *Cont.*

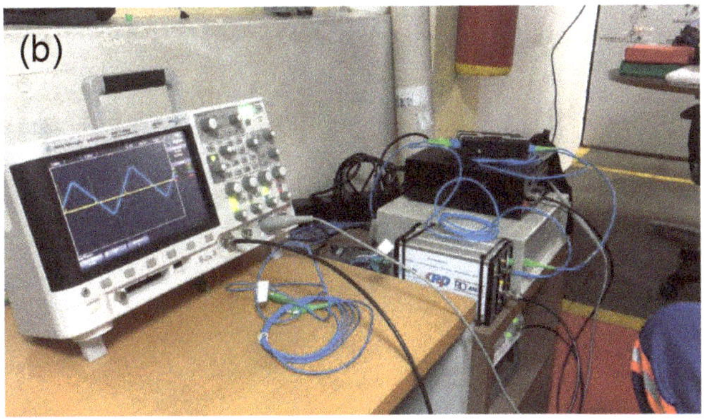

Figure 9. (**a**) Point in the engine where the FBG pressure sensor was installed and (**b**) interrogator and the demodulator tool kit installed in a control room 50 m from the engine under test.

Figure 10 shows the curves of wavelength shifting versus pressure for one FBG sensor obtained previously of the field trial, considering the Bragg wavelength in room temperature that was 1532.90 nm (sensor #1). This sensor was submitted to 3 cycles of static pressure to verify the sensor hysteresis. As we can observe, the sensor curves are linear. The R2 coefficient is 0.9943. The other pressure sensor used in the field test had the Bragg wavelength at room temperature at 1548.12 nm.

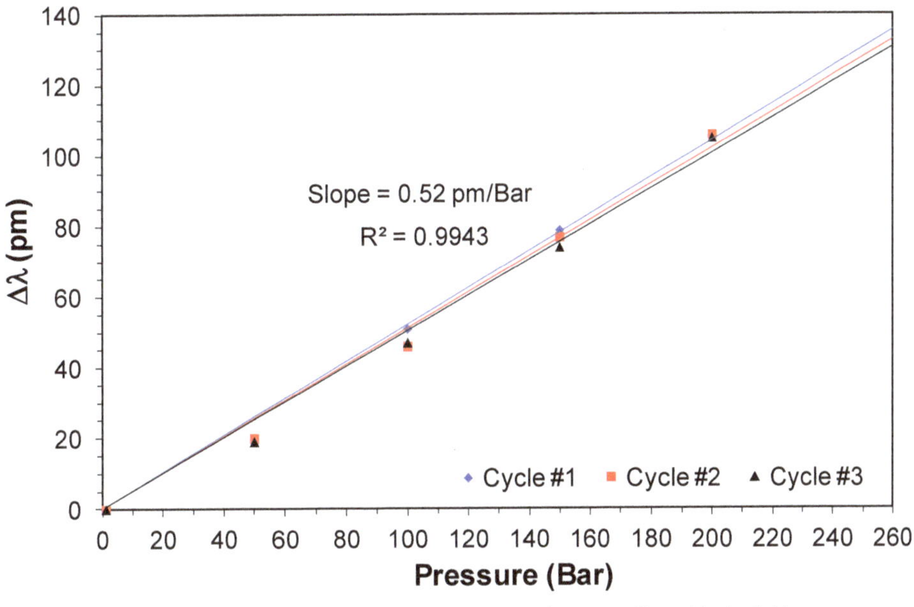

Figure 10. Wavelength shifting versus pressure sensor for sensor #1 used in the field test.

Figure 11 shows de PWM signal versus time of sensor #1 at the engine monitoring point obtained during the temperature stabilization period. The Video S1 (in Supplementary Materials) shows the evolution of this signal seen in an oscilloscope during the stabilization period.

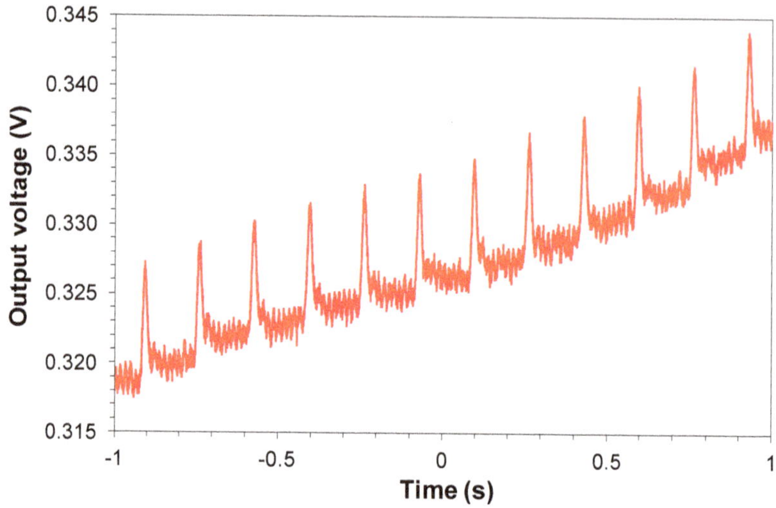

Figure 11. PWM signal versus time of sensor #1 that was obtained during the temperature stabilization period at the engine monitoring point.

Figure 12 shows a qualitative comparison of the dynamic curves of sensor #1 in the monitoring point of engine combustion obtained by the PWM demodulator (with 10 moving average) and the commercial interrogator in terms of wavelength shifting considering the Bragg wavelength in room temperature.

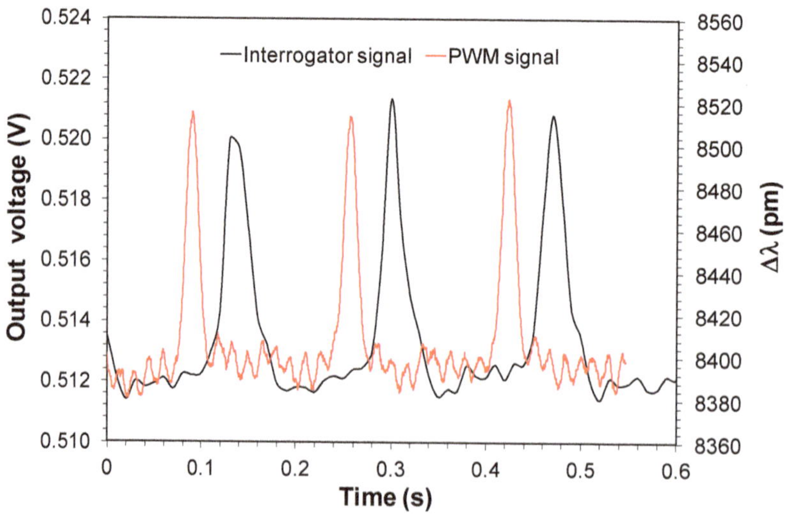

Figure 12. Dynamical curves of FBG pressure sensor in the monitoring point of engine combustion obtained by the PWM demodulator and the interrogator.

As we can observe, the output signal intensity of the demodulator in Figure 12 is lower than the one obtained in laboratory measurements showed in Figure 7a. In Figure 7a the peak-to-peak intensity is ~0.12 V versus ~0.009 V in Figure 12. This fact is partly attributed to the higher sweep range used in field tests (~20 nm) compared with the sweep range used in laboratory measurements (7 nm). This higher sweep range was implemented to measure different Bragg wavelengths of two

FBG sensors. Also, we observed that the PWM signal presents the narrowest linewidth compared with the interrogator signal.

The total wavelength shifting showed in the peak bases in Figure 12 for the interrogator signal (~8380 pm) is attributed to FBG sensor substrate deformation and the temperature of FBG in the monitoring point. Based on the stainless steel FBG substrate dimensions where the FBG was fixed (70 mm), its thermal expansion coefficient 16.10^{-6} °C^{-1}, and the FBG temperature sensitivity of 13 pm°C^{-1} [24], we can estimate that the temperature operation for this sensor was around 200 °C.

Finally, Figure 13 shows a comparison of the PWM signal and the reference sensor. The time scale was synchronized in order to have a better comparison of the temporal characteristics of the sensors. Except for the noise in the PWM signal base, it shows a response similar qualitatively to the reference sensor. The first hypothesis for noise was due to the tunable laser jitter [32], which would originate from the triangular signal source that sweeps the laser. In our experiments, we used the BK Precision model BK4054B waveform generator. This generator features an RMS 300 ps + 0.05 ppm cycle-to-cycle jitter in 1 kHz and 1 Vpp. Considering the sweep frequency of 1 kHz (1 ms cycle), a variation of 300 ps would have little effect on the creation of the observed noise. A second hypothesis raised would be due to the noise margin in the decision threshold of the Flip-Flop D 74LC74 logic gate, which could widen or shorten the PWM pulses due to the variation of the decision point. This hypothesis was discarded since the voltage levels provided by the optical receiver of the demodulator to the Flip-Flop D inputs were designed to work saturated. Finally, the most likely hypothesis is attributed to the residual noise originated from the demodulator's power supply. The noise frequency is close to 60 Hz.

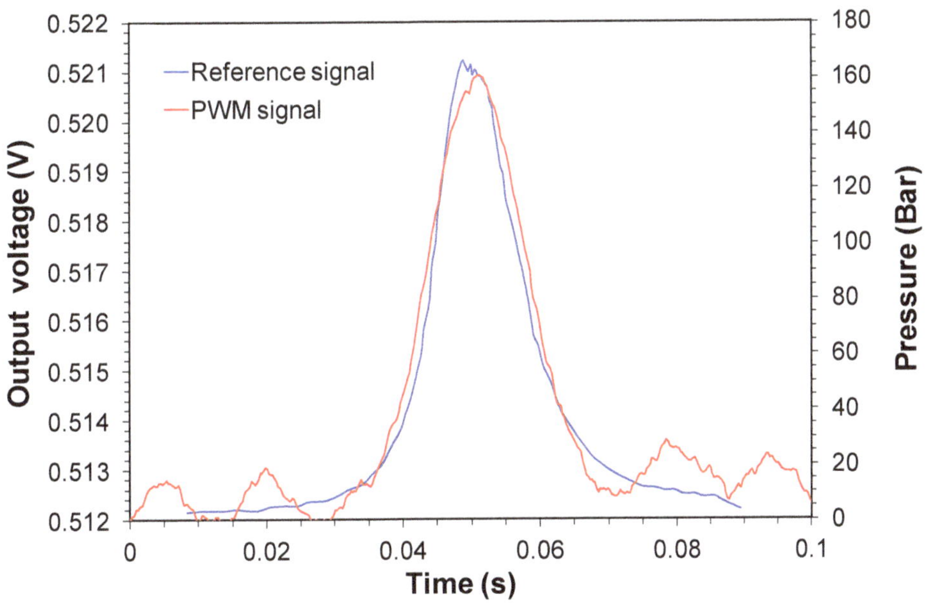

Figure 13. Dynamical curves of FBG pressure sensor in the monitoring point of engine combustion obtained by the PWM demodulator and the reference.

5. Discussion

To obtain a robustness FBG system for instantaneous combustion chamber control, we demonstrated a filterless, multi-point, and temperature-independent FBG dynamical demodulator using PWM, which can be installed close or inside the engine operating environment.

The demodulator uses just electronic components except for the photodetector. All these devices must be selected for operation in high temperatures. Besides, the demodulator works digitally, which reduces the error on the FBG signal conversion. Because the laser output changes in wavelength and time during the sweep, the optical wavelength variation of FBG in the engine became an electrical time variation in the demodulator. When the demodulator receives the analog time variation signal of FBG, it transforms this signal in a digital signal with a pulse width variation. The pulse width variation has the information of dynamical pressure modulated in the FBG. The active electrical filter removes the digital modulation of the PWM signal recovering the original FBG modulated-signal. This recovered signal is sent to the ECU. No synchronization signal is necessary for this system.

The maximization of the demodulator output signal can be obtained by reducing the laser sweep spectral band to the spectral band of the operation temperature of the sensors. The Bragg wavelength (in room temperature) can be the same wavelength for all sensors in the network, but this condition is not obligatory.

The laboratory tests demonstrated that the proposed demodulator reproduced the engine characteristics adequately. Also, the demodulator signal had more resolution than the commercial interrogator. The laser sweep frequency must be 10 times higher than the maximum sensor frequency response to have enough sampled points during the measurements. We also observed some noise in the signal base (in laboratory and field test) that we attribute to the residual noise originated from the demodulator's power supply. Also, we measured the linearity of the demodulator signal that is enough to reproduce the original characteristics of the FBG modulated signal.

In the field test, the demodulator was not tested close to the engine, where we have high-temperatures. The objective of the field test was to verify the performance of the PWM demodulator in terms of signal processing using a dynamical FBG pressure sensor installed in an environment with variable temperature. The design of the electronic-board of the demodulator with high-performance electronic devices is a future project. The demodulator worked well in the field test and, it was proved by the comparison of the PWM signal with the reference pressure sensor.

6. Conclusions

In this work, we investigated an innovative filterless, multi-point, and temperature-independent FBG dynamical demodulator using the PWM technique. The demodulator was developed to monitor the instantaneous combustion chamber pressure in closed-loop control of fuel mass fraction burned of the thermoelectric engines. The demodulator must work integrated with an ECU close to the engine, where the temperature is high.

The system was characterized in the laboratory using an FBG sensor modulated in a frequency of 6 Hz and a tunable laser with a 1 kHz sweeping frequency and wavelength range from 1527 to 1534 nm. A commercial FBG interrogator was used to compare the results obtained in this application. Besides, it was evaluated in a field test in an engine of a thermoelectric power plant. The demodulator worked well in the field test and, it was proved by the comparison of the PWM signal with one reference pressure sensor. Briefly, we will evaluate the demodulator, integrated with an ECU, very close to the engine in closed-loop control of fuel mass fraction burned of the thermoelectric engines.

Supplementary Materials: The following are available online at http://www.mdpi.com/1424-8220/20/20/5825/s1, Video S1: The FBG-PWM signal seen in an oscilloscope during the stabilization period.

Author Contributions: Conceptualization, J.B.R. and M.C.A.; methodology, M.C.A.; software, R.S.P.; validation, M.C.A., R.S.P. and F.R.B.; formal analysis, C.F.; investigation, J.B.R.; resources, R.S.P.; data curation, A.d.A.S.; writing—original draft preparation, J.B.R.; writing—review and editing, R.S.P.; project administration, D.V.; funding acquisition, M.A.R.J. All authors have read and agreed to the published version of the manuscript.

Funding: This research was funded by ANEEL (Agência Nacional de Energia Elétrica), grant PD-7236-0003/2016 and CNPq (Conselho Nacional de Desenvolvimento Científico e Tecnológico) sponsors the author J.B.R. under scholarship DT grant 304626/2017-1.

Conflicts of Interest: The authors declare no conflict of interest.

References

1. Ashok, B.; Ashok, S.D.; Kumar, C.R. A review on control system architecture of a SI engine management system. *Annu. Rev. Control* **2016**, *41*, 94–118. [CrossRef]
2. Johnson, R.; Evans, J.; Jacobsen, P.; Thompson, J.; Christopher, M. The Changing Automotive Environment: High-Temperature Electronics. *IEEE Trans. Electron. Packag. Manuf.* **2004**, *27*, 164–176. [CrossRef]
3. Biba, D.-R.; Musuroi, S.; Svoboda, M. A New Approach to Lifetime and Loading Stress Level Estimation for Multilayer Ceramic Capacitors in Electronic Control Units. In Proceedings of the 2017 International Conference on Optimization of Electrical and Electronic Equipment (OPTIM) & 2017 Intl Aegean Conference on Electrical Machines and Power Electronics (ACEMP), Brasov, Romania, 25–27 May 2017; pp. 37–42.
4. Stryjek, P.; Nikisz, T.; Sykulski, K.; Sowa, M. Electronic control systems of internal combustion engines with respect to the implementation of special functions in military vehicle. *Szybkobieżne Pojazdy Gąsienicowe* **2016**, *1*, 173–182.
5. Pawletko, R.; Polanowski, S. Evaluation of Current Developments and Trends in the Diagnosis of Marine Diesel Engines Based on the Indicator Diagrams Analysis. *J. Kones. Powertrain Transp.* **2014**, *21*, 389–396. [CrossRef]
6. Maroteaux, F.; Saad, C.; Aubertin, F.; Canaud, P. Analysis of Crank Angle Resolved In-Cylinder Combustion Modeling for Real Time Diesel Engine Simulations. *SAE Tech. Pap. Ser.* **2015**, 1. [CrossRef]
7. Campanella, C.E.; Cuccovillo, A.; Campanella, C.; Yurt, A.; Passaro, V. Fibre Bragg Grating Based Strain Sensors: Review of Technology and Applications. *Sensors* **2018**, *18*, 3115. [CrossRef]
8. Tian, S.; Zhao, X.; Zhou, Z.; Ou, J. Application of fiber Bragg grating sensors in civil engineering. *Int. Soc. Opt. Photonics* **2005**, *5851*, 182–189. [CrossRef]
9. Antunes, P.; Varum, H.; André, P. Optical FBG Sensors for Static Structural Health Monitoring. *Procedia Eng.* **2011**, *14*, 1564–1571. [CrossRef]
10. Allil, R.C.S.B.; Werneck, M.M.; Ribeiro, B.A.; De Nazaré, F.V.B. Application of Fiber Bragg Grating Sensors in Power Industry. In *Current Trends in Short- and Long-Period Fiber Gratings*; IntechOpen: London, UK, 2013; pp. 133–166.
11. Sarkar, B.; Koley, C.; Roy, N.; Kumbhakar, P. Condition monitoring of high voltage transformers using Fiber Bragg Grating Sensor. *Measurement* **2015**, *74*, 255–267. [CrossRef]
12. Tam, H.Y.; Liu, S.Y.; Guan, B.-O.; Chung, W.H.; Nguyen, A.; Cheng, L.K. Fiber Bragg grating sensors for structural and railway applications. In Proceedings of the SPIE 5634, Photonics Asia–Advanced Sensor Systems and Applications II, Beijing, China, 8–12 November 2004.
13. Liang, M.; Fang, X. Application of Fiber Bragg Grating Sensing Technology for Bolt Force Status Monitoring in Roadways. *Appl. Sci.* **2018**, *8*, 107. [CrossRef]
14. Leal-Junior, A.G.; Marques, C.; Neto, A.F.; Pontes, M.J. Multi-interface level in oil tanks and applications of optical fiber sensors. *Opt. Fiber Technol.* **2018**, *40*, 82–92. [CrossRef]
15. Melo, L.; Rodrigues, J.; Farinha, A.; Marques, C.; Bilro, L.; Alberto, N.; Tomé, J.; Nogueira, R. Concentration sensor based on a tilted fiber Bragg grating for anions monitoring. *Opt. Fiber Technol.* **2014**, *20*, 422–427. [CrossRef]
16. Al-Fakih, E.; Abu Osman, N.A.; Adikan, F.R.M. The Use of Fiber Bragg Grating Sensors in Biomechanics and Rehabilitation Applications: The State-of-the-Art and Ongoing Research Topics. *Sensors* **2012**, *12*, 12890–12926. [CrossRef] [PubMed]
17. Presti, D.L.; Massaroni, C.; Leitao, C.S.J.; Domingues, M.D.F.; Sypabekova, M.; Barrera, D.; Floris, I.; Massari, L.; Oddo, C.M.; Sales, S.; et al. Fiber Bragg Gratings for Medical Applications and Future Challenges: A Review. *IEEE Access* **2020**, *8*, 156863–156888. [CrossRef]
18. Cavillon, M.; Lancry, M.; Poumellec, B.; Wang, Y.; Canning, J.; Cook, K.; Hawkins, T.W.; Dragic, P.; Ballato, J. Overview of high temperature fibre Bragg gratings and potential improvement using highly doped aluminosilicate glass optical fibres. *J. Phys. Photonics* **2019**, *1*, 42001. [CrossRef]
19. Broadway, C.; Min, R.; Leal-Junior, A.G.; Marques, C.; Caucheteur, C. Toward Commercial Polymer Fiber Bragg Grating Sensors: Review and Applications. *J. Light. Technol.* **2018**, *37*, 2605–2615. [CrossRef]
20. Grobnic, D.; Mihailov, S.; Smelser, C.; Ding, H. Sapphire Fiber Bragg Grating Sensor Made Using Femtosecond Laser Radiation for Ultrahigh Temperature Applications. *IEEE Photon. Technol. Lett.* **2004**, *16*, 2505–2507. [CrossRef]

21. Tosi, D. Review and Analysis of Peak Tracking Techniques for Fiber Bragg Grating Sensors. *Sensors* **2017**, *17*, 2368. [CrossRef]
22. Fukuda, M. Reliability Testing of Semiconductor Optical Devices. In *Materials and Reliability Handbook for Semiconductor Optical and Electron Devices*; Springer Science and Business Media LLC: Berlin/Heidelberg, Germany, 2012; pp. 3–17.
23. Dias, J.S.; Leite, R.; Ferreira, E. Electronic technique for temperature compensation of fibre Bragg gratings sensors. *AEU Int. J. Electron. Commun.* **2008**, *62*, 72–76. [CrossRef]
24. Rosolem, J.B.; Penze, R.S.; Bassan, F.R.; Floridia, C.; Peres, R.; Dini, D.C.; Vasconcelos, D.; Junior, M.A.R. Electroless Nickel-Plating Sealing in FBG Pressure Sensor for Thermoelectric Power Plant Engines Applications. *J. Light. Technol.* **2019**, *37*, 4791–4798. [CrossRef]
25. Stan, N.; Bailey, D.C.; Chadderdon, S.L.; Webb, S.; Zikry, M.; Peters, K.; Selfridge, R.H.; Schultz, S.M. Increasing dynamic range of a fibre Bragg grating edge-filtering interrogator with a proportional control loop. *Meas. Sci. Technol.* **2014**, *25*, 65206. [CrossRef]
26. Li, Z.; Zhou, L.; Sun, W. A Novel Method for the High-Speed Demodulation of FBG Sensor Arrays. In Proceedings of the 2017 5th International ConfereNCE on Computer-Aided Design, Manufacturing, Modeling and Simulation, Busan, Korea, 22–23 April 2017; Aip Publishing: Melville, NY, USA, 2017; Volume 1834, p. 20021.
27. Chen, W.-W.; Chen, J.-F. *Review of the PWM Control. Circuits for Power Converters*; Springer Science and Business Media LLC: Berlin/Heidelberg, Germany, 2017; pp. 37–79.
28. Yamaguchi, T.; Endo, W.; Shinoda, Y. Interrogation System with Automatic Recognition and Delay Correction Functions of Fiber Bragg Gratings by Pulse Modulation with Wavelength-Swept Laser. *IEEE Sens. J.* **2019**, *19*, 10519–10528. [CrossRef]
29. Wu, W.; Xu, R.; Liu, X.; Chen, T. Tunable Laser Using Semiconductor Optical Amplifier and Its Application in Fiber Grating Sensor System. In Proceedings of the Optoelectronic Materials and Devices III, Shanghai, China, 2–6 November 2009.
30. Caldwell, J. Analog Pulse Width Modulation, Texas Instruments Incorporated, Technical Document—Reference Guide SLAU508. June 2013. Available online: https://www.ti.com/lit/ug/slau508/slau508.pdf?ts=1602574479776&ref_url=https%253A%252F%252Fwww.google.com.hk%252F (accessed on 12 October 2020).
31. Rosolem, J.B.; Penze, R.S.; Floridia, C.; Bassan, F.R.; Peres, R.; Da Costa, E.F.; Silva, A.A.; Coral, A.D.; Nogueira, J.R.; Vasconcelos, D.; et al. Dynamic Effects of Temperature Temperature on FBG Pressure Sensors Used in Combustion Engines. *IEEE Sens. J.* **2020**, 1. [CrossRef]
32. Oiwa, M.; Kim, J.; Tsuji, K.; Onodera, N.; Saruwatari, M. Amplitude-Noise and Timing-Jitter Reduction Via Pulsed Injection Locking of SOA Fiber Ring Laser. In Proceedings of the OECC/ACOFT 2008—Joint Conference of the Opto-Electronics and Communications Conference and the Australian Conference on Optical Fibre Technology, Sydney, Australia, 7–10 July 2008; pp. 1–2.

sensors

MDPI

Article

Dual Wavelength Differential Detection of Fiber Bragg Grating Sensors with a Pulsed DFB Laser

François Ouellette *, Zhonghua Ou and Jianfeng Li

State Key Laboratory of Electronic Thin Films and Integrated Devices, School of Optoelectronic Science and Engineering, University of Electronic Science and Technology of China (UESTC), Chengdu 610054, China; ozh@uestc.edu.cn (Z.O.); lijianfeng@uestc.edu.cn (J.L.)

* Correspondence: fouellette58@yahoo.ca

check for updates

Received: 15 July 2020; Accepted: 21 August 2020; Published: 24 August 2020

Abstract: We show how dual wavelength differential detection can be used to measure fiber Bragg grating sensors using nanosecond pulses from a single DFB laser diode, by taking advantage of its dynamic chirp. This can be performed in two ways: by measuring the reflected power from two separate pulses driven by two different currents, or by taking two delayed digitized samples within a single pulse. A prototype instrument using fast digitizing and processing with an FPGA is used to characterize the chirp, from which the performance can be optimized for both measurement schemes.

Keywords: fiber Bragg grating; fiber optic sensor; dynamic chirp; DFB laser

1. Introduction

Fiber Bragg grating (FBG) sensors have been used for three decades now [1], in applications as varied as aircraft wing shape measurements [2], temperature monitoring in medical treatments [3], pressure and temperature sensing in oil wells [4], or strain sensing in biomechanics [5]. FBG sensors have traditionally been measured either with a combination of a broadband optical source and a spectrometer, or a wavelength-swept source and a photodiode [1]. Both techniques involve acquiring a set of data points covering the entire range of wavelengths potentially reached by the sensor. The state of the sensor is determined by estimating the center of the Bragg grating spectrum from the resolution-limited spectral data, using more or less sophisticated algorithms [6,7]. Such spectral techniques allow the use of wavelength-division multiplexing (WDM) of multiple sensors along the same fiber, each sensor being an FBG with a different central wavelength. While FBG sensing systems have been used for many years, it is still felt that their more widespread use would benefit from more inexpensive, and higher resolution instrumentation.

Other interrogation techniques have nevertheless been proposed. For example, Morozov et al. [8] have proposed to probe the FBG with the two frequency sidebands of a modulated source. The state of the FBG can then be extracted from the amplitude and phase of the reflected modulated signal. Kulchin et al. [9] have proposed a scheme using an OTDR instrument, where the convolution of the reflected FBG spectrum with that of two or more reference FBGs can determine the state of the sensor. Those two techniques have some points in common with the technique presented here, such as the use of two frequencies, or the use of time-division multiplexing, though their implementation is quite different.

Recently, another technique to probe the state of an FBG sensor has been proposed and demonstrated [10–17]. It relies on measuring the reflectivity of the grating at two, or sometimes more, fixed wavelengths, all located within the spectrum of the FBG. The computed difference or ratio of the two or more reflectivity values can then be uniquely related to the location of the center of the spectrum. This technique has been dubbed dual-wavelength differential detection (DWDD) [13] and we shall use the same name and acronym here. Because it uses fixed wavelengths, DWDD lends itself more naturally to time-division multiplexing (TDM), where multiple identical sensors are interrogated by short optical pulses, with optical delays between the arrival times of each reflected pulse. Like other TDM schemes, this can be either in a serial configuration, with identical sensors distributed along one fiber, or in a parallel spatial-multiplexing configuration where the signal is split into multiple branches, each branch having a single sensor [18].

So far, most of the previous implementations of DWDD have have involved complex and expensive components and instruments. Wilson et al. [17] originally used a single laser diode, but relied on the diode operating on two longitudinal modes, one on each side of the grating spectrum, a condition not easy to obtain, and prone to mode partition noise. The two wavelengths also had to be separated by a monochromator before detection. Cheng and Xia [13] obtained a DWDD signal by taking the logarithmic ratio of the reflected signals from a tunable laser, manually switched between two wavelengths. To perform an equivalent TDM measurement, they used an electro–optic modulator to do a frequency sweep of the laser light, detecting it with an expensive network vector analyzer. Rohollanehjad et al. [16] used three tunable cw lasers, simultaneously modulated by an electro-optic modulator, and amplified by an EDFA, with three tunable filters to separate the wavelengths at detection. Xia et al. [14] used two pulsed tunable laser diodes with a programmable delay. Li et al. [15] measured both the static and dynamic strain and temperature on a chain of gratings, using a DWDD scheme for the static part. For this, they used two cw DFB laser diodes, amplified by switched semiconductor optical amplifiers, and further amplified by an EDFA, in order to avoid any effect of dynamic chirp. Apart from Wilson et al. [17], these demonstrations calculate the DWDD signal by subtracting the logarithms of the two signals, which does not eliminate noise due to power fluctuations of the source. Such a mathematically complex algorithm is also difficult to compute in real time for each pulse, and measurement rates in those demonstrations remains quite small.

We have recently demonstrated a much simpler and inexpensive DWDD instrument that uses a single, pulsed commercial DFB laser diode [11], and is suitable for TDM measurements of up to 15 sensors. We have also used a similar technique to measure tilted fiber Bragg gratings with very high resolution [12]. While other demonstrations have tried to avoid the dynamic chirp of the laser sources, we use the current and time dependence of the wavelength to our advantage in order to generate the two (or more) required wavelengths. In Reference [12], we have demonstrated that high signal-to-noise ratio detection and high resolution digitization, in conjunction with good wavelength stability of the source, or of an external reference, can achieve an effective spectral resolution better than 0.08 pm, which is a more than 10 times improvement over spectral instruments. These demonstrations show that DWDD instruments can offer an inexpensive, high performance alternative to spectral-based instruments for the interrogation of multiple FBG sensors.

Our DWDD scheme is illustrated in Figure 1. We use measurements of the reflected power at two close wavelengths λ_+ and λ_- separated by a fraction of the grating full-width at half-maximum (FWHM) w_B. As the Bragg wavelength λ_B shifts under the influence of the measured parameter, the difference between the two reflected powers P^+ and P^- changes. For example, as illustrated, as λ_B shifts from $-w_B/2$ to $-w_B/2$ with respect to the average between those two wavelengths λ_{av}, the difference between the two reflected powers is inverted. It is thus essentially related to the slope of the spectrum at λ_{av}, and is a one-to-one function of $\lambda_{av} - \lambda_B$. Dividing the difference by the sum of the reflected powers normalizes the signal and makes it insensitive to the propagation loss between the source and the sensors, as well as to power fluctuations of the source. A Gaussian-shaped spectrum also results in a signal that is linear with respect to the Bragg wavelength, as was demonstrated in [10,12]. In [11], this simple algorithm was implemented in an FPGA chip, and could calculate the status of the sensor from nanosecond pulses in real time, leading to effective measurement speeds of more than 1 MHz for 15 simultaneously measured sensors.

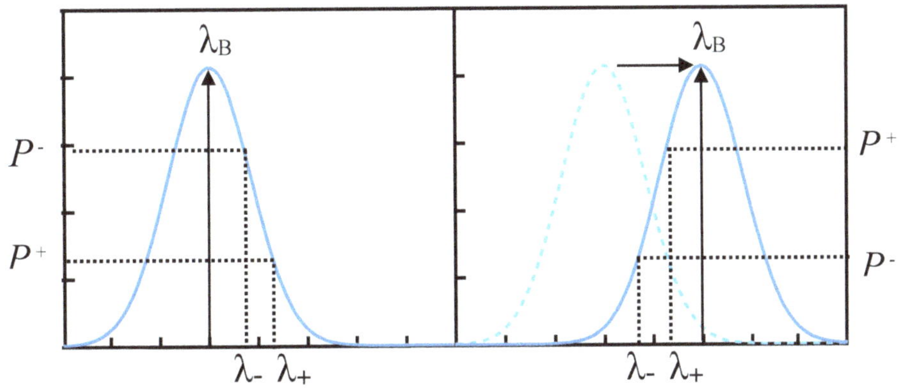

Figure 1. Dual-wavelength differential detection (DWDD) scheme involving the measurement of the reflected power at two closely spaced wavelengths λ_+ and λ_-. As λ_B shifts, the difference between the two reflected powers also changes.

We have previously studied in detail the dynamic chirp for a square-wave modulated DFB laser diode at a low frequency of 5 kHz, and how it could be used to measure tilted FBG sensors [12]. Here we investigate in more detail how knowledge of the dynamic chirp behaviour of a nanosecond-pulsed DFB laser diode can be used to maximize the sensitivity and resolution of the DWDD signal. Using a prototype instrument, we extract the relevant coefficients for the dynamic chirp, and demonstrate two methods by which the dynamic chirp can be used to get a DWDD signal: one using two pulses with different driving currents, and one using two digitized samples at different times within a single laser pulse.

2. DWDD Implementation with a Single DFB Laser

2.1. DWDD Algorithm and Resolution

If the wavelength of the DFB laser source is different at different times, then by digitizing the reflected signals, one can select the digitized samples that correspond to the desired probing wavelengths λ_+ and λ_-, and use them to compute the DWDD response. In our case, we use a simple algorithm involving basic arithmetic operations that can be implemented in an FPGA chip for real time calculation [10].

In order to estimate the resolution of our DWDD measurement, it should be noted that notions normally used for calculating the resolution of spectral scanning schemes, such as the bandwidth of the laser source, do not formally apply when one is dealing with chirped laser pulses. The wavelength of the laser is a time-varying quantity, and by measuring the average reflectivity over a given time interval, one is actually measuring the average reflectivity over a corresponding wavelength interval. For short enough time intervals, the reflectivity is a continuous function of the difference between the average wavelength and the FBG central wavelength. Since the DWDD scheme relies on calculating the difference in reflectivity between two such average wavelengths, the resolution is in fact related to the precision with which one can detect small changes in reflectivity.

We assume that the FBG spectrum has a Gaussian shape with a FWHM w_B. The Gaussian shape is a very good approximation for the central region of a weak uniform grating ($R < 30\%$). In cases where the emitted powers at the two wavelengths are different (as when we use pulses with different driving currents), the reflected signals P^+ and P^- at wavelengths λ_+ and λ_- can be normalized using a reference measurement of the pulse powers $P^{\pm}_{ref}$ as they are launched into the fiber. For example, we performed that by using the signal from the internal photodiode of the DFB laser. For cases where the power is the same at both wavelengths (as when we use two samples within the same square pulse), such normalization is not necessary. Considering the normalized powers $P_n^{\pm} = P^{\pm}/P_{ref}^{\pm}$, the following algorithm provides a sensing signal S that is linear with the Bragg wavelength [10,12]:

$$S = \frac{P_n^+ - P_n^-}{P_n^+ + P_n^-} = \alpha\left(\lambda_{av} - \lambda_B\right), \tag{1}$$

where the sensitivity is given by the slope α:

$$\alpha = 4\ln(2)\frac{\delta}{w_B^2}, \tag{2}$$

with $\delta = \lambda_+ - \lambda_-$ being the wavelength separation.

The linear response is a result of the Gaussian shape of the spectrum and the small wavelength separation. An almost purely linear response ($r^2 > 0.998$) is obtained for $\delta \leq 0.4w_B$. The signal S, which is dimensionless, varies from positive to negative as λ_B shifts from shorter wavelengths to longer ones. When $(\lambda_{av} - \lambda_B)$ becomes too large, the reflected powers become smaller, which increases the measurement noise. Therefore the measurement range is limited by the signal-to-noise ratio (SNR), and is proportional to w_B. Thus it can be tailored by changing the length, and therefore the bandwidth, of the grating, to match a particular sensing application. While we will show that the amount of chirp may place some limit on the bandwidth of the sensors, there are also ways to adapt the range using the packaging of the sensor itself, to make it more or less sensitive to a given parameter.

Because the range can be tailored in such a manner, and made to match the required range of the measured parameter, the performance of a DWDD instrument is best described by the ratio of the range to the resolution, which can be expressed as the number of bits of resolution B_r. This also makes it easier to compare with other sensing technologies. B_r depends on the smallest measurable difference in the powers P^+ and P^-, which depends on the SNR and the resolution of the analog-to-digital converter (ADC). B_r also depends on the slope α which governs how much change in power there is for a change in the Bragg wavelength. The resolution is largest when the average wavelength λ_{av} coincides with the center of the grating spectrum at λ_B, and the reflected power is maximum. It then decreases as λ_B moves away from λ_{av}. Considering a practical limit to the range as one FWHM w_B, the resolution at $\lambda_{av} - \lambda_B = 0.5w_B$ can be expressed by [12]:

$$B_r = B_{pd} - 1 - \log_2\left[1 + \frac{1}{2\ln(2)\delta/w_B}\right]. \tag{3}$$

B_{pd} is the effective number of bits of resolution of the digitized and averaged photodiode signal as obtained for $\lambda_{av} - \lambda_B = 0$, i.e., the maximum reflected power. This depends on both the effective resolution of the ADC, the SNR of the photodiode, and other sources of noise such as that from the laser current driver. Bits are also gained by averaging over N pulses, to the tune of $\log_2\sqrt{N}$. For nanosecond pulses, averaging can be performed over many thousands of pulses while still resulting in a measurement rate in the tens or hundreds of Hz, bringing significant resolution gain. For the case where the powers $P_0^{\pm}$ are not the same and a normalization to a reference pulse has to be done, B_{pd} will be lower because it includes the noise from the reference measurements, and also because the lower of the two $P^{\pm}$ has a smaller SNR.

One bit from B_{pd} is then lost when the signal is at half the maximum possible reflected value. The last term in Equation (3) represents the effect of the sensitivity α, which depends on the ratio δ/w_B. At a maximum practical value of $\delta/w_B = 0.4$, this accounts for 1.5 more bits lost from B_{pd}. For smaller δ/w_B, more bits are lost. Figure 2 shows the total bits lost from B_{pd} as a function of δ/w_B. Aside from maximizing the SNR, using a high value of δ is therefore essential to optimize the resolution of a DWDD instrument.

In Reference [12], we achieved nearly 12 bits of resolution, using a 14-bit ADC, with enough room for improvement to reach at least 13 bits, which would make the DWDD sensing system on par with competing technologies such as resistive strain gauges.

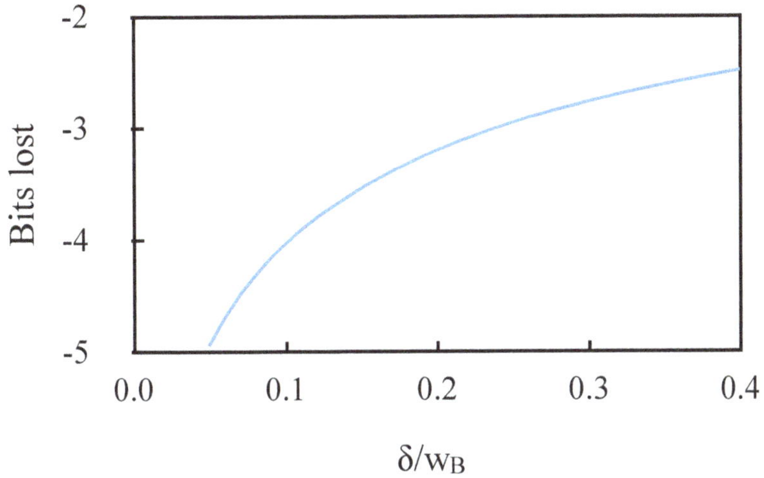

Figure 2. Resolution bits lost from B_{pd} as a function of (δ/w_B).

2.2. Dynamic Chirp of a Nanosecond-Pulsed DFB Laser

To generate two wavelengths with a significant wavelength spacing δ, we use the dynamic chirp of a single DFB laser diode, i.e., the dependence of its emission wavelength on both time and current. The wavelength of a DFB laser diode is directly related to the refractive index of its active region. When it is pulsed, two effects determine its behaviour. First, the injected current creates free carriers that change the refractive index and the wavelength in the negative direction. This change is linearly proportional to the current. On a time scale of nanoseconds, the concentration of free carriers follows the current adiabatically, thus this wavelength shift remains constant over the pulse duration for a square pulse. The second effect is the heating which occurs in the junction, which is also linearly proportional to the current, but has a more complex dynamic behaviour. The heat is generated in the active region, but gradually spreads to the entire diode chip, then its submount, and is ultimately dissipated by a heat sink in an active way if the temperature is controlled by a thermo-electric cooler (TEC). These different levels result in a temporal behaviour with as many as four different time constants, as has been shown by Shalom et al. [19]. Those time constants are all of different order of magnitude and range from about 15 ns to 50 µs. In our case, we use pulses of 5–30 ns, and therefore, following the same model as Shalom et al. [19], but keeping only the adiabatic effect and the nanosecond scale thermal effect, we can describe the chirp behaviour using the following equation:

$$\lambda(t) - \lambda_0 = I\left[A + B\left(1 - e^{-t/\tau}\right)\right], \tag{4}$$

where $t = 0$ is the beginning of the pulse, I is the current, τ is the thermal response time, and A and B are the adiabatic and thermal coefficients. λ_0 is the nominal resonant wavelength of the DFB laser cavity in the absence of current.

Since the dynamic chirp is linearly proportional to the current, one way to exploit it is to use two pulses emitted in succession, with different driving currents, and therefore two different wavelengths. We call this scheme the two-pulse scheme. Enough delay is allowed between the pulses for the reflections from all the sensors to reach the photodiode, as illustrated in Figure 3b. On the other hand, since the wavelength also evolves along a single pulse, one can also use two digitized samples at different times along the pulse, as was done in Reference [12] for a square-wave modulated laser diode. We call this the one-pulse scheme, which is illustrated in Figure 4b. The achievable value of δ for both schemes requires a knowledge of the coefficients in Equation (4).

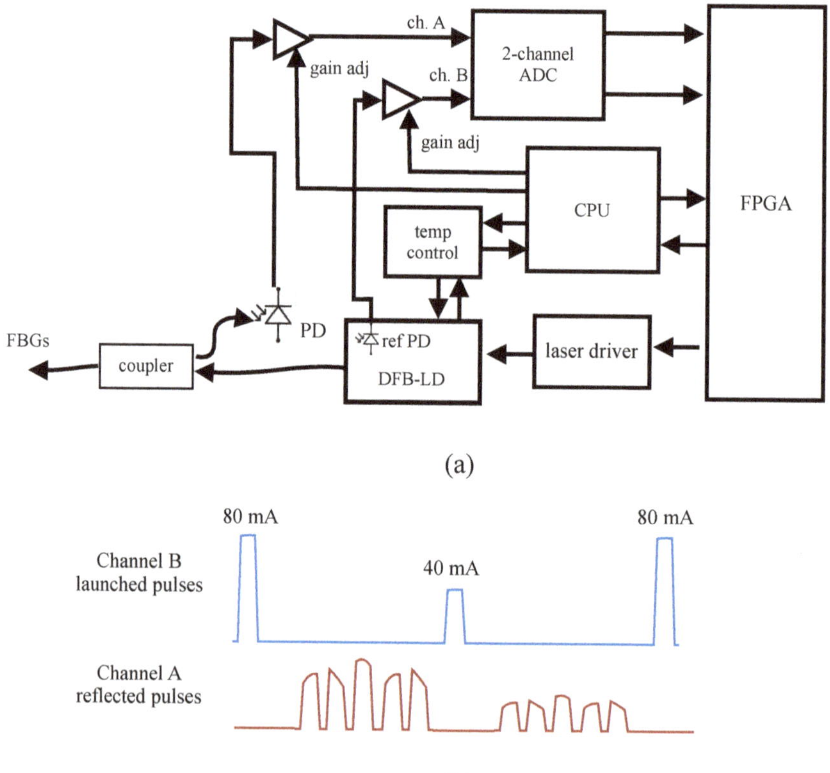

Figure 3. (**a**) Instrument schematic; (**b**) Temporal trace for the emitted pulses detected in channel B of the analog-to-digital converter (ADC), and the reflected pulses detected in channel A (in the case of 5 sensors).

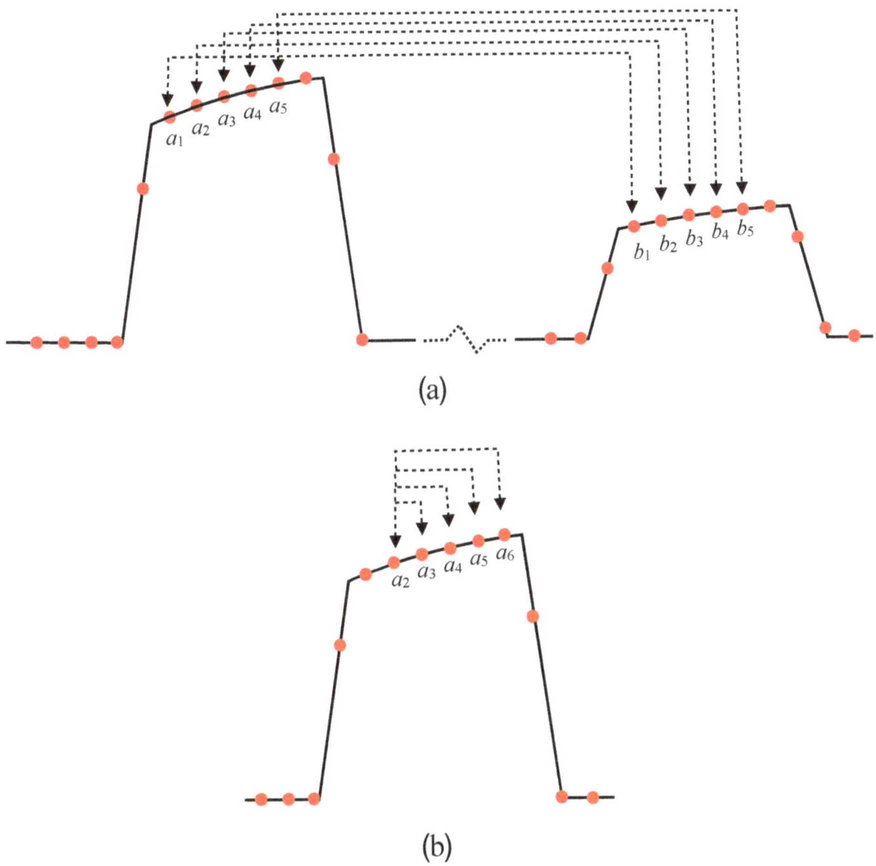

Figure 4. Digitized samples used for the measurements of S vs. the laser wavelength: (**a**) two-pulse scheme and (**b**) one-pulse scheme.

In our prototype instrument, we chose to use pulses with 80 mA and 40 mA driving currents. Such a choice was made to have a large enough value of δ, while not sacrificing too much SNR in the second pulse. Figure 5a illustrates the temporal behaviour of the emission wavelength according to the model of Equation (4), with respect to the nominal wavelength λ_0, for two perfectly square 30 ns pulses with these two driving currents, the wavelength shift for 40 mA being half that for 80 mA. Figure 5b shows the time dependence of the difference δ_{2p} between the wavelength of the pulses for the two-pulse scheme. The values of A, B, and τ used in Figure 5 are those found in our experiments described below, and the dots are the experimental measurements. In practice, the pulse rise and fall times were about 1 ns, which does not affect the model significantly.

According to the model of Equation (4), as the current rapidly rises, the wavelength drops simultaneously. Following this fast drop, the wavelength then rises with an experimentally determined time constant of 17.8 ns. After the end of the pulse, the wavelength rises rapidly because the adiabatic contribution vanishes, and then decreases with the same thermal time constant as the laser junction cools down. This is however not observable because there is no more optical power being emitted, hence it is shown as a dashed line in Figure 5a.

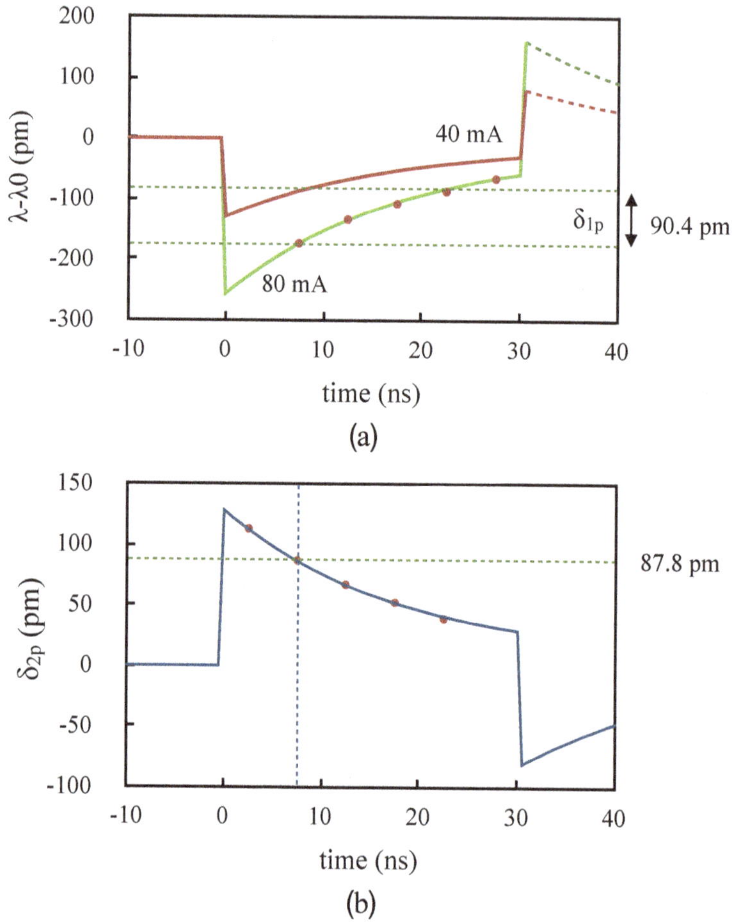

Figure 5. (**a**) Wavelength vs. time for 80 mA and 40 mA, 30 ns pulses; (**b**) Difference δ_{2p} between the wavelengths of the 80 mA and 40 mA pulses. The dots are experimental measurements. The dotted lines in (**a**) represent the expected value of δ_{1p} at 7.5 ns, and in (**b**) that of δ_{2p} for 15 ns delay from the 7.5 ns sample.

In practice, there is also a slower rise due to the thermalization processes on longer time scales, but that is not significant on the nanosecond time scale. This slower thermal effect gives rise to a steady state wavelength shift, which depends on the average driving current and the repetition rate of the pulses. However, we do not need to consider it here.

The relative contributions of the adiabatic and thermal shifts were found to have nearly equal magnitude, though of opposite signs, consistent with the values measured by Shalom et al. [19]. Because of this, we found that the one-pulse and two-pulse schemes can have an equivalent sensitivity, given enough delay between the samples of the one-pulse scheme.

For the two-pulse scheme, the wavelength difference δ_{2p}, and therefore the sensitivity, is higher at the beginning of the pulses, as can be deduced from Figure 5b. Therefore, maximum sensitivity is obtained for shorter pulses. For the one-pulse scheme, we found that the wavelength difference δ_{1p} when using samples at t and $t + 15$ ns, is nearly the same as δ_{2p} obtained using the samples at time t in the 40 mA and 80 mA pulses. This is illustrated by the dashed lines in Figure 5b, which show the value of δ_{2p} for a sample taken 7.5 ns from the start of the pulse (87.8 pm), and in Figure 5a, which show the value of δ_{1p} from the difference in wavelength between a sample taken 15 ns later than the 7.5 ns sample (90.4 pm).

In the following sections, we describe the design of our prototype instrument, and the measurements made to obtain the data, and extract the parameters used in Figure 5.

3. Instrument Design

Our prototype instrument was originally conceived and designed with the two-pulse measurement scheme in mind. As the laser source, we use a commercial DFB laser diode (Jiuzhou EO) mounted in a butterfly package with internal TEC, isolator, and monitor photodiode. It has a rated power of 8 mW for a 90 mA current. The wavelength of the DFB laser diode is also a function of temperature. This feature is useful for fine-tuning the laser to the centre of the desired sensing range. From the spectrum of the laser diode at various temperatures, we obtained a linear response of 97 pm/°C over a 10 °C range, in line with the laser specifications.

In order to implement a TDM scheme with sufficiently short distances between the gratings, pulses in the nanosecond range are required, and the processing circuitry must be accordingly fast. We chose a low cost 10 bit, dual-channel ADC, with 200 Msps per channel (TI ADC10DV200). Thus the signal is sampled in 5 ns intervals. One channel was used to measure the reflected pulses, while the other one measured the reference pulse from the laser diode internal monitor photodiode. The sampled signals are then fed to the FPGA (Xilinx Spartan 6 LX25) for fast calculations of the DWDD signals. Both the laser driver (Maxim 3967AETG+) and the detection and amplification circuits have a response time of about 1 ns.

The instrument schematic is shown in Figure 3a, while Figure 3b illustrates (not actual data) the sequence of lauched pulses detected in channel B of the ADC and the received pulses from the multiple FBGs detected in channel A. A first pulse at 80 mA was launched, and time was allowed for the echoes from up to 15 sensors to be received by the photodiode and digitized by the ADC (though only 5 are shown in the Figure). A second pulse at 40 mA was then launched, and the echoes were similarly detected and digitized. The FPGA then performed calculation of the 15 DWDD signals in real time, after which another pair of pulses is launched. All the components were mounted on a PCB (designed and fabricated by IPCB Systems, Bromont, QC, Canada). A CPU controls the various components, and communicates with a computer via a USB link, where the acquisition parameters can be selected from a user interface.

The user interface also has an oscilloscope mode, where the values of the signals for each sample along the acquisition sequence can be visualized graphically, showing the reflected pulses like an oscilloscope trace. They can also be measured directly by zooming in on the trace, with digitized values from 0 to 512 for positive signals. We used this mode to directly measure the reflected powers in order to calculate the value of S for the one-pulse scheme, without having to reprogram the FPGA.

The instrument was designed with a target ratio of range to resolution of at least 1000. Since S is not expected to be larger than 1, only four significant decimal digits are displayed. In the rest of the paper, we will display the S values multiplied by 10,000, so that they are represented by a dimensionless integer corresponding to those four significant digits.

This instrument showed high performance for measuring FBGs as temperature sensors [11]. As an illustration, Figure 6 shows three sets of measurements of S as a function of temperature for FBGs immersed in water, using the two-pulse scheme with 5 ns pulses. The FBG had 710 pm bandwidth and about 1.5% reflectivity. Over the temperature range, the signal had extremely good linearity ($r^2 > 0.9996$) and the measurements were highly repeatable. Taking into account the noise in the value of S, we estimated the effective resolution as 0.05 °C. More details on the instrument performance in measuring multiple sensors will be given elsewhere.

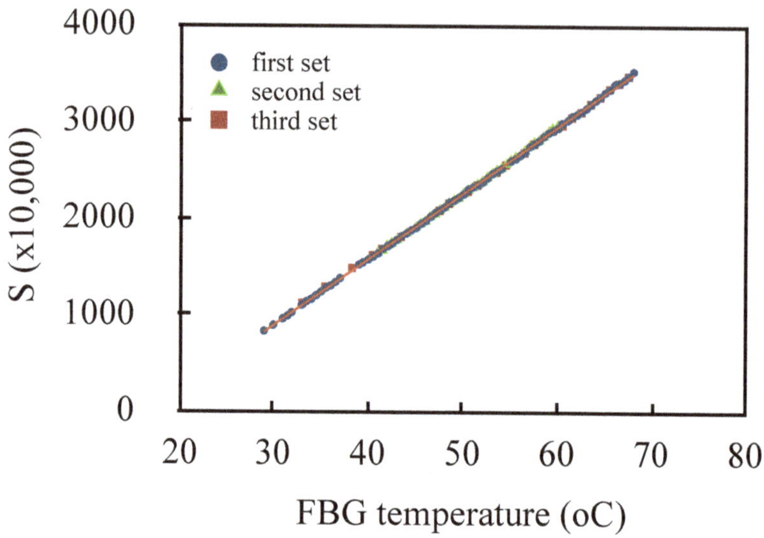

Figure 6. Signal S as a function of the FBG sensor temperature for three measurement sets.

4. Characterization of the Dynamic Chirp

The effect of the dynamic chirp could be readily visualized from the reconstructed pulse shapes in the oscilloscope mode of our instrument, as can be seen in Figure 7. When the laser diode average wavelength is on the side of the FBG spectrum (at about half-maximum), the reflection increases or decreases with time, depending on which side. When the average wavelength is centered on the Bragg wavelength, the reflection is flat across the pulse. This strong reshaping indicates that the amount of chirp is a significant fraction of w_B.

In order to characterize the dynamic chirp with more precision, we measured the value of the slope α of the signal S as a function of the laser wavelength when using different pairs of digitized samples, either between the 80 mA and 40 mA pulses, or within the 80 mA pulse, as shown in Figure 4. Instead of scanning the grating temperature, which can be time consuming, we performed the measurements by scanning the laser diode temperature, knowing that it shifts with a value of 97 pm/°C. The slope of the resulting curves was then used to calculate the value of the wavelength difference δ from Equation (2). w_B had previously been obtained by recording the reflected power as a function of the laser diode wavelength, and fitting it to a Gaussian with $w_B = 710$ pm.

Denoting the samples in the 80 mA pulse as a_n and in the 40 mA pulse as b_n, in the first set of measurements, illustrated in Figure 4a, we used 30 ns pulses and measured S using five corresponding pairs of samples $a_1 - b_1$ to $a_5 - b_5$, starting from about 2.5 ns into the pulse. All the linear fits had a regression coefficient $r^2 > 0.99$, and thus gave an accurate value of δ. The five curves are shown in Figure 8a. According to Equation (4), and for a difference in driving currents of 40 mA, the values of δ_{2p} obtained from the slopes should follow:

$$\delta_{2p}(t) = \lambda_+(t) - \lambda_-(t) = 40\left[A + B\left(1 - e^{-t/\tau}\right)\right]. \quad (5)$$

A second set of measurements was made by calculating S from the reflected power for the 80 mA pulse, using pairs of sample $a_2 - a_3$, $a_2 - a_4$, $a_2 - a_5$ and $a_2 - a_6$, thus effectively measuring

$\lambda_+(t) - \lambda_+(t_0)$ along the pulse relative to the second sample at $t = t_2$. Those curves are shown in Figure 8b. In this case, the measured values $\delta_{1p}(t)$ should be given by:

$$\delta_{1p}(t) = \lambda_+(t) - \lambda_+(0) = 80B\left(1 - e^{-t/\tau}\right). \tag{6}$$

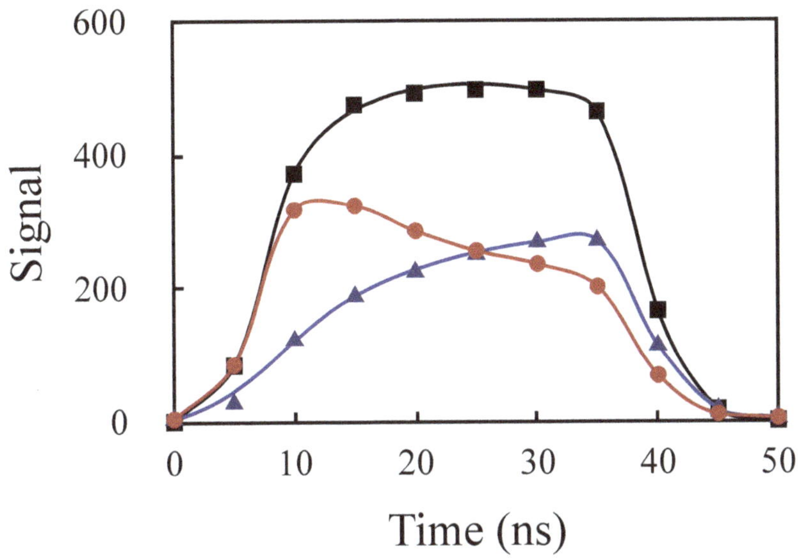

Figure 7. Pulse shape after reflection for three laser diode temperatures corresponding to the center of the grating (squares), and at about half-maximum on both sides (circles and triangles).

From Equations (5) and (6), we find that:

$$\delta_{2p}(t) - \delta_{2p}(0) = \delta_{1p}(t)/2 = 40B\left(1 - e^{-t/\tau}\right) \tag{7}$$

Therefore the two sets of measaurements can be combined to fit Equation (7), which is shown in Figure 9, from which we determined the values of B and τ. Using these values, A could then be estimated from the $\delta_{2p}(t)$, using:

$$A = \frac{\delta_{2p}(t)}{40} - B\left(1 - e^{-t/\tau}\right). \tag{8}$$

The best estimate for A was obtained by averaging the results of Equation (8) from the five experimental values of δ_{2p}. The curves of Figure 5, which include the data points for both sets of measurements, were plotted using the parameters from this fit. The values obtained were $A = -2.81$ pm/mA, $B = 2.64$ pm/ma, and $\tau = 17.8$ ns. These were within the range of those obtained by Shalom et al. [19] for different models of DFB lasers.

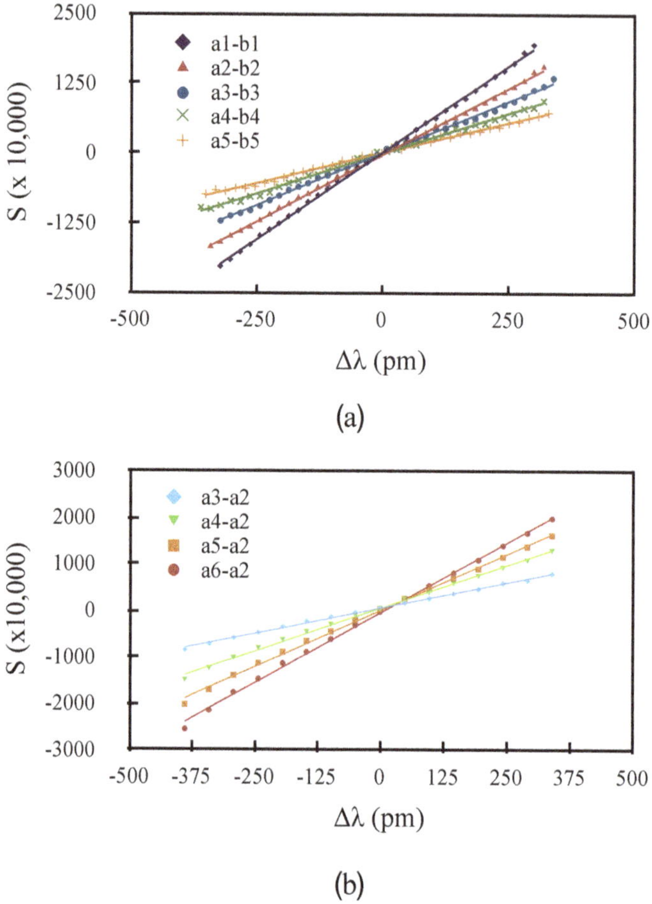

Figure 8. S vs. $\Delta\lambda$ for (**a**) two-pulse scheme, and (**b**) one-pulse scheme illustrated in Figure 4.

With these parameter values in hand, we can now compare the relative sensitivities of the one-pulse and two-pulse schemes, and find how much delay between the samples of the one-pulse scheme gives an equal sensitivity to the two-pulse scheme. Given the time t_0 of the first sample, taken from the onset of the pulse, the delay between samples Δt for equal sensitivity of both schemes is:

$$\Delta t = -\tau \ln \left[\frac{1 + (1 + \frac{A}{B})\, e^{t_0/\tau}}{2} \right]. \quad (9)$$

Because $A \approx -B$, the dependence on t_0 was small, and Δt varied from 13.5 ns to 14.1 ns for t_0 going from 0 to 8 ns. Therefore, a three-sample delay (15 ns) should give about the same sensitivity. To confirm this, we performed one more set of measurements of S using the samples $a_2 - b_2$ of a 40 ns pulse in the two-pulse scheme, and the samples $a_2 - a_5$ for the one-pulse scheme, as shown in Figure 10. An essentially similar slope was obtained for both schemes. For larger Δt, however, the sensitivity of the one-pulse scheme would exceed that of the two-pulse scheme. Whereas, the value of δ_{2p} is limited by $(I_+ - I_-)A = 112$ pm, that of δ_{1p} is limited by $I_+B = 211$ pm. For example, for a 35 ns delay, the one-pulse scheme is 60% more sensitive than the two-pulse scheme.

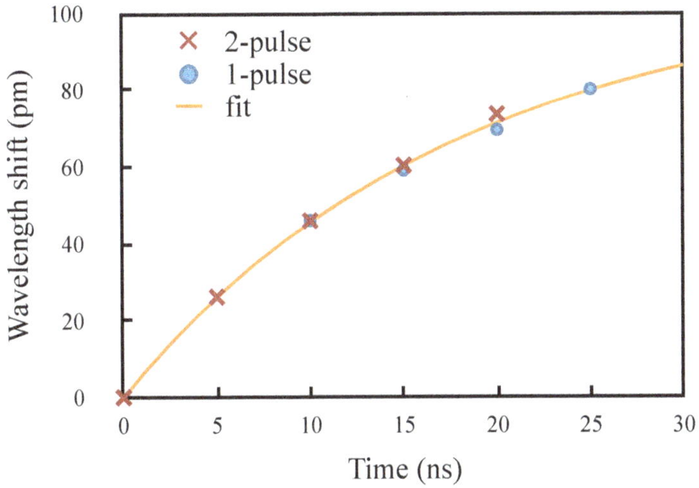

Figure 9. Best fit for B and τ.

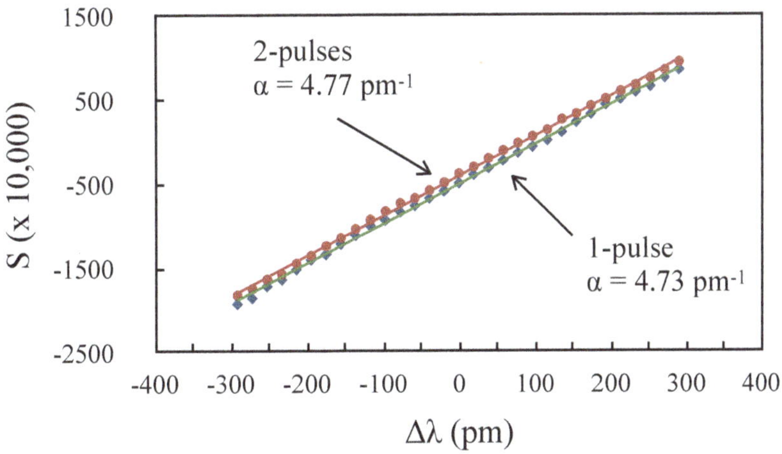

Figure 10. S vs. $\Delta\lambda$ for both schemes, for 15 ns delay between samples in the one-pulse scheme.

5. Discussion and Conclusions

The use of a pair of short pulses has the advantage that a shorter delay between the pulses can be used. For sensors along a single fiber, this means a shorter spacing between the sensors. In a parallel configuration, less fiber is required to impart a delay between each branch. Using 5 ns pulses with 5 ns latency between pulses means that a 1 m spacing can be used.

On the other hand, the one-pulse scheme has many advantages of its own. First of all, for applications that do not require such short spacing between the sensors, the one-pulse scheme can have a larger sensitivity for longer delays between samples. Secondly, for a square pulse, the power is essentially constant during the pulse. Therefore, there is no need to normalize to a reference power. The algorithm for S is then even simpler and faster to compute. Since it involves fewer measured quantities, the compound error is also smaller. Finally, the single pulse can use the maximum driving current allowed by the laser diode, which maximizes the power and thus the SNR. As mentioned in Section 2, the resolution as given by Equation (2) will be smaller when a power normalization is performed, as in the two-pulse scheme. Though the exact number of bits lost depends on the details of the instrument, it should represent at least one bit in a typical case.

In terms of measurement rate, for an equivalent sensitivity, both schemes are pretty much on par. Given a latency time of 5 ns between each pulse, to account for the finite response time of the detection system, the two-pulse scheme requires two such periods for each sensor. Thus the total measurement time for each sensor is twice the sum of the pulse duration plus the latency time, which is 20 ns. The one-pulse scheme requires 20 ns for the pulse, to ensure that the 2 probed samples separated by 15 ns are within the pulse, and 5 ns latency time, for a total of 25 ns. For 15 sensors, this represents basic measurement rates of 3.3 and 2.7 MHz. No application requires such a fast measurement rate, and a faster rate is really only useful because more averaging can be performed. The 30% difference would only provide a fraction of improvement in the final resolution.

One important aspect of the use of DWDD as a commercial system is that it allows, and actually requires, both the sensors and the instruments to be standardized. All instruments, and all sensors, should function around the same standardized wavelength. While this standardization requirement can be seen as a limitation, because it imposes stringent requirements on reproducibility of the components and manufacturing process, it can also be seen as an enabler for widespread deployment. For example, similar standardization has enabled the wide deployment of dense wavelength division multiplexing (DWDM) systems. The DFB laser diode used for this work indeed has a precise wavelength because it is intended for use in such systems. Precisely wavelength-matched FBG filters for DWDM systems are also commonly available. Standardization reduces costs, because it allows components to be mass produced without modifying the production line, and to be stocked in inventory for rapid, on demand delivery. By contrast, current WDM-based FBG sensor systems are practically all custom-designed and custom-made, which makes them economically viable only for niche applications.

As a related point, it is well-known that the characteristics of DFB lasers (central wavelength, chirp) can age and drift over time, and also that individual devices will have different characteristics. Therefore, a DWDD instrument, like any other sensing instrument, requires calibration to ensure accurate measurement, as should the sensors used with it. This is typically done using so-called golden units, that are well-calibrated sensors with properties that can ultimately be traced to national or international standards. This is actually a much more subtle and complicated issue than it might seem at first sight, and is most often avoided in much of the FBG sensor literature. In the case of DWDD, we expect to address it more fully in our future work. As for ensuring long-term accuracy, we previously discussed [12] how a sensible way of ensuring it is to use FBGs themselves as internal references. Properly annealed FBGs are known to have a great degree of long-term stability. A pair of properly-positioned reference FBGs can account not just for the wavelength drift, but also for the change in calibration slope brought about by potential change in the chirp coefficient δ. In a TDM-based instrument, those reference FBGs only take two slots in the measurement sequence, still leaving multiple slots for the actual sensors. Furthermore, the temperature dependence of the FBG wavelength is one tenth that of a typical DFB laser, so temperature-controlled reference FBGs can have extremely high wavelength stability. That way, a DWDD instrument can achieve a resolution unattainable with spectral scanning instruments [12].

In conclusion, we have shown that the dynamic chirp of a single DFB laser diode can be used to perform dual-wavelength differential detection of fiber Bragg grating sensors with nanosecond duration pulses, using a simple algorithm that can be implemented in a fast FPGA chip for real time calculation. Because the adiabatic and thermal contributions to the dynamic chirp have opposite sign but nearly equal magnitude, we have shown that the DWDD measurement can be implemented in two ways with equivalent sensitivity: either use two 5 ns pulses with different driving currents, or use two samples within a single pulse, separated by at least 15 ns.

Because of its simplicity and low cost, the DWDD method implemented in this way opens the way to inexpensive, high-resolution FBG sensing systems, making them competitive with other sensing technologies in both price and performance.

Author Contributions: Conceptualization, F.O., Z.O. and J.L.; methodology, F.O.; software, F.O.; validation, F.O., Z.O., and J.L.; formal analysis, F.O.; investigation, F.O.; resources, Z.O. and J.L.; data curation, F.O.; writing—original draft preparation, F.O.; writing—review and editing, J.L.; visualization, F.O.; supervision, F.O.; project administration, F.O.; funding acquisition, F.O. and J.L. All authors have read and agreed to the published version of the manuscript.

Funding: This work was supported by by the National Natural Science Foundation of China (NSFC) (Grant No. 61722503, 61421002, 61307102, and 61327004), the Fundamental Research Funds for the Central Universities (Grant No. ZYGX2019Z012), the Joint Fund of Ministry of Education for Equipment Pre-research (Grant No. 6141A02033411), and the Field Funding for Equipment Pre-research (Grant No. 61404140106).

Conflicts of Interest: The authors declare no conflict of interest.

Abbreviations

The following abbreviations are used in this manuscript:

DWDD	Dual wavelength differential detection
DFB	Distributed feedback
FBG	Fiber Bragg grating
FPGA	Field programmable gate array
TDM	Time division multiplexing
EDFA	Erbium doped fiber amplifier
FWHM	Full width at half maximum
SNR	Signal to nois ratio
ADC	Analog to digital converter
TEC	Thermoelectric cooler
PCB	Printed circuit board
CPU	Central processing unit
DWDM	Dense wavelength division multiplexing

References

1. Chen, J.; Liu, B.; Zhang, B. Review of fiber Bragg grating sensor technology. *Front. Optoelectron. China* **2011**, *4*, 204–212. [CrossRef]
2. Ma, Z.; Chen, X. Fiber Bragg Gratings Sensors for Aircraft Wing Shape Measurement: Recent Applications and Technical Analysis. *Sensors* **2019**, *19*, 55. [CrossRef] [PubMed]
3. Schena, E.; Tosi, D.; Saccomandi, P.; Lewis, E.; Taesung, K. Fiber Optic Sensors for Temperature Monitoring during Thermal Treatments: An Overview. *Sensors* **2016**, *16*, 1144. [CrossRef] [PubMed]
4. Qiao, X.; Shao, Z.; Bao, W.; Rong, Q. Fiber Bragg Grating Sensors for the Oil Industry. *Sensors* **2017**, *17*, 429. [CrossRef] [PubMed]
5. Al-Fakih, E.; Osman, A.; Azuan, N.; Adikan, M.; Rafiq, F. The Use of Fiber Bragg Grating Sensors in Biomechanics and Rehabilitation Applications: The State-of-the-Art and Ongoing Research Topics. *Sensors* **2012**, *12*, 12890. [CrossRef] [PubMed]
6. Tosi, D. Review and analysis of peak tracking techniques for fiber Bragg grating sensors. *Sensors* **2017**, *17*, 2368. [CrossRef] [PubMed]
7. Dyer, S.D.; Williams, P.A.; Espejo, R.J.; Kofler, J.D.; Etzel, S.M. Fundamental limits in fiber bragg grating peak wavelength measurements. In Proceedings of the 17th International Conference on Optical Fibre Sensors, Bruges, Belgium, 23–27 May 2005.
8. Morozov, O.G.; Natanson, O.G.; Aybatov, D.L.; Talipov, A.A.; Prosvirin, V.P.; Smirnov, A.S. Metrological aspects of symmetric double frequency and multi frequency reflectometry for fiber Bragg structures. In Proceedings of the Optical Technologies for Telecommunications, Ufa, Russian, 26–28 November 2007.
9. Kulchin, Y.N.; Vitrik, O.B.; Dyshlyuk, A.V.; Shalagin, A.M.; Babin, S.A.; Nemov, I.N. Differential reflectometry of FBG sensors in the wide spectral range. *Laser Phys.* **2011**, *21*, 304. [CrossRef]

10. Ouellette, F. Apparatus for Measuring Optical Signals From Multiple Optical Fiber Sensors. U.S. Patent 9,810,556 B2, 7 November 2017.
11. Ouellette, F.; Li, J. Dual-wavelength differential detection of fiber bragg grating sensors with a single dfb laser diode. In Proceedings of the Asia Communications and Photonics Conference, Chengdu, China, 2–5 November 2019.
12. Ouellette, F.; Ou, Z.; Li, J.; Albert, J. High-resolution interrogation of tilted fiber Bragg gratings using an extended range dual wavelength differential detection. *Opt. Express* **2020**, *28*, 14662–14676. [CrossRef] [PubMed]
13. Cheng, R.; Xia, L. Interrogation of weak bragg grating sensors based on dual-wavelength differential detection. *Opt. Lett.* **2016**, *41*, 5254–5257. [CrossRef] [PubMed]
14. Xia, L.; Wu, Y.; Rahubadde, U.; Li, W. Tdm interrogation of identical weak fbgs network based on delayed laser pulses differential detection. *IEEE Photonics J.* **2018**, *10*, 1–8. [CrossRef]
15. Li, Z.; Tong, Y.; Fu, X.; Wang, J.; Guo, Q.; Yu, H.; Bao, X. Simultaneous distributed static and dynamic sensing based on ultra-short fiber bragg gratings. *Opt. Express* **2018**, *26*, 17437–17446. [CrossRef] [PubMed]
16. Rohollahnejad, J.; Xia, L.; Cheng, R.; Ran, Y.; Rahubadde, U.; Zhou, J.; Zhu, L. Tdm interrogation of intensity-modulated usfbgs network based on multichannel lasers. *Opt. Express* **2017**, *25*, 670–680. [CrossRef] [PubMed]
17. Wilson, A.; James, S.W.; Tatam, R.P. Time-division-multiplexed interrogation of fibre bragg grating sensors using laser diodes. *Meas. Sci. Technol.* **2001**, *12*, 181–187. [CrossRef]
18. Rao, Y.J.; Lobo Ribeiro, A.B.; Jackson, D.A.; Zhang, L.; Bennion, I. Combined spatial- and time-division-multiplexing scheme for fiber grating sensors with drift-compensated phase-sensitive detection. *Opt. Lett.* **1995**, *20*, 2149–2151. [CrossRef] [PubMed]
19. Shalom, H.; Zadok, A.; Tur, M.; Legg, P.; Cornwell, W.; Andonovic, I. On the various time constants of wavelength changes of a dfb laser under direct modulation. *IEEE J. Quantum Electron.* **1998**, *34*, 1816–1822. [CrossRef]

MDPI
St. Alban-Anlage 66
4052 Basel
Switzerland
Tel. +41 61 683 77 34
Fax +41 61 302 89 18
www.mdpi.com

Sensors Editorial Office
E-mail: sensors@mdpi.com
www.mdpi.com/journal/sensors

www.ingramcontent.com/pod-product-compliance
Lightning Source LLC
Chambersburg PA
CBHW040749250926
51850CB00007BB/1452
9783036512860